Water Management for Irrigation

Water Management for Irrigation

Edited by **Keith Wheatley**

R **CALLISTO**
REFERENCE

New York

Published by Callisto Reference,
106 Park Avenue, Suite 200,
New York, NY 10016, USA
www.callistoreference.com

Water Management for Irrigation
Edited by Keith Wheatley

International Standard Book Number: 978-1-63239-768-3 (Hardback)

The publisher's policy is to use permanent paper from mills that operate a sustainable forestry policy. Furthermore, the publisher ensures that the text paper and cover boards used have met acceptable environmental accreditation standards.

Trademark Notice: Registered trademark of products or corporate names are used only for explanation and identification without intent to infringe.

Printed in the United States of America.

Contents

Preface

This book provides comprehensive insights into the field of water management with respect to irrigation. Irrigation refers to the practice of using and distributing water via artificial sources for maximum healthy produce. It is especially done in the areas which witness low rainfall and have scarce water facilities to ensure better crop production. Fresh water being a scant resource makes water management an essential practice. Therefore, water management while irrigating crops becomes a necessity. This book will discuss in detail the various techniques applied to minimize water wastage and pollution while irrigation. It will also give detailed information about the varied theories related to this field. Those with an interest and an eye for detail, will find this text full of crucial information. It will serve as a beneficial guide for students, researchers, farmers, environmentalists and others interested in this subject.

The information shared in this book is based on empirical researches made by veterans in this field of study. The elaborative information provided in this book will help the readers further their scope of knowledge leading to advancements in this field.

Finally, I would like to thank my fellow researchers who gave constructive feedback and my family members who supported me at every step of my research.

Editor

Rainfall Distribution Functions for Irrigation Scheduling: Calculation Procedures Following Site of Olive (*Olea europaea* L.) Cultivation and Growing Periods

Chiraz Masmoudi-Charfi[1*], Hamadi Habaieb[2]

[1]Institution of Agriculture Research & High Education, Tunis, Tunisia
[2]National Institute of Agronomy, Tunis, Tunisia
Email: [*]masmoudi.chiraz@yahoo.fr

Abstract

In Tunisia (36.5°N, 10.2°E, Alt. 10 m), rainfall is the major factor governing olive production. It is characterized by large variability in time and space, making yields of olive trees highly dependent on the amount of water received and timing. Thus, improvement of olive productivity by irrigation is necessary. This study aimed to determine the crop water needs of olive orchards and the rainfall frequencies at which they are covered following age and sites of olive production. For this purpose, the rainfall distribution functions are established for different cities of Tunisia (Tunis, Bizerte, Béja, Nabeul, Sidi Bouzid, Gabes and Sousse). For all sites and growing periods, the reference evapotranspiration (ET_0) was computed by using several formulas. Their performance against the Penman Monteith (PM) method was evaluated graphically and statistically in all considered cities in order to evaluate their accuracy for better adapting them to the existing environmental conditions, particularly when data are missing to compute ET_0-PM. Results presented herein show that the estimated ET_0 values strongly correlate with ET_0-PM at all sites and formulas with r values up to 0.88. Particularly, the methods of Turc and Ivanov appropriately predict the ET_0-PM in all climatic regions of Tunisia and may constitute an appropriate alternative for ET_0 estimation when data are missing to compute ET_0-PM. However, although the Turc method performs well with all climatic zones, arid and semi-arid, in western, northern and coastal areas of Tunisia, the Ivanov method appears to be more appropriate to the northern areas (Béja and Bizerte) characterized by semi arid climate and having annual rainfall of up to 450 mm, though a poorer agreement was found when using the Eagleman formula. Estimates of ET_0 by using the Hargreave-Samani (HS) formula for the east-southern area (Gabes) characterised by arid climate show satisfactory agreement with ET_0-PM estimates, corroborating previous findings reporting

---

that the HS method performs well in most climatic regions, with the exception of humid areas where it tends to overestimate ET_0. It appears also that at a given site, the most appropriate method for ET_0 estimation at annual scale may be different from that giving the best value of ET_0 when considering the growing stages of the olive tree. The formula of Turc, although it was appropriate when estimating the annual ET_0 value for Sousse, it wasn't adequate at seasonal scale. Adversely, although the method of BC is suitable for stages 1, 2, 4 and 5 at Sousse, the appropriate method for the overall cycle is that of Turc. Results also show that the average annual value of ET_0-PM calculated by using full datasets from January to December is well correlated to the maximum and minimum daily estimates with r values ranging between 0.70 and 0.83. These variations indicate that there is no weather-based evapotranspiration equation that can be expected to predict evapotranspiration perfectly under every climatic situation due to simplification in formulation and errors in data measurement. Nevertheless we can say that when data are missing, ET_0 can be estimated with a specific formula, suggesting that of Turc for Tunis, Sidi Bouzid, Sousse and Béja at annual scale despite of their appartenance to different climatic regions, while the method of Ivanov is quite valuable for Bizerte and Nabeul. Determination of the crop evapotranspiration (ETc) on the basis of ET_0-PM computations following age and the growing periods show positive values of (ETc-P) at annual scale for Tunis, Nabeul, Sousse Bizerte and Beja when young olive plantations are considered, but for old trees, values are positive only for Tunis, Bizerte and Béja. Seasonal differences between ETc and P (rainfall) recorded during the irrigation period are negative even for young plantations. The lowest and highest deficits are observed at Béja and Gabes cities, respectively. The rainiest periods are December-February for Béja, Tunis and Bizerte and September-November for the other sites with similar trends for rainfall frequency (F). The driest period is that of July-August for all sites, with F values exceeding 0.9 in most cases. Only 10% of water needs are supplied by rainfall during this period of fruit development even for one year old orchards. Therefore, irrigation is needed all time for adult trees even at the rainiest locations. For young plantations, irrigation becomes necessary beginning from the second period of tree development, *i.e.* April-June for Bizerte, Béja, Nabeul and Tunis since the early spring period for both young and old plants for Gabes and Sidi-Bouzid. It appears from this analysis based on the seasonal rainfall frequencies and water needs computed with the PM-formula, that there is a need for irrigating olive plantations aging more than 5 years in most case studies and especially when olive is cultivated in the western areas of Tunisia. Irrigation is needed during the growing fruit period but also during the other seasons, when shoots grow. Results also indicate that the use of no adequate method to estimate ET_0 allowed overestimating or underestimating of water requirements. So it is desirable to have a method that estimates ET consistently well and future research is needed to reconcile which should be the standard method of calculating the change in the crop coefficient over time. However, despite a quite good performance of the PM-equation in most applications, particularly when it is used for irrigation scheduling purposes, some problems may appear because of lack of local information on values and determination of the effective rainfall. In conclusion we can say that on the basis of the results produced, we can decide for each region and growing period if complementary irrigation is needed or not. Indicative amounts are given for each case study. Also, it appears to our best knowledge that this work with that of Nasr (2002) [1] is the only one that has estimated ET_0 for the whole country using PM-ET_0 compared to other empirical methods, but we didn't test the possible advantages in using calibrated values for the radiation adjustment coefficient or temperature adjustment for dew point temperature estimation as proposed by Allen (1996) [2]. This calibration is therefore a line to be explored. Additional research is needed on developing crop coefficients that use the Penman-Monteith equation when calculating ET and a standardized method of calculating the time base for the crop coefficients preferably based on a growing degree day concept.

Keywords

Methods of ET_0 Computation, FAO-PM Method, Climatic Water Deficit, Irrigation Application, Rainfall Frequency

1. Crop Evapotranspiration and Methods of ET_0 Computation: Brief Review

Most water absorbed by roots is lost through leaf stomata. This process is driven by the climatic demand and involves the need of measuring many biological and environmental variables like evapotranspiration. A large number of formulas have been developed and/or improved from 1942 to 2005 to compute the monthly or the seasonal "consumptive water use" later termed reference evapotranspiration (ET_0), starting with the Blaney-Criddle formula (BC) method and ending with the Penman-Monteith equation (PM) [3]-[6].

Empirical methods were primarily based on solar radiation (Turc, 1961; Christiansen-Hargreaves, 1969; Hargreaves-Samani, 1985), temperature (Blaney-Criddle, original and modified, 1950) and relative humidity (Ivanov, 1954; Eagleman, 1967). The formulas of Penman (Penman Original, 1963) and FAO-PM [7] [8] are based on combination theory and necessitate intermediary calculi, involving simultaneous measurements of radiation, air temperature, humidity and wind speed. Calculation procedures are described in Rana and Katerji (2000) and Habaieb and Masmoudi-Charfi (2003) [6]-[9] and recently summarized by Sammis *et al.* (2011) [10].

Many of these formulas have been used for irrigation and water rights management and yield prediction, but some of them provided errors and large uncertainities because most evapotranspiration available data are site specific and are difficult to apply to other situations. For this reason a consultation of experts and researchers, with the *International Commission for Irrigation and Drainage* and the *World Meteorological Organization*, have been organized by the FAO in the 90th in order to review the methodologies and procedures adopted for determinating the crop water requirements (*ETc*). This consultation considered the PM-model as the most advanced method for *ETc* determination in commercial orchards. It is aimed to define the grass reference evapotranspiration, *i.e.*, the rate of evapotranspiration from an hypothetical crop approximately resembling the evapotranspiration from an extensive surface of a disease-free green well-watered grass cover of uniform height, actively growing, completely shading the ground, and with adequate water and nutrient supply. Though, it is selected as the method by which the evapotranspiration of this reference surface (ET_0) can be unambiguously determined, providing consistent ET0 values in all regions and climates. Indeed, as the methodology of ET_0-PM estimation was successfully applied at different time scales in various climatic regions of the world [11]-[24], it was adopted since the year 1998 as the referenced method [8].

As redefined by *The Food and Agricultural Organization of United Nations* [8], the PM-equation overcomes shortcomings of the previous FAO Penman method, providing values of ET_0 more consistent with actual crop water use data worldwide and was found to be more consistent over a wider range of climatic conditions than other equations [16]. It became the *American Society of Civil Engineers (ASCE) Standardized Reference ET equation* and the current recommended equation to calculate the evapotranspiration rate from non-water stressed crop, *i.e.*, the evaporation power of the atmosphere which is expressed by the reference crop evapotranspiration (ET_0).

The PM-equation has two components: the radiation component controls *ET* by providing energy to drive the ET process [20] and the wind and vapor pressure deficit component which "controls the rate of transport of water vapor from the plant and soil surface and the capacity of the air to absorb water vapor". But estimating all these parameters needs auxiliary sub-models locally calibrated or parameterizations for conductances [13]. This implies many uncertainties which are introduced through the parameterization of the other variables [25]. In addition, as most available evapotranspiration data are site specific and are difficult to apply to other situations, the panel of experts developed a simplified way to deal with evapotranspiration dependence on climate by relating it to standard reference evapotranspiration (ET_0) by the crop coefficient, K_c, which is defined as the ratio of the crop evapotranspiration to the reference evapotranspiration as: $K_c = crop\ ETc/grass\ ET$, where *ET* is the maximum daily, monthly or seasonal evapotranspiration determined under non-stressed conditions. It is an empirical seasonal factor relating the seasonal plant water usage for a specific crop to the total seasonal consumptive water use factor generated under experimental conditions. K_c depends on the development rate of the crop and varies following year, season, the amount of water available to roots and the effective transpiring leaf area which was used in the PM-model to estimate daily tree transpiration of whole trees [8] [12] [26]-[28]. But as optimal conditions are difficult to realize, crop coefficients were taken as the ratio of evapotranspiration of well-watered orchards.

However, although the PM-method is considered as the most rational and elaborated approach, it could not be used for reliable estimation of ET_0 with incomplete series of climatic data. The lack of full weather datasets in many parts of the world, particularly in remote areas, limits the application of ET_0-PM, which is a major problem. Indeed, computation of ET_0 with the PM-equation requires a large number of climatic data, which are often very scanty, forcing the user to choose some other alternatives. Alternative accurate approaches requiring limited data has led to a huge number of related studies focusing various climates [15]-[32]. These procedures have been tested by Nandagiri and Kovoor (2005) [33].

Basing upon former studies to compare the performance of ET temperature methods, mainly the study made by Jensen et al. (1990 and 1997) [27]-[34], when full weather data are lacking, ET_0 can be estimated either using the empirical Hargreaves-Samani (HS) equation [4], or empirically estimating R_n, VPD and U_2 for using in the ET_0-PM equation, including using data from neighbour weather stations [16]. Various ET temperature methods were then excluded, particularly the ET climatic equation of Thornthwaite (1948) [35], that largely underestimates ET_0 comparatively to ET_0-PM [36]. In both aforementioned methods the minimum set of data required consists of T_{max} and T_{min}. The latter approach for using the ET_0-PM with only T_{max} and T_{min} is called PM temperature (PMT) method and is also referred in literature as reduced set PM method [32]. Both the HS and PMT methods have received a continuous attention from research contrarily to the use of neighbour weather data. The recent methodology reported by Martí and Zarzo (2012) [37] based on principal component analysis to estimate ET_0 when no local climatic inputs are available may provide new developments in this domain.

The ET_0-PM is widely used to manage irrigation of orchards characterized by incomplete soil cover. Irrigation water supplies are computed on the basis of the crop evapotranspiration (ETc) and rainfall amounts (R). In Tunisia (36.5°N, 10.2°E, Alt. 10 m), rainfall is the main fresh water resource providing annually 36 Km^3 of water in average, of which only 3 km^3 could be potentially collected as runoff water in large dams [38]. Large spatio-temporal variability characterise rainfall distribution with annual amounts ranging between 1500 mm in the extreme north of the country (Mogods mountains) and less than 80 mm in the southern area (Sahara). Non-arid area is estimated only at 37,000 km^2 (24%), arid area at 55,000 km^2 (35%) and desert at 63,000 km^2 (41%) [39] [40]. Renewable groundwater resources are estimated at 1.7 km^3 and mostly used (83%) for agriculture [41]. About 500,000 ha of annual and perennial crops are irrigated permanently, amongst 66,000 ha of olives beneficiate regularly of complementary irrigation.

Olive (Olea europaea L.) is the main component of our agricultural system covering one third of the agricultural area which approximates 1400,000 ha. Most plantations are rainfed, receiving annually no more than 250 mm of rainfall. A major part of these falls is received during the autumn and winter (quiescent period) seasons and only 25% of this amount and perhaps less is profitable to the growing fruits (May to October), that's why fruit size and olive production remain highly dependent on water provided during that period and particularly in the centre and southern areas [42]-[44]. Yields fluctuate from year to year between 60,000 and 300,000 tons as well as the amount of extracted oil (16% to 26%). This situation is very uncomfortable for both orchadists and the government because olive benefits are the main financial resource for agriculture, and Tunisia has to provide obviously thousand tons of olive oil each year for exportation to the EU, which is a strategic partner.

Supplemental irrigation [40] of olive orchards has emerged during the last few decades as an appropriate practice that has the potential to improve their productivity and stabilize yields by reducing spatial and temporal production variability. It has been adopted in different parts of Tunisia, and even institutionalized as a major strategy to respond to growing water scarcity in the irrigated areas and at the same time to increase crop productivity in traditionally rainfed plantations. Furthermore, the Tunisian government has taken since the 90th through its financial and development programs to develop many water conservation strategies as building important hydraulic catchments [39]-[41]. Also, the government encourages farmers to densify the traditional plantations in order to increase their benefits and oil production and to plant olive trees at higher densities, by distributing in some cases free cuttings and N-fertilizers. All kind of waters, fresh, brakish [45] and even margines and waste waters [46] are used. With appropriate horticultural practices, olive production has risen significantly during this last decade, but crops still suffer from a large gap between applying the correct water needs and the optimal production because of summer water shortage and cyclic droughts [42]-[40] [47]-[49]. That's why, other methods and technologies like assisted drip irrigation [50]-[52], strategies of deficit irrigation [51] [53]-[55] and partial root irrigation [56] [57], fertigation, high yielding cultivars grown with high levels of input [58] [59], but also innovative techniques that ensure the best use of natural precipitation, should be promoted to improve the performance of the used practices. However, due to the limited available water resources, development of techniques and methods should target at the most profitable areas where the allocation of irrigation water could be optimized.

Consequently, as olive farming becomes more intensive, crops require greater economical inputs and the changes accompanying the new cultivation systems have revealed a lack of knowledge about management of these orchards. Many problems erase at field level particularly when irrigating them. A large number of orchardists practice irrigation during the high fruited years independently of tree age, LAI and soil coverage increases. Thus, the amount of water distributed to the orchard may not meet its seasonal water needs and sometimes, lead to a loss of large amounts of water, low water use efficiencies, low growth rates and even dratistic yields. Negative effects appear when water shortage occurs during the critical stages of flower differenciation (February) and fruit growth (June-September), but also during the period of flower induction, occurring during the previous summer. So, because fruits are produced on one year old shoots which elongates during the last year [60]-[62], water should be available during all these stages over two growing seasons to cover water needs with regard to the importance of each one of the physiological process involved during fruit development [63].

In many studies carried out in Tunisia [64]-[69] [52] [70] and elsewhere around the Mediterramean countries [14] [71]-[77], irrigation is applied as a compulsory practice when the tree is the most responsive to water. These works sought to evaluate the potential of using different irrigation amounts to complement rainfall. Results show that in environments characterized by alternating wet and dry seasons, adding small amounts of water during the growing season can increase water productivity many-fold. This potential of supplemental irrigation must be explored to make better use of the limited resources available. Other results show that olive production was almost unchanged when water supply increased up to 300 mm. This is well traduced by some quantitative relationships established between production and water supplies [47] [69] [57] and well described in the recent paper "*Manuel d'Irrigation de l'Olivier, Techniques et Applications*", published in Tunisia by Masmoudi-Charfi *et al.* (2012) [78].

Adequate irrigation management requires an accurate estimation of the whole orchard water losses [6] [13] [79]-[82]. This has been measured in olive plantations by using the water balance method [77] [80] [81] [83] and lysimeters [84] providing *in situ* measurements of the actual water used by plants cultivated under field conditions. This involves estimation of water used at specific time and location as well as the relationship between water consumption and biomass production [57] [69] [85] and its dependency with climatic data and soil coverage [69] [77]. But, when these values are applied at field level, some problems may arise inherent to the spatial representativeness of measurements. Necessarily, the field should be cultivated under the same conditions.

Nowadays, the general approach used to calculate the water requirements of olive orchards is the crop evapotranspiration ET [11] [13] [29] [68] [76] [77] [81] [86], a term more descriptive of the water sources involved, which are the amounts of water evaporated from the soil (E) and transpired by the plant (T) per unit area. These processes occur simultaneously and there is no easy way of distinguishing between them [8] [12] [13]. Plant evaporation measurements named transpiration (T) are difficult to perform and many parameters should be taken into account [11] [49] [83] [84] [87] [88]. Errors or/and consistent changes may arise [16] [25] following age, the anatomy of trees and their vigor, soil water availability and the evaporative demand. If water is not limiting factor, transpiration will be conditioned by leaf area. This may result in a problem of representativeness, since well developed olive trees, shading a large area may provide within the same orchard higher transpiration rates. Moreno *et al.*, (1996), Fernandez *et al.*, (2001) and Abid-Karray (2006) [14] [89] [90] measured this component by the heat pulse technique of Granier (1985) [91]. In Tunisia, Masmoudi-Charfi *et al.* (2013) [55], showed that tree transpiration of young olive trees cv., Chétoui, approximates 50% *ETc*. For soil evaporation, many alternative methods and models were proposed since the 70th [92]-[94]. However, although this component is theoretically well defined, it is difficult to measure for orchards and isolated trees. Most models used for this purpose are based on the fraction of solar radiation reaching the soil surface, which decreases over the growing period as the crop develops. Some attempts should be made to take into account the special inhomogeneity of the environment under the tree [81]. Works carried out in Centre Tunisia on ten years old trees of cultivars Picholine and Meski have shown that soil evaporation equals to 53% *ETc* [51]. In other studies made in Spain these components were measured under olive orchards with microlysimeters [94], while the approach developed by Villalobos *et al.* (2000) [12] is based on the PM-equation [8] [16] with the use of crop coefficients, taken equal to 0.3 - 0.4 for young trees and ranging between 0.5 and 0.7 for adult plants [95]. Specific values were recently determined in Tunisia for adult trees by Braham and Boussadia (2013) [96] for the period of flowering and fruit set.

This study was carried out in several geographical sites of Tunisia covering the most common regions of olive cultivation, which are characterized by different climates. The main objectives are: 1) to evaluate the potential and accuracy of different ET_0 calculation formulas for better adapting them to the existing environmental conditions;

2) to define for each location the most appropriate formula to determine ET_0 and 3) to determine the amount of irrigation application to olive orchards depending on seasonal rainfall amounts, tree age and the growing stages through the establishment of the rainfall distribution functions for each growing period and location. These objectives are set to support a wide range of irrigation management and water resources applications for use in regions where weather data are missing, incomplete or of questionable quality. For each site, the comparative study between ET_0-PM and the other estimates, allowed us to choose the formula that gives the most valuable values for a specific city using a limited number of climatic data.

2. Materials and Methods

2.1. Areas of Olive Cultivation and Sites of Study

In Tunisia olive trees are mostly present in the center and southern regions of the country, contributing with 72% in number and 88% in area (**Table 1**).

Olive is cultivated under different growing environments, systems (rainfed, irrigated) and densities (17 to 1250 trees/ha). Soil is generally clay-loamy in the north, where Mediterranean climate prevail, and salty in the centre and southern regions, characterized by hard and dry conditions with absolute temperature exceeding 40°C during the summer months. Fresh water used for irrigation is available in the north and becomes scarcer and brackish elsewhere. Deep resources are found in the southern areas. Water is supplied to olive orchards during the dry season from May to October. Drip irrigation concerns less than 20% of plantations. Cities concerned with this study and their coordinates (Lat. 36° to 38°, Long. 8° to 11°, Alt. 2 m to 314 m) are shown in **Table 2** and **Figure 1**.

Table 1. Distribution of olive trees in Tunisia.

Area	% of total number	% of total area
North	28	12
Center	60	68
South	12	20
Total	57,000,000 trees	1,400,000 ha

Table 2. Coordinates of the studied cities.

Site	Localization	Main crops	% Vegetables	agriculture Fruit trees	area Cereals	(%) Olive tree area
Nabeul	LAT: 36°85'N LON: 11°08'E, ALT: 30 m	Fruit trees, vine, vegetables	53.1	32.3		9.3
Sousse	LAT: 35°76'N LON: 10°75'E, ALT: 2 m	Fruit trees and vegetables	83.2			9.0
Sidi Bouzid	LAT: 34°41'N LON: 8°81'E, ALT: 314 m	Fruit trees, vine, wheat, vegetables		48.2	19.6	16.2
Gabes	LAT: 33°88'N LON: 10°10'E, ALT: 5 m	Fruit trees, vegetables	29.6	47.2		3.2
Tunis	LAT: 36°83'N LON: 10°23'E, ALT: 4 m	Fruit trees, vine, vegetables	48.9	35.3		10.4
Béjà	LAT: 36°48'N LON: 8°8'E, ALT: 144 m	Wheat/Annual crops/fruit trees		27.1	36.9	18.7
Bizerte	LAT: 37°25'N LON: 09°8'E, ALT: 3 m	Fruit trees, vine, annual crops	48.3	24.0		10.5
Tunisia		**Olive trees and Cereals**	**30.8**	**40.0**	**14.2**	**1,400,000 ha**

NB: fruit trees including olive trees. Remarks: 1. Although Sfax is the main region of olive cultivation it was not considered in the results because the climatic data are incomplete. 2. Climatic data used for the city of Sousse were recorded with an automatic climatic station located in Monastir, which is a locality of Sousse.

Figure 1. Cities of Tunisia.

Cities of Béja, Tunis, Bizerte and Nabeul are all situated in the North of the country, while Sousse and Sidi Bouzid are located in the Center, and Sfax and Gabes are both coastal sites of South Tunisia.

2.2. Metdology

2.2.1. Steps of Work

The following steps were followed in this study:

1) Computation of the monthly reference evapotranspiration (ET_0, mm) by using several formulas, comparing estimates to ET_0-PM computations;

2) Estimation of the crop evapotranspiration (*ETc*, mm) following age and the growing periods;

3) Establishment of the rainfall distribution functions for all growing periods and cites;

4) Visualizing *ETc* values on the rainfall distribution graphs;

5) Computing the climatic deficit (P-*ETc*) and determination of the irrigation water amounts (I, mm) following tree age and site.

On the basis of the results produced, we have to decide for each region and period of growth if complementary irrigation is needed or not. Indicative amounts are given for each case study.

2.2.2. Climatic Variables

Climatic data were provided by the *National Institute of Meteorology* (INM) website [92], covering many years (**Table 3**). Data used for this study are maximum and minimum temperatures (T_{max} and T_{min}, °C), maximum and minimum relative humidity (RH_{max} and RH_{min}, %), maximum (N, hours/day) and actual sunshine durations (n, hours/month), wind speed measured at 2 m height (U_2, m/s) and atmospheric radiation (R_a, Mj/m^2/day). The majority of the selected cities have the longest and nearly complete data records for the requested period.

Rainfall records are averages of 99 years-long-period (1901-2000). Values are presented in **Figure 2** following the regions and the growing periods of olive trees.

Annual rainfall amounts ranged approximately between 200 mm at Gabes, an east-southern area of Tunisia and 600 mm at Béja and Bizerte which are continental and coastal areas located in the western and northern areas of

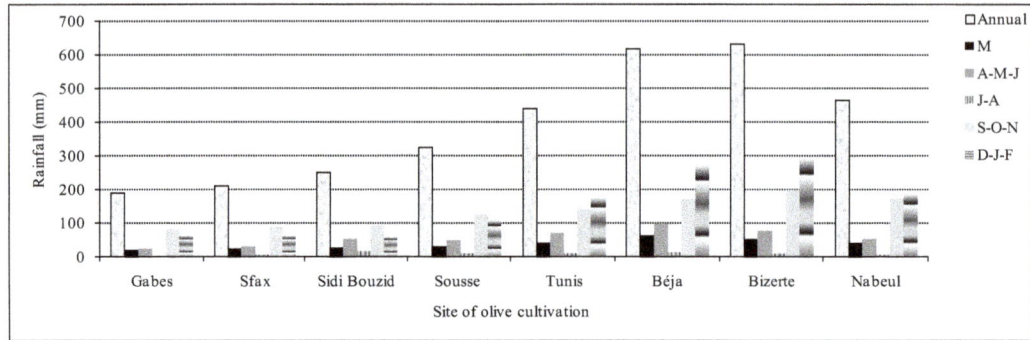

Figure 2. Rainfall amounts (P, mm) recorded in Tunisia following the growing period and sites of olive cultivation. Data are averages computed over 99 years-long-period between years 1901 and 2000.

Table 3. Climatic data recorded at the studied cities (INM, 2013) [97].

	Tunis	Nabeul	SB	Béja	Gabes	Bizerte	Monastir
T_{min} (°C)	13.4	15.1	12.5	10.8	162	13.1	15.2
T_{max}°C	23.5	22.7	25.3	239	243	22.6	24.0
U_2 (m/s)	4.2	3.0	2.8	42	34	3.9	4.5
es (mb)	2.3	2.4	2.5	23	26	2.2	2.5
R_a (Mj/m²/day)	29.3	29.3	30.3	296	303	29.2	29.6
N (h/day)	5.0	12.0	12.0	120	12.0	12.0	12.0
HR (%)	68.9	72.4	59.0	650	63.0	70.8	66.0
n (h/month)	238.3	213.5	229.2	2048	265.0	242.2	223.5
Rainfall (P, mm/year)	449.7	468.1	251.8	628.3	190.1	634.6	328.1

the country, respectively (**Figure 1**). For Sousse and Sidi Bouzid, annual rainfall varied between 200 mm and 300 mm; all of them are located in the Centre of Tunisia. For Tunis and Nabeul, annual precipitations ranged between 400 mm and 500 mm. The driest month is July and the wettest is January.

Monthly average rainfall amounts (mm) recorded between 1901 and 2000 and their ecartypes are presented in **Figure 3** for the different cities. Largest variations between years were recorded during the rainiest months, *i.e.* December-February and September-November following the site. Lower variations between averge values were recorded during the summer months for all sites.

2.2.3. Climatic Characterization of the Studied Cities

The UNEP aridity index, which is adopted by the FAO and used worldwide, consists of the ratio of mean annual precipitation (P) to mean annual potential evapotranspiration ET_0 computed with the Thornthwaite method [35] [86] [98]. Values are presented in **Table 4** to climatically characterize the different cities on the basis of the following classification: *Hyper-arid*: 0 - 0.08, *Arid*: 0.08 - 0.2, *Semi-arid*: 0.2 - 0.5, *Dry sub-humid*: 0.5 - 0.65, *Moist sub humid*: 0.65 - 1, *Humid*: 1 - 2.

Values of aridity index obtained herein points to:

1) Tunisia is not concerned with humid and sub-humid climates.

2) Arid and semi-arid climate dominate the central and southern landscape.

3) Although Bizerte, a coastal region, has a relatively high annual precipitation accompanied with cold winters, it is not considered as a sub-humid region. Its AI is close to that of Béja, a continental area of north Tunisia.

4) Arid climate refers to the areas of west (Sidi Bouzid) and south (Gabes) Tunisia, although Gabes is a coastal site.

The Aridity Index AI ranges between 0.11 and 0.43, making Tunisia mostly concerned with the arid (Gabes and Sidi Bouid) and semi-arid climates (other cites). This agrees partially with Kassas (2005) [39], for which

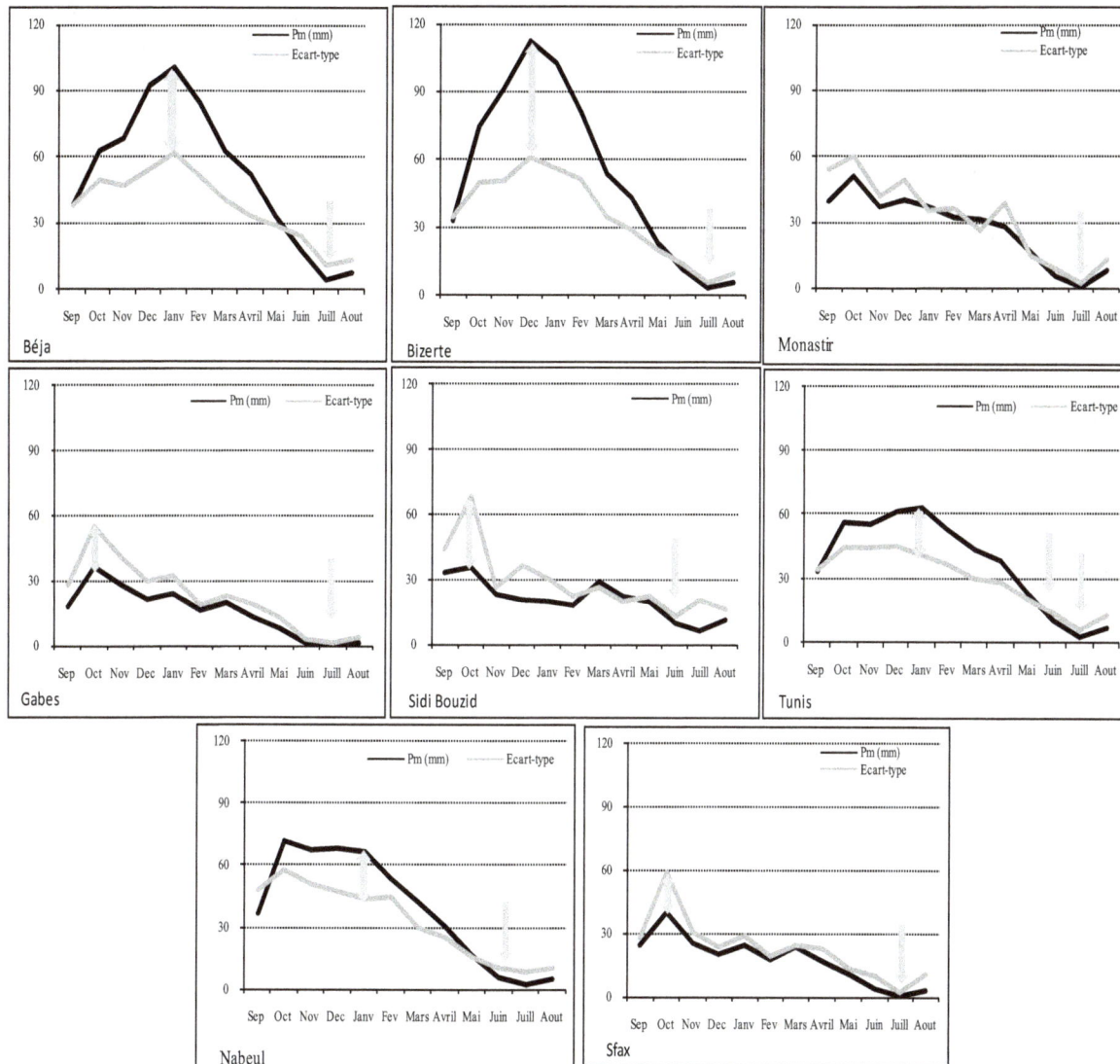

Figure 3. Monthly average rainfall amounts (mm) recorded between years 1901 and 2000 and the corresponding ecartypes.

Table 4. Mean annual values of Thornth waite (TW) potential evapotranspiration (mm), precipitation (P, mm) and index of aridity (IA) calculated for the studied cites.

	ET_0-TW (mm/year)	Rainfall (P, mm/year)	Index of aridity	Type of Climate
Tunis	1556.2	449.7	0.29	Semi-arid
Béja	1471.2	628.3	0.43	Semi-arid
Nabeul	1592.9	468.1	0.29	Semi-arid
Monastir	1679.5	328.1	0.20	Semi-arid
Gabes	**1762.0**	**190.1**	**0.11**	**Arid**
SB	**1634.9**	**251.8**	**0.15**	**Arid**
Bizerte	1494.4	634.6	0.42	Semi-arid

*Thornthwaite (TW) classification based on the UNEP aridity index [86]. SB: Sidi Bouzid city.

Tunisia is concerned with the semi-arid and arid climates in most areas and sub-humid in the extreme north of the country dominated by the mountains of *Kroumirie*, where the aridity index varies between 0.50 and 0.65.

The climate sub-regions of Tunisia produced herein based on the spatial variability of the aridity index, are in relatively good agreement with the results of "The precipitation based regionalization studies for Tunisia" illustrating the main structure of the Tunisian climate [41] but different from others produced by using other index to climatically characterise the different regions of the country. According to the FAO classification based on the length of the growing period, Tunisia is found out of the dry lands because crops can grow over a long period, exceeding 120 days (4 months).

2.2.4. Methods of ET_0 Calculation

Empirical formula used for ET_0 calculation is detailed in a previous paper published by Habaieb and Masmoudi-Charfi (2003) [9], and summarized in **Table 5**. The parameterization scheme for ET_0 (L ground $m^{-2} \cdot d^{-1}$) estimation is reported in **Table 6**. This schema adapted the PM-equation to be used with daily total net radiation (R_n, MJ ground $m^{-2} \cdot d^{-1}$) and the corresponding daily averages for air temperature (T_{mean}, °C), wind speed (U_2, $m \cdot s^{-1}$), vapor pressure deficit (VPD, kPa) and conductances [8] [13] [99]. The equation uses standard records of solar radiation (sunshine), air temperature, humidity and wind speed. Currently the climate data to calculate ET_0-PM are readily available from automated climate stations [1]. To ensure the integrity of computations, the weather measurements should be made at 2 m (or converted to that height, FAO 56) above an extensive surface of green grass, shading the ground and not short of water.

To appropriately compute the parameters of ET_0-PM (mm/day), the procedures proposed by Allen *et al.*, (1998) [8] should be followed as indicated above. Gavilán *et al.*, (2007) [20] reported that the methods proposed by Allen *et al.* (1998) [8] for estimating R_n and G are appropriate for ET_0 estimation for both daily and hourly

Table 5. Methods for ET_0 calculation and corresponding formulas. Parameters used in the formula are defined in the list of abbreviations below.

Method of ET_0 computation	Climatic variable	Formula
Blaney-Criddle modified (1950) by Doorenbos and Pruitt (1977)	$T_{average}$, n, N	$ET_o = (8 + 0.46\ T_{average}) \times n/N/i$
Hargreaves-Samani (1985)	$T_{average}$, R_a, T_{max}, T_{min}, R_a	Original: $ET_o = 0.0023\ R_a\ \Delta T^{0.5} \times (T_{average} + 17.8)$ Modified: $ET_o = 0.0035\ R_a\ \Delta T^{0.5} \times (T_{average} + 12.54)$
Christiansen-Hargreaves (1969)	R_s or R_a, U_2, $T_{average}$, $RH_{average}$, E	Original: $ET_o = 0.492\ R_s \times C_{TT} \times C_{WT} \times C_{HT}$ Modified: $ET_o = 0.324\ R_a \times C_{TT} \times C_{WT} \times C_{HT} \times C_{ST} \times C_R$
(1) Penman Original (1963) (2) Penman Monteith (1998)	$T_{average}$, e_o, U_2, n, N, R_a, $RH_{average}$,	$ET_o = [\Delta/(\Delta + \gamma) \times (R_n + G)] + [\gamma/(\Delta + \gamma) \times 15.36 \times (1 + 0,0062 \times U_2 \times (e^o - e))]$ $ET_o = \dfrac{\left[0.408 \times \Delta \times (R_n - G)\right] + \left[900 \times \gamma/(T + 273) \times (e^o - e) \times U_2\right]}{\Delta + \left[\gamma \times (1 + 0.34 \times U_2)\right]}$
Ivanov (1954)	T_{max}, T_{min}, T_m, e_o, $RH_{average}$, RH_{max}, RH_{min}	$ET_o = 0.0018 \times (T + 25)^2 \times (100 - e/e^o\ 100)$
Eagleman (1967)	T_{max}, T_{min}, $T_{average}$, e_o, $RH_{average}$, RH_{max}, RH_{min}	$ET_o = 0.035 \times e^o \times (100 - RH_{average})^{0.5}$
Stephens and Stewart (1965)	$T_{average}$, R_a, n, N	$ET_o = (0.014\ T_{average} - 0.37) \times R_s/1500/0.039$ with $R_s = (0.25 + 0.5\ n/N) \times R_a$
Turc (1961-1965).	$T_{average}$, $RH_{average}$, R_a, n, N	RH > 50%: $ET_o = 0.40\ (R_s + 50)\ T_{average}/(T_{average} + 15)$ RH < 50%: $ET_o = 0.40\ (R_s + 50)\ T_{average}/(T_{average} + 15)\ (1 + 50 - RH_{average})/70$ With $R_s = (0.25 + 0.5\ n/N) \times R_a$

where: T_{max}: Maximum air temperature (°C), T_{min}: Minimum air temperature (°C), $T_{average}$: Mean daily air temperature (°C), T_{month}: Average monthly air temperature (°C), ΔT: $T_{max} - T_{min}$ (°C), Δ: Slope of the saturated vapor pressure curve (KPa $°C^{-1}$), γ: Psychrometric constant (KPa·$°C^{-1}$), was set constant and equal to 0.066 (KPa/°C), U_2: Wind speed measured at 2 m height (m/s), RH_{max} and RH_{min}: Maximum and minimum relative humidity of the air (%), e_s: Saturated vapor pressure (KPa), $e_s = 0.5[e_o(T_{max}) + e_o(T_{min})]$, e^o (T_{max}): Saturation vapor pressure (KPa) at T_{max}, e^o (T_{min}): Saturation vapor pressure (KPa) at T_{min}, e_a: Average value of vapor pressure or actual vapor pressure (kPa), $e_a = [e_{oTmax}\ RH_{min} + e_{oTmin}\ H_{max}]/200$. $e_o - e_a$: Saturated vapor pressure deficit of the air (kPa), VPD, R_a and R_s: atmospheric and solar radiation (Mj/m^2/day). R_{ns}: short radiations (Mj/m^2/day). R_{nl}: long radiations (Mj/m^2/day). N: Maximum (Hours/day) sunshine duration, n: Actual sunshine duration (Hours/month), n/N = p: daylight hours monthly/annual daylight hours, i: number of days/month. R_n: net radiation at the crop surface (MJ·m^{-2}·day^{-1}), R_n is computed as the algebraic sum of the net short and long short radiations. K: Constant of Boltzman. ET_0: Reference evapotranspiration (mm/day). G: Soil heat flux density (MJ·m^{-2}·day^{-1}), flux of heat into the soil, set equal to zero to represent the condition of an isolated tree. G_{month}: Monthly soil heat flux density (MJ·m^{-2}), PM: Penman-Monteith equation.

Table 6. Daily ET_0 (mm) estimation procedure according to Penman-Monteith (PM) formula.

Variable	Unit	Formula	
T_{max}	°C		
T_{min}	°C	$T_{average} = (T_{max} + T_{min})/2$	°C
$T_{average}$	°C	Δ	
Altitude	m	γ	
U_2	m/s	$(1 + 0.34\ U_2)$	
		$\Delta/(\Delta + \gamma\,(1 + 0.34\ U_2))$	
		$\gamma/(\Delta + \gamma\,(1 + 0.34\ U_2))$	
		$(900/(T_{average} + 273)) \times U_2$	
Vapor Pressure Deficit (VPD)			
T_{max}	°C	$e^\circ\,(T_{max})$	KPa
T_{min}	°C	$e^\circ\,(T_{min})$	KPa
Saturation vapor pressure: $e_s = [e^\circ\,(T_{max}) + (e^\circ\,(T_{min})]/2$			KPa
e_a derived from air moisture			
RH_{max}	%	$e^\circ\,(T_{min}) \times RH_{max}/100$	KPa
RH_{min}	%	$e^\circ\,(T_{max}) \times RH_{min}/100$	KPa
e_a: average value			KPa
Saturation vapour pressure deficit $(e_s - e_a)$ KPa			
Radiation			
Latitude			
Day		R_a	MJ·m⁻²·day⁻¹
Month		N	Hour
n	Hour	n/N	
if R_s is not given: $R_s = (0.25 + 0.5\ n/N) \times R_a$			MJ·m⁻²·day⁻¹
$R_{so} = [0.75 + 2\ (\text{altitude})/100000] \times R_a$			MJ·m⁻²·day⁻¹
R_s/R_{so}			MJ·m⁻²·day⁻¹
$R_{ns} = 0.77\ R_s$			MJ·m⁻²·day⁻¹
T_{max}	MJ·m⁻²·day⁻¹	$\Delta\ T_{max}.\ K^4$	MJ.m⁻².day⁻¹
T_{min}	MJ·m⁻²·day⁻¹	$\Delta\ T_{min}.\ K^4$	MJ.m⁻².day⁻¹
$(\Delta\ T_{max}.\ K^4 + \Delta\ T_{min}.\ K^4)/2$			MJ·m⁻²·day⁻¹
e_a	KPa	$(0.34 - 0.14\ e_a^{1/2})$	
R_s/R_{so}		$(1.35\ R_s/R_{so} - 0.35)$	
$R_{nl} = (\Delta\ T_{max}.\ K^4 + \Delta\ T_{min}.\ K^4)/2 \times (0.34 - 0.14\ e_a^{1/2}) \times (1.35\ R_s/R_{so} - 0.35)$			MJ·m⁻²·day⁻¹
$R_n = R_{ns} - R_{nl}$			MJ·m⁻²·day⁻¹
T_{mois}	MJ·m⁻²·day⁻¹	G	MJ·m⁻²·day⁻¹
T_{mois}^{-1}	MJ·m⁻²·day⁻¹	$G_{month} = 0.14\ (T_{month} - T_{month} - 1)$	MJ·m⁻²·day⁻¹
$R_n - G$			MJ·m⁻²·day⁻¹
$0.408\ (R_n - G)$			MJ·m⁻²·day⁻¹
Reference evapotranspiration			
$\Delta/(\Delta + \gamma)\,(1 + 0.34\ U_2) \times 0.408\ (R_n - G)$			mm/day
$[\gamma/(\Delta + \gamma)\,(1 + 0.34\ U_2)] \times 900/(T_{average} + 273)) \times U_2 \times (e_s\text{-}e_a)$			mm/day
$ET_0 = \dfrac{0.408\Delta\left(R_n - G\right) + \left[900\,\gamma/\left(T + 273\right) \times U_2 \times \left(e_s - e_a\right)\right]}{\Delta + \gamma\left(1 + 0.34 U_2\right)}$			mm/day

NB: Representative meanings of the variables indicated in **Table 6** are defined in the previous page.

time scales. Gong *et al.*, (2006) [17] performed a sensitivity analysis of ET_0-PM parameters and pointed to the very high influence of solar radiation and relative humidity in accurate estimation of ET_0. Recently, Allen *et al.*, (2011) [25] published a paper in which they present the factors governing measurement accuracy, while Popova *et al.*, (2006) [30] validate the FAO methodology for computing ET_0 with missing climatic data.

2.2.5. Growing Periods, Kc Values, Crop Water Needs and Irrigation Amounts

1) Growing periods

Olive is grown in the Mediterranean region over 270 days-long-period, beginning from March. Flower differenciation occurs from 15 February to 15 March while early fruit growth and pit hardening were observed from end of May to end of June. The ultimate fruit growth and oil synthesis were always observed beginning from 15 September. In order to adapt these stages to the available rainfall data, five growing periods were considered in this study, slightly different from the subdivisions made for the bisannual growing cycle [60] [63]:

    1: March: shoot growth and flower development;

    2: April-June: flowering and early fruit growth;

    3: July and August: fruit development;

    4: September-November: shoot growth, fruit enlargement, oil synthesis and olive maturation;

    5: December-February: quiescence;

2) Crop coefficient (K_c) values

Values of K_c recommended by Allen *et al.*, (1998) [8] for adult olive trees ranged between 0.5 and 0.7. For young trees, Lebourdelles (1977) [65] recommended the use of values of 0.3 for trees aged 1 year and 2 years, 0.4 for trees of 3 - 5 years and 0.5 - 0.7 for adult plants. Recently, Braham and Boussadia (2013) [96] found for Tunisia values ranging between 0.46 and 0.51 (**Table 7**) which were determined by using the sap flow technique of Granier (1985) [91], with an average value of 0.48 in April and 0.47 for May.

3) Crop water needs and irrigation amounts

Crop water needs were determined following the FAO method where $ETc = ET_0 \times K_c \times K_r$. The coefficient K_r was introduced to take into account the soil coverage.

Irrigation water requirements were determined by subtracting the rainfall (P) that contributes to the evapotranspiration process from the estimated ETc.

3. Results

3.1. Spatial Pattern of ET$_0$-PM

The range of annual ET_0-PM varies from 1321.6 mm up to 1570.1 mm (**Table 8**), with maximum value observed in the arid area of Gabes (south-east of Tunisia) characterized by high temperature and radiation levels. Lowest annual ET_0 is recorded at Béja, a continental area of north-western Tunisia. Inversely, the highest seasonal ET_0-PM value is recorded in the continental area of Centre-western Tunisia (Sidi Bouzid), a mountainous area situated at 314 m height, while the lowest value is observed at Bizerte, a coastal and windy town of North Tunisia.

Daily maximum values ranged between 6.5 mm (Bizerte) and 7.1 mm (Sidi Bouzid), while minimums were

Table 7. Values of K_c obtained for olive trees cultivated in the Centre of Tunisia under adequate watering conditions [96].

Mois	Avril				Mai		
Stades phénologiques							
Kc	0,51	0,47	0,47	0,47	0,47	0,46	0,46

recorded in December and January and varied between 1.2 mm (Béja) and 2.2 mm (Sousse) (**Table 9**). Highest daily values of ET_0-PM are those recorded for the arid regions of Gabes and Sidi Bouzid, while the lowest are observed for the East (Bizerte) and West (Béja) northern areas.

The spatial pattern of ET_0 estimated using full datasets show a gradual increase of ET_0 to peak in July (**Figure 4**). Average annual value is well correlated to the maximum and minimum daily estimates, providing a poly

Table 8. Annual and seasonal ET_0-PM (mm).

Site	ET_0-PM (mm/year)	ET_0-PM (mm/season)
Tunis	1423.6	896.5
Nabeul	1484.7	871.1
SB	1485.0	907.1
Sousse	1527.1	879.8
Bizerte	1334.1	839.6
Gabes	1570.1	894.0
Béja	1321.6	885.7

Table 9. Average daily estimates of ET_0 (mm/day) computed with the PM-method.

PM (mm/day)	JAN	FEB	MAR	AVP	MAY	JUN	JULY	AUG	SEP	OCT	NOV	DEC
Tunis	1.65	2.13	2.76	3.59	4.86	6.08	6.99	6.49	4.85	3.30	2.27	1.70
Sidi Bouzid	1.91	2.51	3.15	4.02	5.06	6.18	7.10	6.38	4.91	3.34	2.34	1.81
Sousse	2.18	2.72	3.35	4.02	4.87	5.75	6.64	6.24	5.23	4.01	2.94	2.16
Nabeul	2.05	2.65	3.25	3.96	4.62	5.90	6.63	6.19	5.12	3.54	2.72	2.10
Bizerte	1.60	1.98	2.60	3.38	4.38	5.65	6.54	6.17	4.67	3.10	2.05	1.62
Béja	1.19	1.78	2.32	3.28	4.59	5.93	7.01	6.56	4.83	2.94	1.72	1.16
Gabes	2.37	2.89	3.50	4.31	5.11	5.93	6.55	6.27	5.35	4.05	2.98	2.21
Max	2.37	2.89	3.50	4.31	5.11	6.18	7.10	6.56	5.35	4.05	2.98	2.21
Min	1.19	1.78	2.32	3.28	4.38	5.65	6.54	6.17	4.67	2.94	1.72	1.16

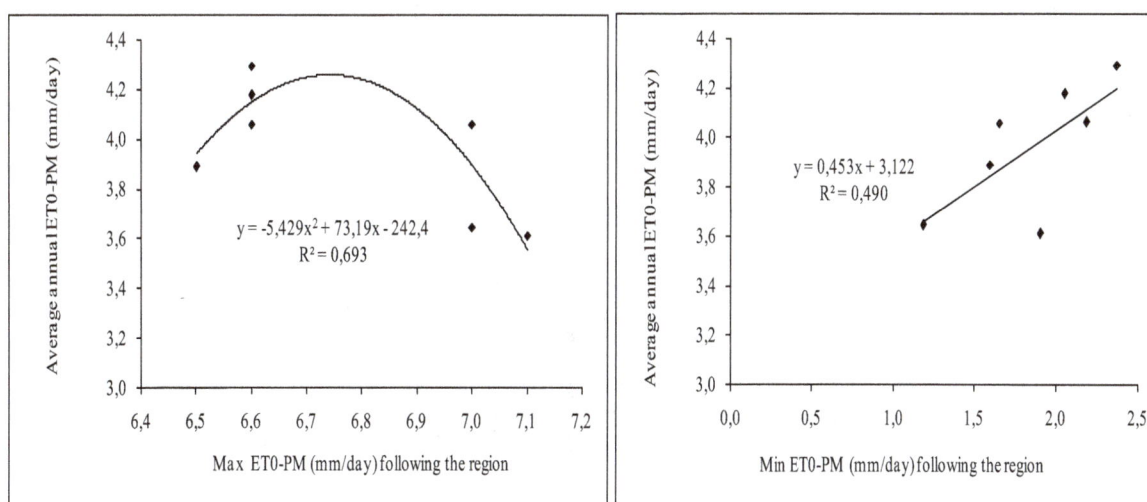

Figure 4. Statistic relationships between the average annual ET_0-PM (mm/day) and the maximum and minimum daily estimates computed for the studied cities ($p = 0.05$).

nomial curve in the first case (r = 0.83) and a linear positive correlation in the second case (r = 0.70). For regions where maximum ET_0-PM (July) exceeds 6.6 mm/day (Tunis = 7 mm; SB = 7.1 mm; Béja = 7.0 mm), minimum values recorded in winter are lower, ranging between 1.2 and 1.9 mm/day (Tunis = 1.7 mm; Sidi Bouzid = 1.9 mm; Béja = 1.2 mm). For the case of Sousse, Nabeul and Gabes, all coastal areas, maximum ET_0-PM is 6.6 mm (July) while minimum values reached 2.2 mm; 2.1 mm and 2.4 mm respectively. This is the result of the proximity of the sea which temperates the climate of the surrounding areas by decreasing the summer ET_0-PM values and increasing the winter ones.

Seasonal values computed following the growing periods show lower and higher estimates for west Tunisia (Beja) during winter-early spring and summer periods, respectively (**Figure 5**). ET_0-PM varies between 50 mm and 70 mm during period 5 (December-February) and reaches 100 mm (Gabes) in March (period 1). Estimates of ET_0 ranged between 100 mm and 180 mm during period 2 (April-June). Lower values recorded during period 1 and 2 are those of Béja and Bizerte. Maximums are recorded in July, exceeding 200 mm/month for all stations. Those recorded in August are slightly lower. During period 5 (September-November), ET_0-PM ranges between 50 mm and 150 mm.

The spatial distribution of ET_0 didn't follow the typical variability stipulating their increase southward and westward due to the decrease in latitude and the increase of altitude, respectively as reported by Razieia and Pereira (2013) [86]. However the spatial pattern of ET_0 computed at Bizerte resemble to that of Tunis, both located in North of Tunisia. Highest daily values were recorded for the arid regions of Gabes and Sidi Bouzid, while the lowest are observed for both the East (Bizerte) and West (Béja) northern areas. During summer months, Tunis and Béja present the highest values although they are situated at different latitudes and altitudes.

3.2. ET_0 Estimates Following the Empirical Formulas Compared to PM-Computations

The performance of empirical methods against the PM-ET_0 estimates in all considered sites are evaluated graphically (**Figure 6**) and statistically (**Table 11**). Significant positive correlations are observed with r values exceeding 0.88.

To assess the performance of these methods with respect to ET_0-PM, relative to all values for each station, the r values were determined. When the coefficient of correlation r is close to 1.0, most of the variation of the observed values can be explained by the linear model.

Variations between values are due to site characteristics and the formula used for ET_0 estimation. Values of r ranged between 0.880 and 0.999 (**Table 10**). The Eagleman method gave the highest values out of the range of those given by all the other formulas, while the lowest were provided by the HS formula. The lowest r coefficients were obtained for the cities of Sousse (r = 0.888) and Gabes (r = 0.885) when ET_0 is estimated by using the PO and Ivanov methods, respectively. These statistical coefficients are site specific even for the same species (e.g., olives) and the function for one orchard could not be used for the other.

Results show that when the series of climatic data is incomplete, ET_0 can be estimated by another empirical method depending on the available climatic variables. Formula that gives values of ET_0 approximating ET_0-PM—the universal reference estimating ET_0 method for each site are reported in **Table 7** and **Table 8**. Results show that methods of ST and SW are suitable for all stations. For Tunis, Nabeul and Sidi Bouzid, the method of Chr.

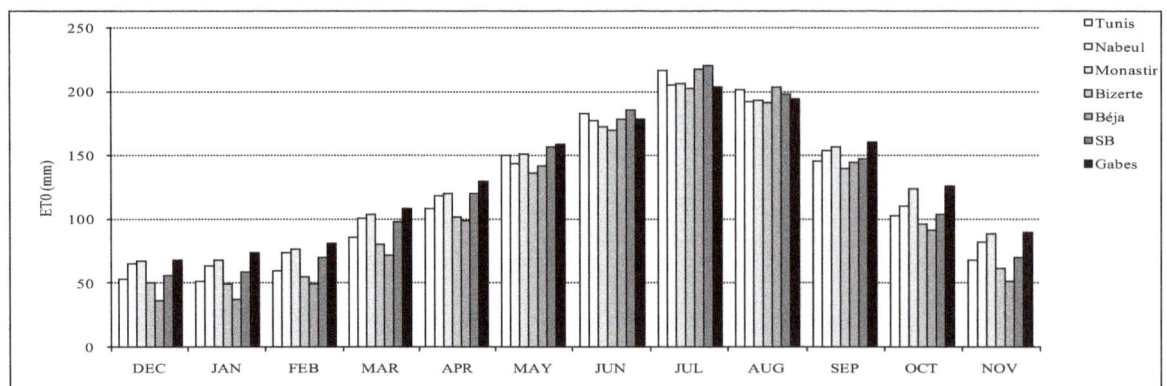

Figure 5. Values of ET_0 following the growing period and site of olive cultivation.

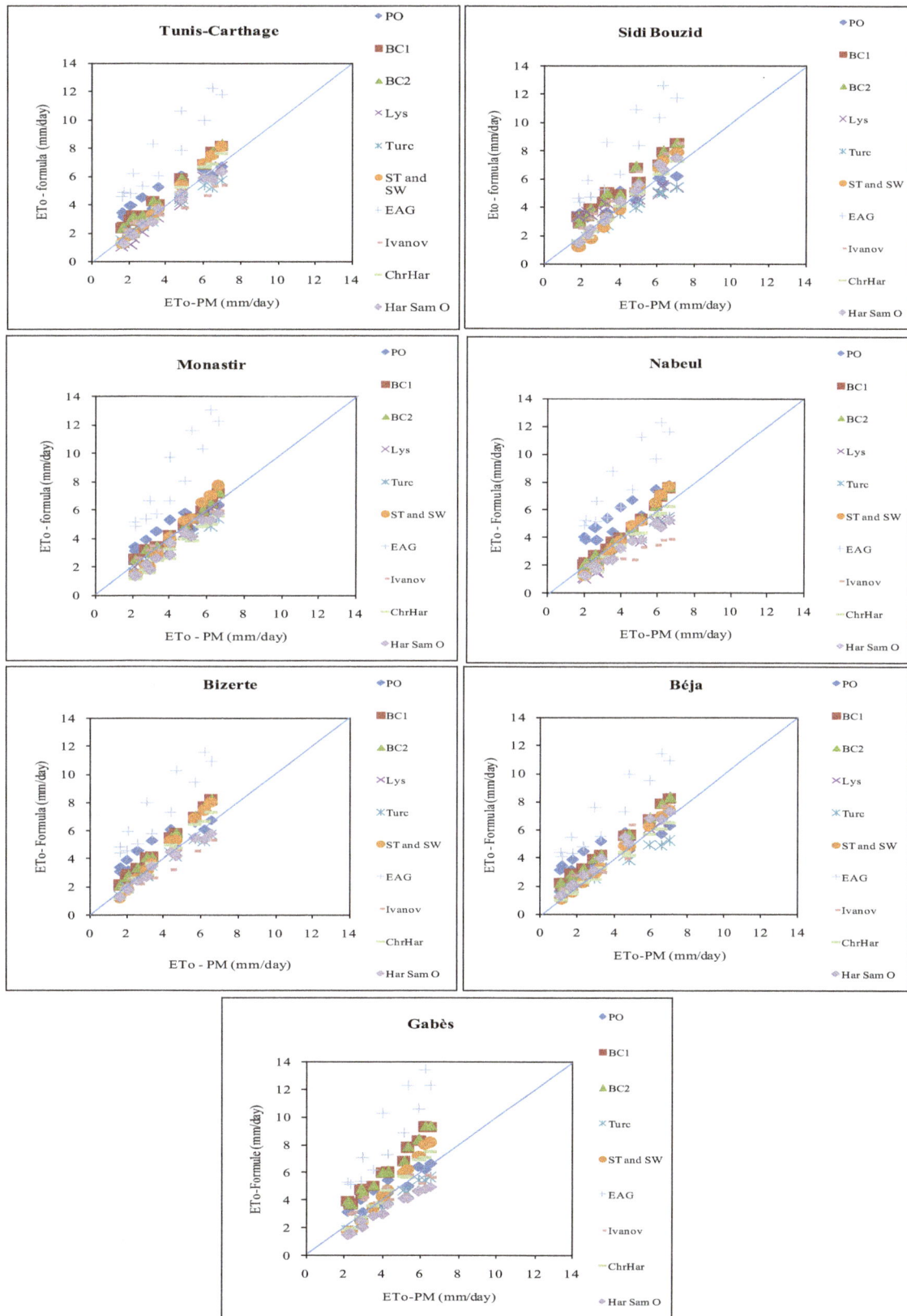

Figure 6. Relationship between ET_0-PM estimates and values of ET_0 determined by different calculation methods (mm/day) for all studied sites (p = 0.05).

Table 10. Coefficients of correlation obtained for the different sites. ET_0 estimated by several methods are compared to ET_0-PM computations.

	PO	BC1	BC2	Turc	ST and SW	EAG	Ivanov	Chr Harg	Harg Sam
Bizerte	0.918	0.999	0.999	0.988	0.999	0.935	0.961	0.996	0.988
Tunis	0.926	0.994	0.994	0.992	0.999	0.939	0.962	0.997	0.990
Béja	0.900	0.997	0.997	0.990	0.999	0.949	0.952	0.996	0.992
Gabes	0.926	0.991	0.991	0.979	0.995	0.909	0.885	0.991	0.970
Sousse	0.888	0.993	0.993	0.981	0.996	0.932	0.958	0.989	0.983
Nabeul	0.927	0.993	0.993	0.990	0.997	0.907	0.950	0.995	0.990
Sidi Bouzid	0.921	0.980	0.980	0.993	0.997	0.922	0.973	0.999	0.995

Harg. gave valuable results as well as that of BC for Bizerte, Béja, Gabes and Sousse, which are coastal regions.

However, the method of calculating the best ET_0 values changes following the growing period. The formula that gives values of ET_0 approximating ET_0-PM for each city (**Table 11** and **Table 12**) indicates:

For Nabeul and Monastir, the BC formula is valuable for most periods of growth.

For Sidi Bouzid, the HS is valuable for stages 1, 4 and 5 and that of Chr. Harg. for stages 2 and 3 and the overall growing cycle.

For Bizerte, the method of ST and SW gives values of ET_0 approximating those of ET_0-PM for the overall growing cycle and stages 1 and 4.

The regions of Tunis, Sousse, Nabeul, Bizerte and Gabes are localized on the coast but a different 'response' is observed for a given stage of development.

For Bizerte, Tunis, Béja, Gabes, Sousse and Nabeul, the ST and SW method is noted as the best one for the overall cycle and for stages 4 for the first stations and 2 for Nabeul.

Linearly with ET_0-PM, the Ivanov method appropriately predicts ET_0 in all sites of Tunisia situated in the North and costal areas as it gives the lowest relative variation of ET_0. **Table 12** suggests that the Ivanov method particularly performs well with the cities of Tunis, Nabeul and Bizerte characterized by semi arid climate, having annual rainfall of up to 450 mm. But when considering the lowest seasonal climatic deficit (P-ETc), it appears that the Turc formula is most appropriate for Tunis but also for Sousse, a coastal area of centre Tunisia. The Turc method also performs reasonably well in western areas, *i.e.* at Sidi Bouzid (arid climate) and Béja (semi-arid climate). Low climatic deficits were recorded at seasonal scale at both sites. For Gabes, the most appropriate method is that of Harg. Sam.

It appears from these results that the most appropriate method for ET_0 estimation at annual scale may be different from that providing the best value of ET_0 when considering the growing periods. Though, it is important to well define the researched objective. Indeed, the formula of Turc although it gave the best response for Sousse when estimating the annual ET_0 value, it wasn't the appropriate method when considering the growing periods. Adversely, although the method of BC is suitable for periods 1, 2, 4 and 5 at Sousse, the appropriate method for the overall cycle is that of Turc. This formula (Turc) is appropriate for Tunis, Sidi Bouzid, Sousse and Béja at annual scale despite of their appartenance to different climatic regions, while the method of Ivanov is quite valuable for Bizerte and Nabeul. At seasonal scale, the formula of BC is appropriate for stages 1, 4 and 5 for Tunis, Nabeul and Monsatir. That of Turc is valuable for stages 1 and 5 at Béja.

3.3. Crop Water Needs

Average *ETc* values are calculated for each site following age and the growing period (**Table 13**). The annual course of *ETc* computed by using ET_0-PM values shows minimum estimates close to 300 mm per year for young trees and maximum values ranging between 800 mm and 1000 mm for the older ones. Highest annual values of *ETc* are those recorded for Gabes, ranging between 335 mm and 894 mm depending on age. Minimum yearly values are those of Béja varying from 282 mm and 752 mm, due to ET_0 variations.

Water requirements (**Table 13**) increases during the first period of growth (March) from 15 mm to 40 mm for

Table 11. Appropriate methods for ET_0 calculation following the region and the growing period. Costal and northern cities.

	Tunis	Nabeul	Sousse	Bizerte
Rainfall amount (mm/year) and Ecartype	449.7/122.4	468.1/160.4	328.1/137.8	634.6/153.3
ET_0-PM (mm/year)	1424	1485	1527	1334
Frequency of water recovery following the stage of development	March: 40% April-June: 1% July-August: 0% Sept-Nov.: 33% Dec-Feb.: 91% Annual: 1%	March: 28% April-June: 0% July-August: 0% Sept-Nov.: 39% Dec-Feb.: 80% Annual: 14%	March: 17% April-June: 0% July-August: 0% Sept-Nov.: 24% Dec-Feb.: 36% Annual: 0%	March: 56% April-June: 0% July-August: 0% Sept-Nov.: 59% Dec-Feb.: 99% Annual: 19%
Lower relative variation of ET_0 in comparison with PM	Ivanov: +0.3% Turc: +12.4%	BC: −4.2% Ivanov: +28%	Ivanov: −1.8% Turc: +20.8%	Chr. Harg. = +1.6% Ivanov: +3%
Formula giving the lowest seasonal deficit	Ivanov Turc	Ivanov Harg Sam	Chr Harg = Turc	Ivanov Turc-Harg Sam
P-ETc > 0 for 1 - 2 years (annual)	All methods except EAG	All methods except PO and EAG	All methods except EAG	All methods
P-ETc > 0 for 3 - 5 years (annual)	All methods except EAG–PO-BC	Turc-Ivanov Chr Harg Harg Sam	Turc Chr Harg Harg Sam	All methods except EAG
P-ETc > 0 for 6 - 10 years (annual)	Turc Chr Harg	Ivanov	-	All methods except EAG, PO et BC
Adopted formula	**Turc**	**Ivanov**	**Turc**	**Ivanov**
r value correlation with PM	0.992	0.950	0.981	0.961
Best correlation with PM	ST and SW r = 0.999	ST and SW r = 0.997	ST and SW r = 0.996	ST and SW et BC r = 0.999
Adopted formula following the stage of development (low relative variability)	Stage 1: Chr Harg. Stage 2: Harg Sam O. Stage 3: PO Stage 4: ST and SW Stage 5: Turc	Stage 1: BC Stage 2: ST and SW Stage 3: Chr Harg Stage 4: BC Stage 5: BC	Stage 1: BC Stage 2: BC Stage 3: PO Stage 4: BC Stage 5: BC	Stage 1: ST and SW Stage 2: Harg Sam Stage 3: PO Stage 4: ST and SW Stage 5: Turc
Adopted formula	**Turc** HR > 50% $ET_o = \dfrac{0.40(R_s+50)T_m}{(T_m+15)}$ HR < 50% $ET_o = \dfrac{0.40(R_s+50)T_m}{(T_m+15)(1+50-HR)/70}$ $R_s = (0,.25 + 0.5\, n/N) \times R_a$	**Ivanov** $ET_0 = 0.0018 \times (T + 25)^2$ $\times (100 \times e/e^o\ 100)$	**Turc** HR > 50% $ET_o = \dfrac{0.40(R_s+50)T_m}{(T_m+15)}$ HR < 50% $ET_o = \dfrac{0.40(R_s+50)T_m}{(T_m+15)(1+50-HR)/70}$ $R_s = (0.25 + 0.5\, n/N) \times R_a$	**Ivanov** $ET_0 = 0.0018 \times (T_m + 25)^2$ $\times (100 - e/e^o\ 100)$
Variables involved	T_m, $RH_{average}$, R_a, n, N	T, e, e° $e^o(T) = 0.6108 \exp(17.27 \times T)/(T + 237.3)$ $e = e^o \times RH_{average}/100$	T_m, $RH_{average}$, R_a, n, N	T, e, e°

Table 12. Appropriate methods for ET_0 calculation following the region and the growing period. Continental and southern cities.

	Sidi Bouzid	Gabes	Béja
Rainfall amount (mm/year) and Ecartype	251.8/121.4	190.1/98.7	628.9/154.8
ET_0-PM (mm/year)	1485	1570	1322
Frequency of water recovery following the stage of development	March: 17% April-June: 0% July-August: 0% Sept-Nov: 8% Dec-Feb: 5% Annual: 7%	March: 5% April-June: 0% July-August: 0% Sept-Nov: 4% Dec-Feb: 14% Annual: 7%	March: 73% April-June: 4% July-August: 0% Sept-Nov: 48% Dec-Feb: 99% Annual: 18%
Lower relative variation of ET_0 in comparaison with PM	Harg Sam O.: +12.5% Turc: +20.4%	Chr Harg: −1.4% Harg. Sam: +23.9%	ST and SW: +2.4% Turc: +9.3%
Formula giving the lowest seasonal deficit	**Turc**	**Harg Sam** **Turc**	**Turc** **Chr Harg**
P-$ETc > 0$, 1 - 2 years (annual)	-	-	All formula
P-$ETc > 0$, 3 - 5 years (annual)	-	-	All formula except EAG
P-$ETc > 0$, 6 - 10 years (annual)	-	-	All formula except EAG, PO et BC
Adopted formula	**Turc**	**Harg Sam**	**Turc**
r value correlation with PM	0.993	0.970	0.990
Meilleure corrélation avec PM	Chr Harg r = 0.999	ST and SW r = 0.995	ST and SW r = 0.999
Adopted formula following the stage of development (low relative variability)	Stage 1: Harg Sam Stage 2: Chr Harg Stage 3: Chr Harg Stage 4: Harg Sam Stage 5: Harg Sam	Stage 1: Chr Harg Stage 2: Turc Stage 3: PO Stage 4: ST and SW Stage 5: Chr Harg	Stage 1: **Turc** Stage 2: Chr Harg Stage 3: Harg Sam Stage 4: ST and SW Stage 5: **Turc**
Adopted formula	**Turc** HR > 50% $ET_o = \dfrac{0.40(R_s + 50)T_m}{(T_m + 15)}$ HR < 50%: $ET_o = \dfrac{0.40(R_s + 50)T_m}{(T_m + 15)(1 + 50 - HR)/70}$ $R_s = (0.25 + 0.5\, n/N) \times R_a$	**Hargreaves-Samani** Original: $ET_o = 0.0023\, R_a\, \Delta T^{0,5} \times (T_m + 17.8)$ Modified: $ET_o = 0.0035\, R_a\, \Delta T^{0,5} \times (T_m + 12.54)$	**Turc** HR > 50% $ET_o = \dfrac{0.40(R_s + 50)T_m}{(T_m + 15)}$ HR < 50%: $ET_o = \dfrac{0.40(R_s + 50)T_m}{(T_m + 15)(1 + 50 - HR)/70}$ $R_s = (0.25 + 0.5\, n/N) \times R_a$
Variables involved	T_m, $RH_{average}$, R_a, n, N	T_m, R_a, T_{max}, T_{min}, R_a	T_m, $RH_{average}$, R_a, n, N

Béja following age, from 21 mm to 55 mm for Sidi Bouzid, from 18 mm to 48 mm for Tunis, from 21 mm to 46 mm for Nabeul, from 22 mm to 52 mm for Sousse, from 17 mm to 45 mm for Bizerte and from 23 mm and 61 mm for Gabes. For the other periods (April to November, 2 to 4), water needs increased consistently with values ranging between 87 mm and 254 mm for the coastal areas and from 90 mm to 266 mm for the continental locations.

Minimum *ETc* values were recorded when the formulas of Turc, Chr. Harg, Harg.Sam and Ivanov are used.

Table 13. Water requirements of olive trees (*ETc*, mm) following age, site, growing periods and methods of ET_0 computation.

Bizerte	1	2	3	4	5	PM	PO	BC	Turc	STand SW	EAG	Ivanov	Chr-Harg	Harg-Sam O.
1 - 2 years	17	87	83	64	34	285	374	366	262	315	575	259	295	261
3 - 5 years	23	116	110	85	45	380	499	488	350	419	766	346	394	347
6 - 10 years	30	156	148	114	60	508	668	653	468	562	1026	463	527	465
>10 years	45	233	221	170	90	759	997	975	699	839	1532	691	787	695
Sousse	**1**	**2**	**3**	**4**	**5**	PM	PO	BC	Turc	STand SW	EAG	Ivanov	Chr-Harg	Harg-Sam O.
1 - 2 years	22	95	84	79	46	326	363	349	262	322	647	320	243	265
3 - 5 years	29	127	112	106	61	435	483	465	349	429	863	426	324	353
6 - 10 years	39	170	150	142	82	582	648	623	467	574	1155	571	434	473
>10 years	58	254	224	211	122	870	967	931	698	858	1725	852	649	707
Béja	**1**	**2**	**3**	**4**	**5**	PM	PO	BC	Turc	STand SW	EAG	Ivanov	Chr-Harg	Harg-Sam O.
1 - 2 years	15	90	88	62	27	282	360	362	239	285	556	315	260	312
3 - 5 years	20	120	118	82	36	376	481	483	319	380	741	420	347	416
6 - 10 years	27	160	158	110	48	504	644	647	428	509	992	562	464	558
>10 years	40	240	236	165	72	752	961	966	639	760	1482	839	693	833
Nabeul	**1**	**2**	**3**	**4**	**5**	PM	PO	BC	Turc	STand SW	EAG	Ivanov	Chr-Harg	Harg-Sam O.
1 - 2 years	21	94	83	74	44	317	378	497	291	366	678	353	353	258
3 - 5 years	28	126	111	99	59	423	505	662	389	488	904	471	470	344
6 - 10 years	38	168	149	132	79	566	676	887	521	654	1211	631	630	461
>10 years	56	251	223	198	118	846	1009	1325	777	977	1809	942	941	688
Tunis	**1**	**2**	**3**	**4**	**5**	PM	PO	BC	Turc	STand SW	EAG	Ivanov	Chr-Harg	Harg-Sam O.
1 - 2 years	18	95	88	68	36	304	370	371	263	320	602	279	307	277
3 - 5 years	24	126	117	90	48	405	494	495	351	426	803	373	410	369
6 - 10 years	32	169	157	121	64	543	661	663	470	571	1075	499	549	494
>10 years	48	252	234	181	95	810	987	989	701	852	1606	745	819	738
Sidi Bouzid	**1**	**2**	**3**	**4**	**5**	PM	PO	BC	Turc	STand SW	EAG	Ivanov	Chr-Harg	Harg-Sam O.
1 - 2 years	21	99	88	69	41	317	362	420	254	312	611	316	302	291
3 - 5 years	27	132	117	92	54	423	483	560	338	417	815	422	402	389
6 - 10 years	37	177	157	123	72	566	647	749	453	558	1092	565	539	520
>10 years	55	265	234	184	108	845	966	1119	676	833	1630	843	805	777
Gabes	**1**	**2**	**3**	**4**	**5**	PM	PO	BC	Turc	STand SW	EAG	Ivanov	Chr-Harg	Harg-Sam O.
1 - 2 years	23	100	83	81	49	335	378	497	291	366	678	353	353	258
3 - 5 years	30	133	111	107	65	447	505	662	389	488	904	471	470	344
6 - 10 years	41	178	149	144	87	599	676	887	521	654	1211	631	630	461
>10 years	61	266	223	215	130	894	1009	1325	777	977	1809	942	941	688

Maximum estimates are those obtained by the Eagleman formula.

Lysimetric values determined by Nasr (2002) [1] representing the effective need of water are significantly lower than ETc computed with the PM-formula (ETc-PM) for Tunis, Nabeul, Sousse and Gabes, which are coastal areas. Annual ratios between the lysimetric values and ETc ranged between 0.74 and 0.96 for these areas and approximate the unit for both Béja and Sidi Bouzid as shown in **Table 14**.

3.4. Water Deficit and Irrigation Amounts Following Location

Annual and seasonal values of (P-ETc) computed for all sites are reported in **Tables 15-21**. Negative values represent the amount of water needed by trees over the year or the irrigation period (May-September) to complement rainfall. Amounts of water available for the crop are designed by the sign (+). Details for all sites are decribed as follows:

Tunis: At annual scale all methods allow recovery of the crop water needs of trees aged one to five years except that of Eagleman. But this result is not suitable for such plantations at seasonal scale. Indeed, all values of (P-ETc) are negative even for the youngest orchards. The lowest difference between rainfall R and ETc is recorded with the Ivanov method (-72 mm). Thus the seasonal rainfall amounts are not suffisant to meet the crop water needs for all tranches of age. The method of Turc, recorded previously as the most appropriate for this region allow recovery of the crop water needs of olive plantations aged one to ten years at annual scale and provide low differences between R and ETc at seasonal scale. The amount of water needed at seasonal scale varies from 76 mm to 344 mm depending on age.

Nabeul: At annual scale all methods allow recovery of the crop water needs of trees aged one to two years except those of Eagleman and PO. All seasonal values of (P-ETc) are negative. The Ivanov method provides the lowest water deficit. Irrigation is needed for all kind of olive plantations from May to September. The amount of

Table 14. Ratio between the lysimetric values and ETc (mm) following the site and the growing period.

	M	A-J	Jt-A	S-N	D-F	Annual
	1	2	3	4	5	Cycle
Tunis	0.74	0.96	0.92	0.80	0.64	0.86
Nabeul	0.83	0.86	0.82	0.70	0.56	0.77
Sousse	0.83	0.90	0.93	0.93	0.77	0.89
Gabes	0.85	0.92	0.92	0.77	0.68	0.84

Table 15. Annual and Seasonal water deficits (P-ETc, mm) at the site of Tunis.

Annual water needs (mm)									
Method Age (Year)	PM	PO	BC	Turc	ST and SW	EAG	Ivanov	Chr Harg	Har Sam O.	Lysimeter
1 - 2	+178.2	+111.8	+111.0	+219.0	+162.5	−120.2	+202.5	+174.7	+205.1	+220.6
3 - 5	+76.9	−11.6	−12.7	+131.3	+56.0	−321.0	+109.3	+72.3	+112.8	+133.5
6 - 10	−60.5	−179.1	−180.6	+12.4	−88.5	−593.4	−17.1	+66.8	−12.5	+15.3
>10	−382.2	−505.2	−507.4	−219.3	−370.0	−1123.9	−263.3	−337.5	−256.4	−215.0
Seasonal water needs (mm)									
Method Age (Year)	PM	PO	BC	Turc	ST and SW	EAG	Ivanov	Chr Harg	Har Sam O.	Lysimeter
1 - 2	−105.6	−109.3	−138.9	−76.2	−131.4	−256.5	−71.9	−118.9	−91.0	−92.2
3 - 5	−169.1	−174.1	−213.6	−129.9	−203.5	−370.3	−124.2	−186.9	−149.7	−151.3
6 - 10	−255.3	−262.1	−314.9	−202.8	−301.3	−524.8	−195.2	−279.1	−229.3	−231.4
>10	−423.1	−433.3	−512.1	−344.8	−491.9	−825.7	−333.4	−458.7	−384.4	−387.5

Table 16. Annual and Seasonal water deficit (P-*ETc*, mm) at the site of Nabeul.

Annual water needs (mm)										
Method Age (Year)	PM	PO	BC	Turc	ST and SW	EAG	Ivanov	Chr Harg	Har Sam O.	Lysimeter
1 - 2	**+71**	−46	**+52**	**+136**	**+83**	−227	**+175**	**+123**	+150	**+145**
3 - 5	−35	−191	−60	**+52**	−19	−432	**+103**	**+35**	+70	**+64**
6 - 10	−178	−387	−211	−62	−157	−710	+7	−85	−37	**−46**
>10	−458	−770	−507	−285	−425	−1251	−181	−318	−247	**−260**
Seasonal water needs (mm)										
Method Age (Year)	PM	PO	BC	Turc	ST and SW	EAG	Ivanov	Chr Harg	Har Sam O.	Lysimeter
1 - 2	−115	−154	−133	−83	−133	−270	−40	−101	−77	−81
3 - 5	−177	−229	−201	−134	−201	−383	−77	−157	−126	−132
6 - 10	−261	−330	−292	−203	−293	−537	−127	−234	−192	−200
>10	−424	−528	−471	−337	−472	−836	−224	−385	−321	−334

Table 17. Annual and Seasonal water deficit (P-*ETc*, mm) at the site of Sidi Bouzid.

Annual water needs (mm)										
Method Age (Year)	PM	PO	BC	**Turc**	ST and SW	EAG	Ivanov	Chr Harg	Harg Sam O.	Lysimeter
1 - 2	−71	−116	−174	−8	−66	−365	−138	−63	−83	−87
3 - 5	−177	−237	−314	−92	−171	−569	−266	−166	−193	−198
6 - 10	−320	−401	−503	−207	−312	−846	−440	−306	−341	−348
>10	−599	−720	−873	−430	−587	−1384	−778	−579	−631	−641
Seasonal water needs (mm)										
Method Age (Year)	PM	PO	BC	Turc	ST and SW	EAG	Ivanov	Chr Harg	Har Sam O.	Lysimeter
1 - 2	−113	−103	−157	−74	−131	−272	−148	−116	−129	−86
3 - 5	−177	−164	−236	−125	−201	−389	−224	−181	−199	−141
6 - 10	−264	−247	−343	−195	−297	−548	−327	−269	−294	−216
>10	−434	−409	−552	−330	−483	−858	−527	−442	−478	−362

water needed at seasonal scale varies from 40 mm to 224 mm depending on age.

Sidi Bouzid: The deficit of water is recorded at both annual and seasonal scales. Rainfall amounts are not suffisant to cover the crop water needs even those of one and two years old plantations. Irrigation is thus needed for all kinds of olive orchards. The method giving the lowest value of (P-*ETc*) is that of Turc. The amount of water needed at seasonal scale varies from 74 mm to 330 mm depending on age.

Sousse: At annual scale all methods allow recovery of the crop water needs of trees aged one and two years except that of Eagleman. At seasonal scale, the deficit is present for all tranches of age with lowest differences recorded for both methods: Turc and Chr. Harg. Seasonal rainfall is not suffisant to cover the crop water needs and irrigation is needed from May to September with amounts ranging between 95 mm and 351 mm depending on age.

Gabès: Water deficit is present at seasonal and annual scale. The rainfall amounts are not high enougth to meet the crop water needs even for young plantations. The most valuable method giving the lowest deficit is that of Harg. Sam. Irrigation is requested at seasonal scale with amounts ranging between 128 mm and 374 mm depending on tree age.

Table 18. Annual and Seasonal water deficit (P-*ETc*, mm) at the site of Soussse.

Annual water needs (mm)									
Method Age (Year)	PM	PO	BC	Turc	ST and SW	EAG	Ivanov	Chr Harg	Harg Sam O.	Lysimeter
1 - 2	+50	+13	+27	+114	+54	−271	+56	+133	+111	+85
3 - 5	−59	−107	−89	+27	−53	−487	−50	+52	+23	−12
6 - 10	−206	−272	−247	−91	−198	−779	−195	−58	−97	−144
>10	−494	−591	−555	−322	−482	−1349	−476	−273	−331	−401
Seasonal water needs (mm)										
Method Age (Year)	PM	PO	BC	Turc	ST and SW	EAG	Ivanov	Chr Harg	Harg Sam O.	Lysimeter
1 - 2	−130	−130	−140	−96	−151	−303	−114	−95	−105	−120
3 - 5	−192	−193	−205	−147	−220	−423	−170	−145	−159	−179
6 - 10	−277	−277	−294	−216	−314	−585	−247	−214	−232	−259
>10	−442	−442	−468	−351	−497	−902	−398	−347	−375	−415

Table 19. Annual and Seasonal water deficit (P-*ETc*, mm) at the site of Gabes.

Annual water needs (mm)									
Method Age (Year)	PM	PO	BC	Turc	ST and SW	EAG	Ivanov	Chr Harg	Har Sam O.	Lysimeter
1 - 2	−134	−177	−296	−90	−165	−477	−152	−152	−57	−82
3 - 5	−246	−304	−461	−188	−287	−703	−270	−269	−143	−176
6 - 10	−398	−475	−686	−320	−453	−1010	−430	−429	−260	−304
>10	−693	−808	−1124	−576	−776	−1608	−741	−740	−487	−553
Seasonal water needs (mm)										
Method Age (Year)	PM	PO	BC	Turc	ST and SW	EAG	Ivanov	Chr Harg	Har Sam O.	Lysimeter
1 - 2	−171	−178	−254	−148	−212	−355	−157	−196	−128	−151
3 - 5	−235	−244	−346	−204	−289	−480	−216	−267	−177	−208
6 - 10	−321	−333	−469	−279	−394	−649	−295	−364	−244	−285
>10	−488	−507	−710	−427	−598	−978	−450	−553	−374	−435

Bizerte: At annual scale all methods allow recovery of the crop water needs of trees aged one to five years except that of Eagleman. Other methods like that of Turc allow also recovery of water needs of older trees, to ten years old. Thus, irrigation is needed during the fruit growth season with amounts ranging between 72 mm and 323 mm depending on age, computed with the method of Ivanov wich provid the lowest deficits.

Béja: At annual scale all methods allow recovery of the crop water needs of trees aged one to five years except that of Eagleman. The most appropriate formula is that of Turc providing the lowest deficit. Irrigation should be applied during summer months from May to September with amounts of water ranging between 49 mm and 296 mm depending on tree age.

The method of Turc appears as the most appropriate for Tunis, Sousse, Sidi Bouzid and Béja while the method of Ivanov is adaquate for Bizerte and Nabeul. For Gabes the method of Harg Sam gave the most adequate values. The main climatic data requested are temperature, humidity and insolation, which are available in most stations. These formulas can be used specifically for these stations when the climatic data are not available to compute ET_0-PM, particularly the solar radiation.

Table 20. Annual and Seasonal water deficit (P-*ETc*, mm) at the site of Bizerte.

Annual water needs (mm)										
Method Age (Year)	PM	PO	BC	Turc	ST and SW	EAG	Ivanov	Chr Harg	Har Sam O.	Lysimeter
1 - 2	+360	+271	+279	+383	+330	+70	+386	+350	+384	
3 - 5	+256	+146	+157	+295	+226	−121	+299	+251	+298	
6 - 10	+137	−23	−8	+177	+83	−381	+182	+118	+180	
>10	−114	−352	−330	−54	−194	−887	−46	−142	−50	
Seasonal water needs (mm)										
Method Age (Year)	PM	PO	BC	Turc	ST and SW	EAG	Ivanov	Chr Harg	Har Sam O.	Lysimeter
1 - 2	−100	−119	−147	−84	−136	−245	−72	−119	−86	
3 - 5	−160	−185	−222	−139	−208	−352	−123	−184	−141	
6 - 10	−241	−274	−324	−212	−305	−498	−191	−273	−216	
>10	−398	−448	−523	−355	−494	−783	−323	−446	−361	

Table 21. Annual and Seasonal water deficit (P-*ETc*, mm) at the site of Béja.

Annual water needs (mm)										
Method Age (Year)	PM	PO	BC	Turc	ST and SW	EAG	Ivanov	Chr Harg	Har Sam O.	Lysimeter
1 - 2	+296	+218	+216	+339	+293	+22	+263	+318	+266	+288
3 - 5	+202	+97	+95	+259	+198	−163	+158	+231	+162	+192
6 - 10	+74	−66	−69	+150	+69	−414	+16	+114	+20	+61
>10	−174	−383	−388	−61	−182	−904	−261	−115	**−255**	**−195**
Seasonal water needs (mm)										
Method Age (Year)	PM	PO	BC	Turc	ST and SW	EAG	Ivanov	Chr Harg	Har Sam O.	Lysimeter
1 - 2	−88	−86	−124	−49	−97	−220	−101	−75	**−105**	−92
3 - 5	−151	−148	−199	−98	−162	−326	−168	−133	**−173**	−156
6 - 10	−236	−232	−300	−165	−251	−471	−258	−212	**−265**	−243
>10	−402	−395	−498	−296	−424	−753	−435	−367	**−445**	−413

Values of (*ETc*-P) computed for the irrigation season, from May to September, are negatives. The deficit of water is present even for young plantations. Rainfall amounts were insuffisant to meet the crop water needs of this species during the period of fruit growth. These amounts should be supplied by irrigation. The lowest deficit is observed at Béja and the highest at Gabes. At annual scale, values of (*ETc*-P) are positive for Tunis, Nabeul, Sousse, Bizerte and Beja when young olive plantations are considered. For older trees, values are positive only for the northern areas of Tunis, Bizerte and Béja. This last location is the only case where water needs are covered for olive trees aged 6 to 10 years. So there is a need for irrigating olive plantations aging more than 5 years and especially when olive is cultivated in the western areas. Irrigation is needed during the growing fruit period but also during the other seasons, when shoots grow.

3.5. Rainfall Distribution Functions and Recovery of Crop Water Needs

Rainfall distribution functions were established for each city following the growing periods (**Figure 7**).

Rainfall distribution functions present different evolutionary. For rainy regions like Béja, Nabeul and Bizerte,

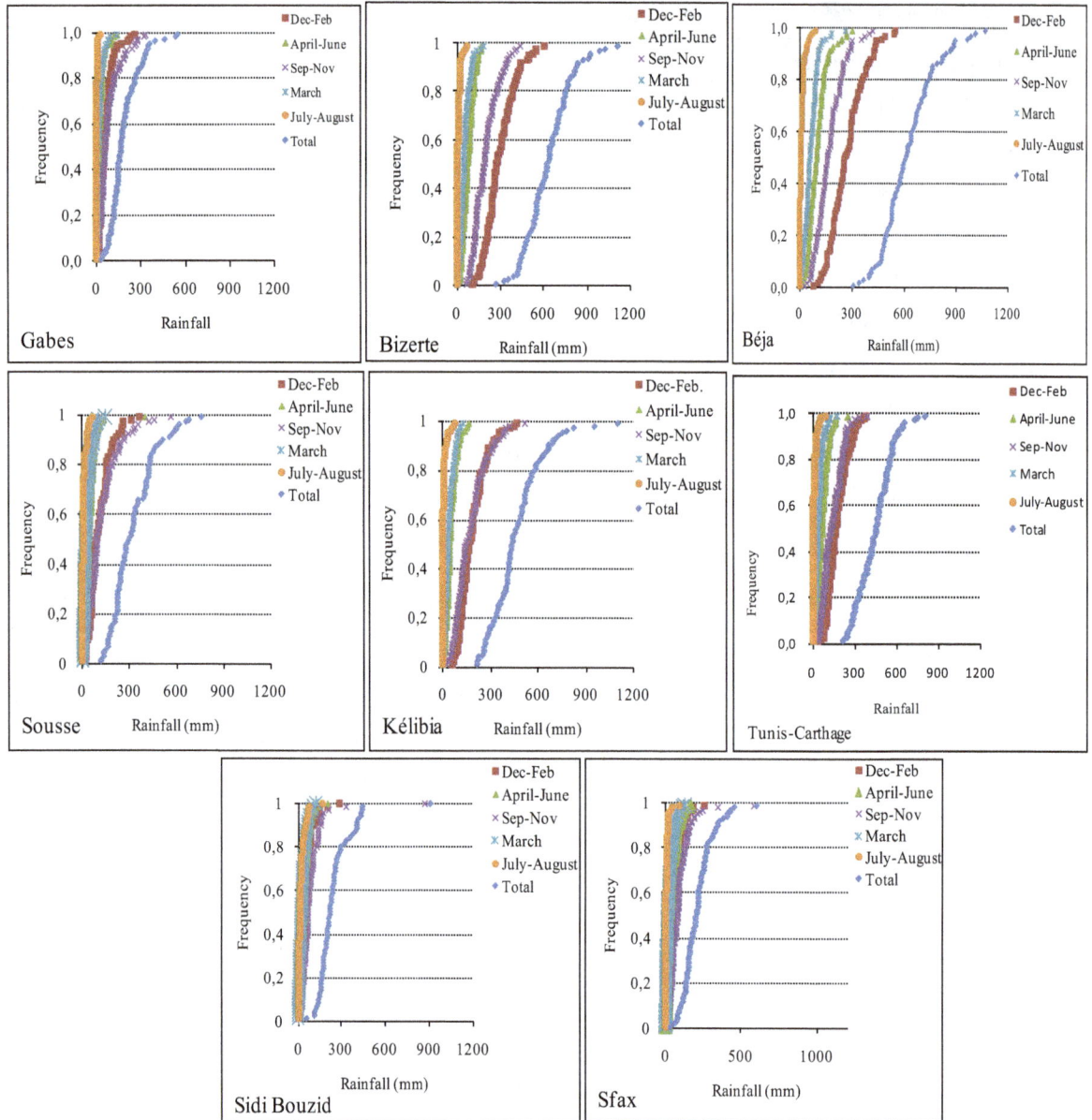

Figure7. Rainfall distribution functions following site of olive cultivation and the growing periods.

the curves looks separate, while those of Gabes and Sidi Bouzid are close one another. On the basis of these seasonal and annual distributions of rainfall, the frequencies of non-satisfactory were determined following tree age and location. Values are reported in **Table 22**. As shown in the following table, the frequencies of non-satisfactory (F) vary consistently from location to another and following the growing period. During July-August, F values exceed 0.9 in all cases and only 10% of water needs are supplied by rainfall even for the rainiest areas like Bizerte. For young trees aged one to five years, rainfall amounts meet excactly the crop water needs at Béja and Bizerte. Rainfall amounts cover 60% to 85% *ETc* of trees aged 6 to 10 years at Bizerte and Béja and trees aged one to five years at Nabeul and Tunis. About 50% *ETc* of one year old trees are covered by rainfall at Sousse. Less than 20% of water needs are covered at Bizerte and Béja for old trees aged more than 10 years, at Nabeul and Tunis for trees aged 6 to 10 years, at Sousse for trees aging 3 to 5 years and at Sidi Bouzid for of one year and two years old trees.

It appears from these results based on the seasonal rainfall frequencies and water needs computed with the PM formula that irrigation supply is necessary all time for trees aging more than 10 years even for the rainiest

Table 22. Frequencies of non-satisfactory of water needs following age and location.

Bizerte	1	F	2	F	3	F	4	F	5	F	T	F
1 - 2 years	17	0.101	87	0.611	83	0.917	64	0.009	34	0.008	285	0.011
3 - 5 years	23	0.148	116	0.828	110	0.999	85	0.050	45	0.005	380	0.038
6 - 10 years	30	0.216	156	0.945	148	0.999	114	0.150	60	0.004	508	0.216
>10 years	45	0.441	233	0.990	221	0.999	170	0.409	90	0.003	759	0.815
Sousse	1	F	2	F	3	F	4	F	5	F	T	F
1 - 2 years	22	0.440	95	0.905	84	0.999	79	0.353	46	0.15	326	0.541
3 - 5 years	29	0.590	127	0.958	112	0.999	106	0.541	61	0.29	435	0.818
6 - 10 years	39	0.740	170	0.999	150	0.999	142	0.673	82	0.471	582	0.924
>10 years	58	0.840	254	0.999	224	0.999	211	0.860	122	0.64	870	0.990
Béja	1	F	2	F	3	F	4	F	5	F	T	F
1 - 2 years	15	0.083	90	0.446	88	0.990	62	0.039	27	0.003	282	0.009
3 - 5 years	20	0.104	120	0.687	118	0.999	82	0.134	36	0.005	376	0.031
6 - 10 years	27	0.193	160	0.879	158	0.999	110	0.299	48	0.006	504	0.218
>10 years	40	0.268	240	0.960	236	0.999	165	0.521	72	0.009	752	0.822
Nabeul	1	F	2	F	3	F	4	F	5	F	T	F
1 - 2 years	21	0.282	94	0.884	83	0.990	74	0.117	44	0.000	317	0.175
3 - 5 years	28	0.376	126	0.886	111	0.999	99	0.248	59	0.009	423	0.427
6 - 10 years	38	0.547	168	0.990	149	0.999	132	0.411	79	0.048	566	0.782
>10 years	56	0.723	251	0.999	223	0.999	198	0.615	118	0.203	846	0.960
Tunis	1	F	2	F	3	F	4	F	5	F	T	F
1 - 2 years	18	0.163	95	0.761	88	0.945	68	0.150	36	0.011	304	0.171
3 - 5 years	24	0.284	126	0.850	117	0.999	90	0.275	48	0.015	405	0.370
6 - 10 years	32	0.398	169	0.980	157	0.999	121	0.456	64	0.021	543	0.768
>10 years	48	0.598	252	0.990	234	0.999	181	0.668	95	0.089	810	0.990
Sidi Bouzid	1	F	2	F	3	F	4	F	5	F	T	F
1 - 2 years	21	0.479	99	0.909	88	0.999	69	0.499	41	0.37	317	0.790
3 - 5 years	27	0.593	132	0.846	117	0.999	92	0.666	54	0.56	423	0.930
6 - 10 years	37	0.732	177	0.860	157	0.999	123	0.764	72	0.74	566	0.930
>10 years	55	0.832	265	0.999	234	0.999	184	0.984	108	0.953	845	0.934
Gabes	1	F	2	F	3	F	4	F	5	F	T	F
1 - 2 years	23	0.694	100	0.843	83	0.999	81	0.653	49	0.496	335	0.916
3 - 5 years	30	0.782	133	0.929	111	0.999	107	0.713	65	0.605	447	0.942
6 - 10 years	41	0.880	178	0.999	149	0.999	144	0.814	87	0.780	599	0.999
>10 years	61	0.952	266	0.999	223	0.999	215	0.826	130	0.860	894	0.999

NB: Period 1: March; Period 2: April-June; Period 3: July-August; Period 4: September-November and Period 5: December-February. Only the PM formula is adopted for ETc estimation.

locations as Bizerte and Béja where 20% only of the water needs are satisfied by rainfall. For younger plantations, irrigation becomes necessary beginning from the second period of development, *i.e.* April-June for Bizerte,
Béja, Nabeul and Tunis. For the other stations, and particularly for Gabes and Sidi-Bouzid, irrigation is necessary for both young and old trees during the early spring period.

Figures 8-11 present the annual and seasonal distribution of rainfall expressed in terms of frequencies with

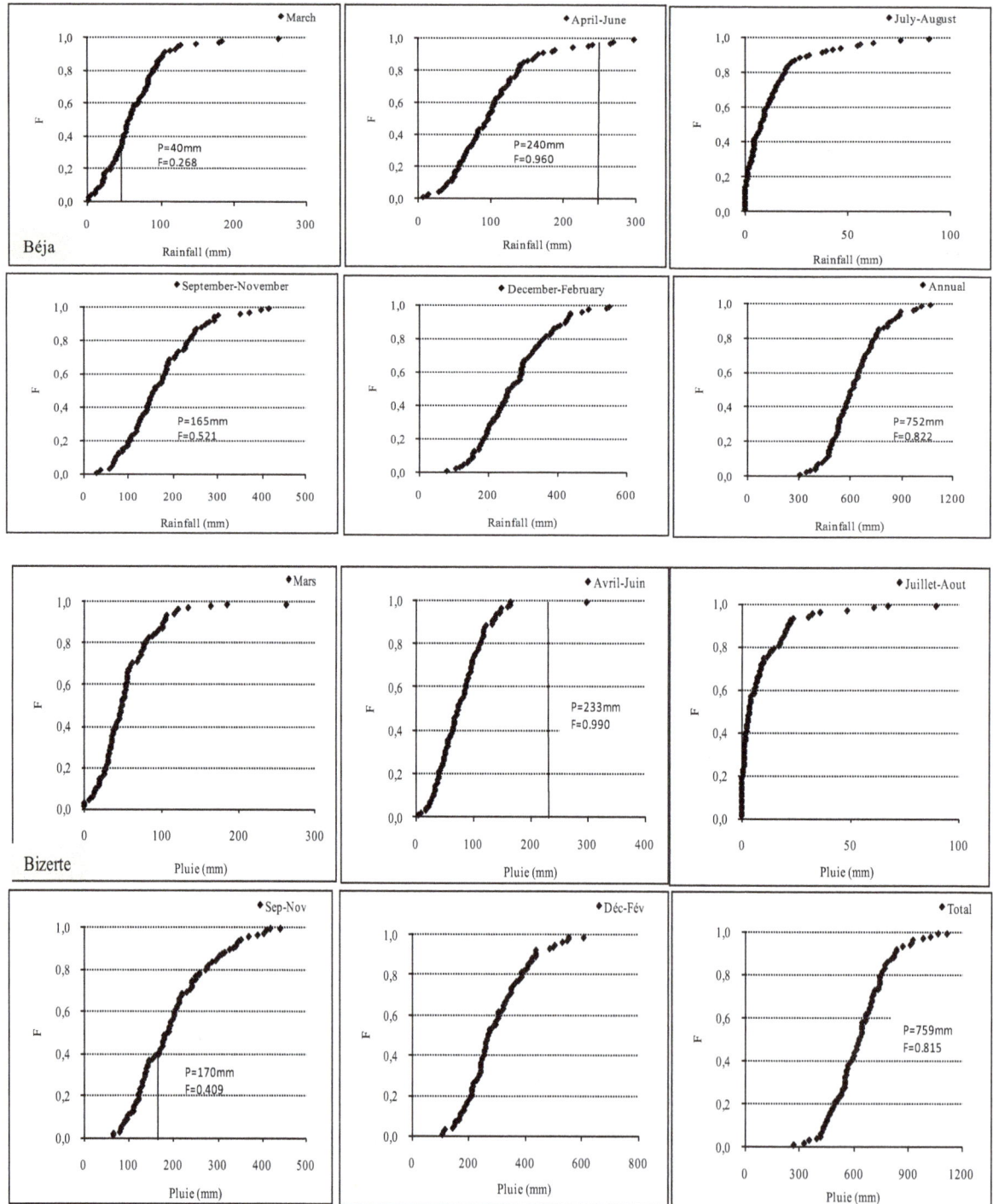

Figure 8. Rainfall distribution functions established for the sites of Béja and Bizerte with series of data recorded during 90
years-long-period and following the growing periods. Bars represent the amount of water needed (*ETc*, mm) for each period
computed with the ET_0-PM formula for 10 years old trees.

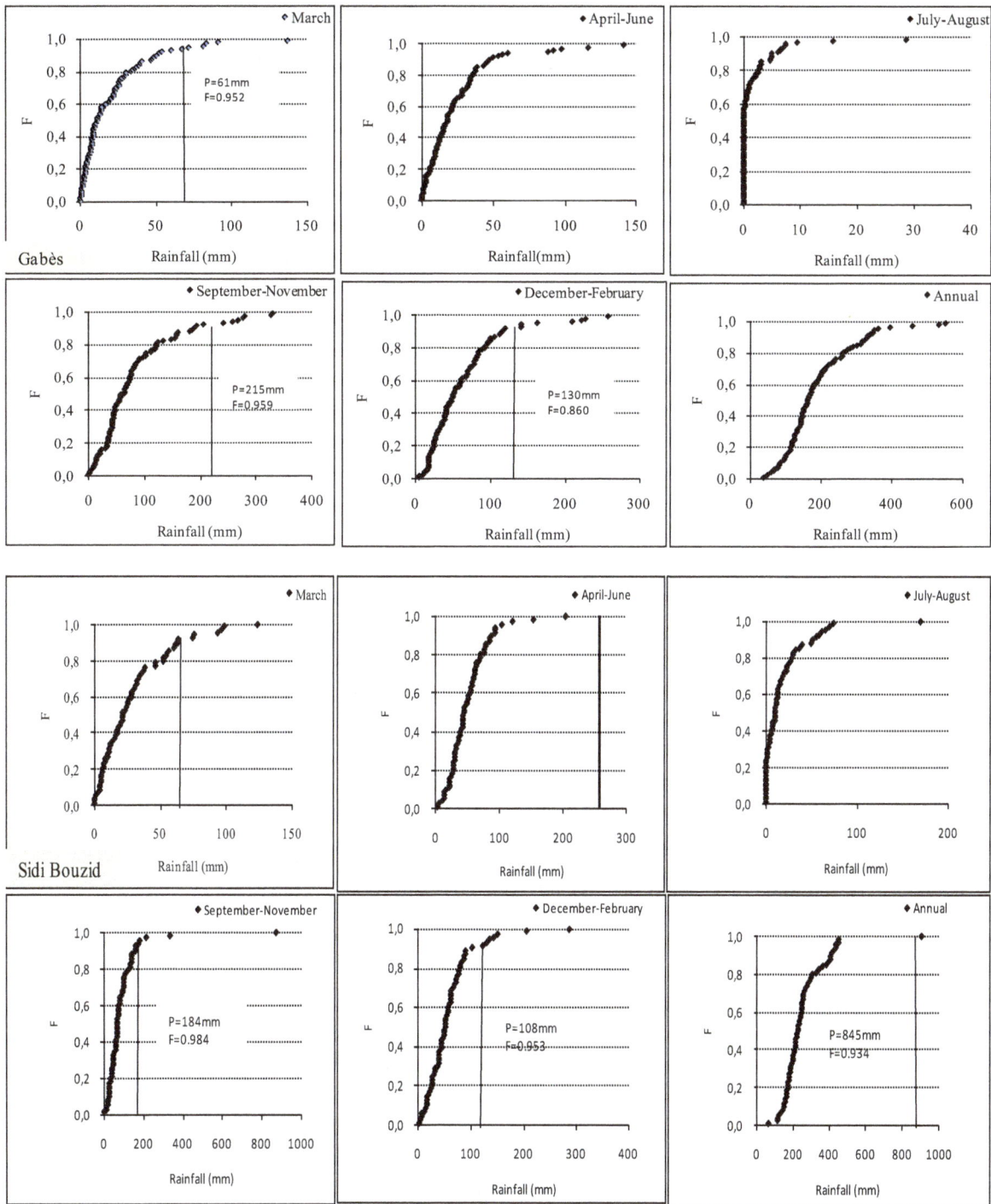

Figure 9. Rainfall distribution functions established for the sites of Gabes and Sidi Bouzid with series of data recorded during 90 years-long-period and following the growing periods. Bars represent the amount of water needed (*ETc*, mm) for each period computed with the ET_0-PM formula for 10 years old trees.

bars representing the amount of water needed for each period of growth following the site.

Rainfall distribution functions established for the first growing period (March), for period 4 (September-November) and period 5 (December-February) showed that an important fraction of water needs, ranging between 60% and 90% is covered by rainfall supplies at Bizerte and Béja. For Nabeul and Tunis, Water amounts needed during stages 1, 4 and 5 for plants aged one to 10 years are correctly covered by rainfall. A similar situation

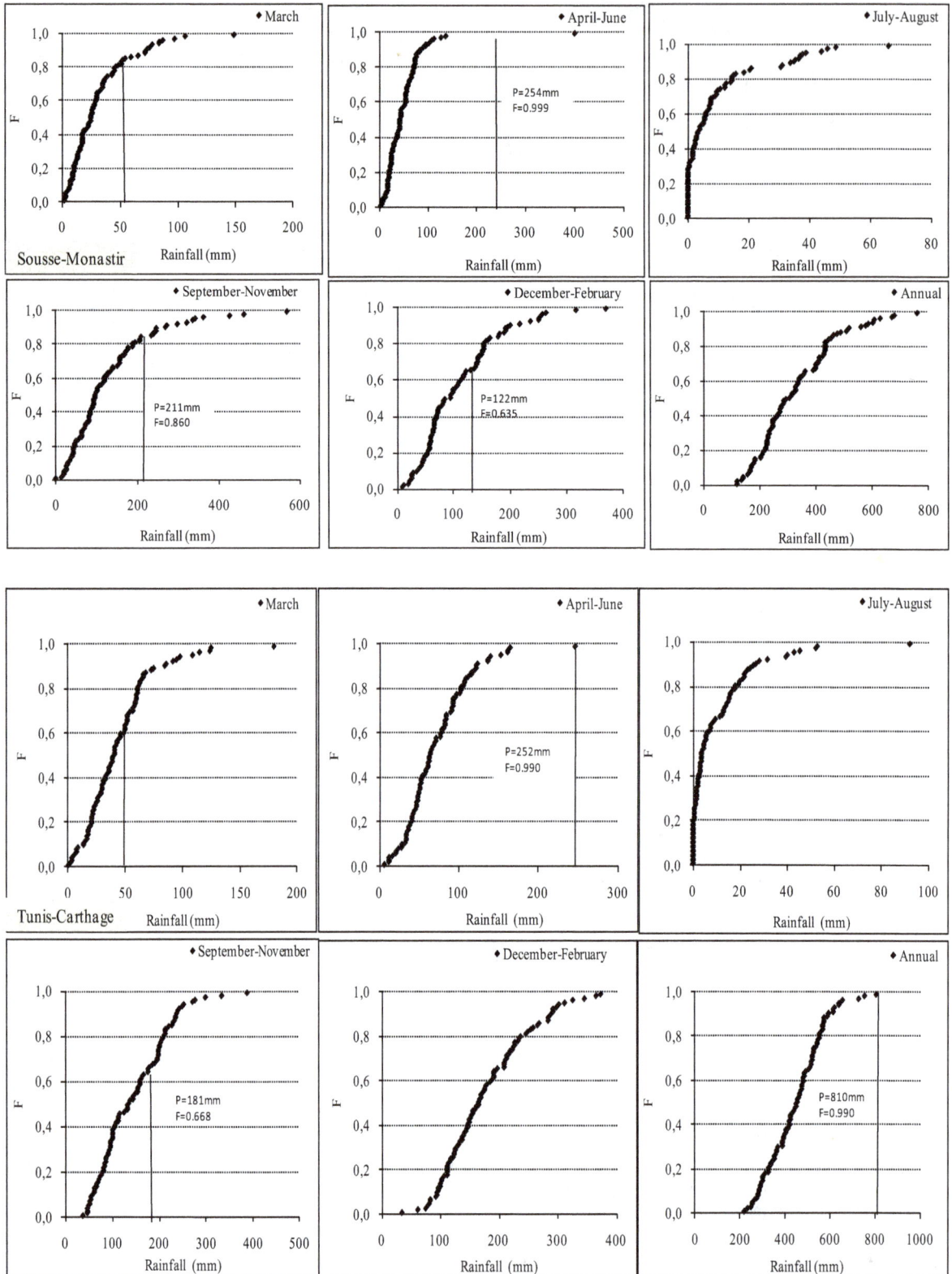

Figure 10. Rainfall distribution functions established for the sites of Sousse and Tunis with series of data recorded during 90 years-long-period and following the growing periods. Bars represent the amount of water needed (*ETc*, mm) for each period computed with the ET_0-PM formula for 10 years old trees.

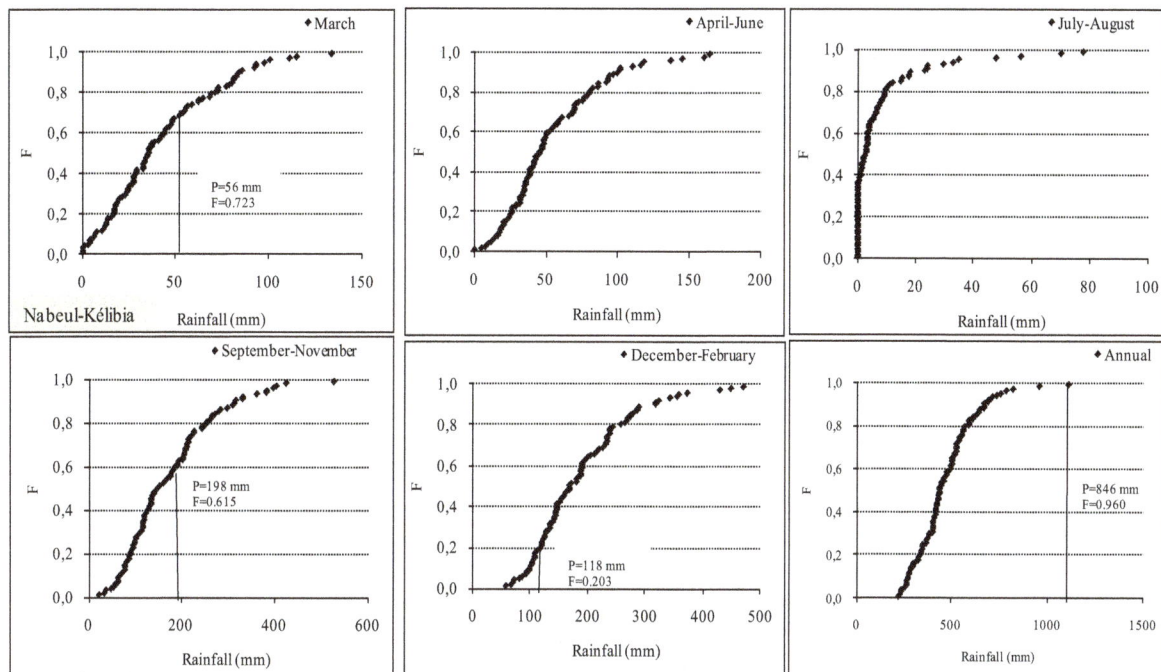

Figure 11. Rainfall distribution functions established for the site of Nabeul-Kélibia with series of data recorded during 90 years-long-period and following the growing periods. Bars represent the amount of water needed (ETc, mm) for each period computed with the ET_0-PM formula for 10 years old trees.

is observed for Sousse where plantations aged one to five years are correctly supplied by rainfall received during stages 1, 4 and 5. For the other locations, less than 50% of water needs are covered by rainfall from September to March.

For the period of April-June, *i.e.*, the period of flowering and early fruit development (period 2), rainfall amounts meet the crop water needs of young trees partially with a fraction ranging between 10% and 55% (40% at Bizerte, 10% at Monastir, 55% at Béja, 15% at Nabeul, 24% at Tunis, 10% at Sidi Bouzid and 16% at Gabes). For adult plants amounts of rainfall are insufficient to cover their water requirements. Therefore irrigation is needed in most cases.

Rainfall amounts received during the summer period (period 3) are very low, not exceeding 10% of the crop water needs for all stations, and irrigation is needed even for one year old plants.

For the period of September-November (period 4), *i.e.*, that of oil synthesis and harvest, annual water needs ranged between 62 and 215 mm following the age. Irrigation is necessary and particularly during the first weeks of this period particularly for Monastir, sidi Bouzid and Gabes where less than 60% of ETc of adult trees are covered.

Rainfall amounts received during winter (December-February, period 5) when flowers differenciate and fruits maturate cover more than 60% of young olive water needs even for the driest areas of Gabes and Sidi Bouzid. For Tunis, Béja, Bizerte and Nabeul, water needs are normally satisfied for all tranches of age.

4. Discussion

Aiming at a comparative analysis of the behaviour of the different methods of ET_0 computation, their performance against the ET_0-PM was evaluated in all considered cities through a set of commonly used statistics.

Results obtained from ET_0 estimation by using different formulas show that the response differs following site, season and the calculation procedure. Analyses of data on differences between ET_0 estimated with different formulas have shown that maximum estimates are provided by Eagleman and PO methods. The BC estimates are more often significantly superior to the other values. Minimum values are provided by ST and SW, Ivanov, Turc and Chr. Harg formulas depending on season and site. Significant differences were found between PO and PM-ET_0 estimates.

Values of ET_0 computed by different empirical formulas and compared to the PM-estimates using full datasets, show that the estimated ET_0 values strongly correlate with ET_0-PM at all stations with all formulas. However, the most appropriate methods are those of Turc and Ivanov, implying that these methods appropriately predict the ET_0 in all climatic regions of Tunisia. A visual inspection of data suggests that the Turc method performs well for all climatic zones, arid and semi-arid, in western, northern and coastal areas of Tunisia. The Ivanov method appears to be more appropriate to the northen areas *i.e.* Béja and Bizerte. Estimates of ET_0 by using the Hargreave-Samani (HS) equation for the east-southern area (Gabes) characterised by arid climate, show satisfactory agreement with PM-ET_0 value, though a poorer agreement was found when using the Eagleman formula.

Results presented herein are coherent with previous findings [5] [82] [100] [101] comparing several evapotranspiration methods, particularly those made in Iran [24] [30] [32] [86], reporting that the HS method performs well in most climatic regions, with the exception of humid areas where it tends to overestimate ET_0. The research works carried out around the Mediterranean basin show different responses. In Northern and Central Tunisia, Jabloun and Sahli (2008) [31] found better estimates for PM than for HS. Paredes and Rodrigues (2010) [102] found small errors with PM relative to HS, thus adopting it to estimate ET_0 in Portugal for irrigation scheduling purposes; they found that estimation errors were larger in humid locations comparative to dryer ones. Temesgen et al., (2005) [100] stated that the HS method underestimates ET_0 for dry and windy locations due to not considering a wind factor and concluded that it is more accurate when applied for 5 or 7-day averages than for daily time scales. Application of PM to different climates of South Africa showed to be superior to HS and that better results were obtained when applied to 5-day rather than daily time cales [29]. Differently, Martinez and Thepadia (2010) [32] found that the HS produced smaller overestimation errors than PM in a humid climate, while Gocic and Trajkovic (2010) [22] proposed software to estimate ET_0 for minimizing computation errors using an adjusted HS equation when weather data are missing. Hargreaves and Allen (2003) and Razieia and Pereira (2013) [5]-[86] tend to explain these differences. Their explanation is based on a careful analysis of the history and applications of the HS equation, considering that as the HS method was developed initially for arid to sub-humid climates and thus it may not fit well to conditions of humid climates, necessitating different calibration with regard to the particularities of these climates. Droogers and Allen (2002) and Trajkovic (2007) [15] [103] attempted to recalibrate the HS coefficients in order to improve its performance. A large number of versions emerged to locally adapt the HS equation, amongst the PM which was lately adapted for using daily forecasted weather data for irrigation scheduling purposes [19]. However, Hargreaves and Allen (2003) [5] concluded that recalibrating the exponents and coefficients of the HS equation only increased the complexity of the equation. The HS method is usually preferred with respective to other more complicated equations since it requires only maximum and minimum air temperatures, which are available in most agro-climatic stations and can be measured with less error and by less trained individuals than the other climate variables used in combination equations. This is very important when solar radiation and wind speed data are lacking.

The consistent deviations of ET_0 estimations from the PM-ET_0 are supposed to be related to the effect of some climatic variables amongst the wind speed, which affect greately ET_0 computation. Discrepancies between HS and PM-ET_0 for example at Bizerte and Nabeul, is attributed to the high wind speed, particularly during spring when ET_0 is higher. In our case, the observed disagreement between estimations may be partially due to not considering wind speed in the HS method. In other case studies, errors may be due to the lack of data and the use of empirical formula for estimating wind speed [89]. Raziei and Pereira (2013) [86] showed that using seasonal regional values as default wind speed values decreased the variance of residuals and the heteroscedasticity of regressions but increased the regression coefficients and therefore ET_0 became over-estimated in the hot and windy locations (case of Bizerte and Nabeul). These authors conclude that, considering the need to minimize errors in estimating ET_0 when only temperature data are available in areas where wind may play a major role, remains an open problem to further research developments.

The BC formula correlates well with PM at all Tunisian sites with r values ranging between 0.888 and 0.927. However at seasonal scale it appears to be not appropriate for all growing periods. It performs well with PM during stages 1, 4 and 5 only for Nabeul and Sousse, which are coastal areas characterized by moderate rainfall amounts. These responses are perhaps due to the fact that the BC formula derived initially in farmers' fields under water stress conditions, and calculates an ET that is most closely related to the average county yields during the years the measurements were taken [104]. Therefore the values obtained in our case did not represent anon-water stressed condition. Razieia and Pereira (2013) [86] consider that the empirical relationship and the originally derived coefficients are nowadays outdated and invalid for today's agriculture production systems even with the subsequent changes that improved the BC-formula by adding more weather and crop variables, like that of Doorenbos and Pruitt (1974) [26] which is generally referred to as the FAO Blaney-Criddle (EtBC) including adjustment factors based on minimum relative humidity, sunshine, and daytime wind speed estimates. This modification to the BC formula was made to make it compatible with the modified Penman's reference ET equation.

Variations recorded when comparing the different methods of ET_0 estimation indicate that there is no weather-based evapotranspiration equation that can be expected to predict evapotranspiration perfectly under every climatic situation due to simplification in formulation and errors in data measurement. According to Razieia and Pereira (2013) [86] it is probable that precision instruments under excellent environmental and biological management conditions will show the FAO Penman-Monteith equation to deviate at times from true measurements of grass ET_0. However, the Expert Consultation agreed to use the hypothetical reference definition of the FAO Penman-Monteith equation as the definition for grass ET_0 when deriving and expressing crop coefficients.

Local information on K_c is scarce for olive, and mainly obtained from ET measurements by using the soil water balance method [76]. Orgaz *et al.*, (2006) [18] reported crop coefficients for olive varying from 0.45 to 0.75 in different locations which are far below the values of annual crops, typically varying from 1.0 to 1.2 [26]. For central Tunisia, Braham and Boussadia (2013) [96] found values of 0.46 during the flowering and fruit set periods. Villalobos *et al.*, (2000) [13] found an average annual crop coefficient of 0.62 which is considered rather low due to the low ground cover and to the enhanced control of canopy conductance by stomatal responses to VPD. These authors showed that the crop coefficient will vary among locations and even among years and seasons depending on soil humectation (rainfall, irrigation), solar radiation and reference evapotranspiration (ET), because this coefficient was derived under non-stressed conditions where water, fertilizer, insects, and salinity do not limit crop growth and production. However, in many cases these non-stressed conditions were not met when deriving the crop coefficients and consequently, the K_c values have been adjusted upward as new measurements of Et have occurred under better irrigation scheduling conditions in the research fields [27] [37].

The variability of K_c makes difficult to apply the FAO method to locations where no experimental information exists. So it is desirable to have a method that estimates directly the actual ET consistently well. In addition, the crop coefficient concept has not only the problem of it being unique to the reference ET formula use but the change in the crop coefficient over time can be a function of days since planting, percent cover or growing degree-days. Variations in K_c and *ETc* estimates are also inherent to differences in resistance to transpiration, crop height, crop roughness, ground cover and crop rooting characteristics which explain the different ET levels in different types of environmental conditions for identical crop. Thus the crop variety and development stage should be considered when assessing the evapotranspiration from crops grown in large, well-managed fields. Additional consideration should be given to the range of management practices that act on the climatic and crop factors affecting the ET process. Cultivation practices and the type of irrigation method can affect the crop characteristics and the wetting of the soil. These results indicate that the use of no adequate method to estimate reference evapotranspiration for *ETc* estimation allowed overestimating or underestimating the water requirements; future research is needed to reconcile which should be the standard method of calculating the change in the crop coefficient over time. Recent works on plant water use showed that integrating daily sap flow is a very good way to establish the connection between the reference evapotranspiration and the individual plant water use [51] [99]. The method was proven to be valuable and valid for canopies with different sizes, exposure and environments. Masmoudi-Charfi *et al.*, (2012) [105], by using the sap flow technique of Granier (1985) [91], found values of K_cxK_r of 0.53 for six-years old trees cultivated in North Tunisia, while Braham and Boussadia (2013) [96] adopting the same technique found specific values ranging between 0.46 and 0.51 for the period of flowing and early fruit development (April-May) for olive orchards cultivated in central Tunisia.This approach eliminates the need of the traditional crop coefficient scheme and future research is needed to reconcile which should be the standard method of calculating the change in the crop coefficient over time. It is desirable to have a method that estimates the (*ETc*) consistently well.

Results relative to rainfall distribution show large annual and seasonal variability. Precipitation occurs sporadically, particularly in the centre and southern areas. Annual precipitations computed, from 1901 to 2000 over 99 years long period varied from less than 200 mm at Gabes to more than 600 mm at Bizerte and Béja. The lowest amount was obtained for the regions of Sidi-Bouzid located at the Center of Tunisia in the western area.

At a seasonal scale, the rainiest periods are December-February for Béja-Tunis-Bizerte and September-November for the sites of Gabes, Sidi Bouzid, Sousse and Nabeul with similar trends for rainfall frequency. The autumnal and winter amounts are equivalent for Nabeul. The driest period is that of July-August for all sites. Minimum variations were observed during the summer period, in June (Nabeul, Sidi Bouzid) or July (Tunis, Sousse, Gabes) depending on site specificities.

On the basis of annual rainfall amounts and ET_0-Thornthwaite estimates, cities of study were classified by using the UNEP aridity index. Values show that Béja and Bizerte as well as Tunis, Nabeul and Sousse have semi-arid climate with IA ranging between 0.20 and 0.43, while Gabes and Sidi-Bouzid are characterized by typical arid-climate with IA of 0.11 and 0.15 respectively, in spite of their location, since Gabes is a coastal town and Sidi Bouzid a western area located at 314 m of altitude. These results show that Tunisia is not concerned with humid and sub-humid climates when the IA is considered to climatically classify these regions, although some published papers considered the region of Kroumirie, extreme NW of Tunisia as humid area [41], while arid and semi-arid climates dominate the central and southern landscape. Also, although Bizerte, a coastal region, has a relatively high annual precipitation accompanied with cold winters, it is not considered as a sub-humid region. Its AI is close to that provided for Béja, a continental area of NW Tunisia, situated at 144 m of altitude.

These conditions make yields of olive orchards highly dependent on rainfall amounts and specially their distribution, resulting in a recurring deficiency of soil moisture and variable yields. Indeed, olive production varies from one year to another by a factor of five depending on the amount of rainfall and fall's timing. This situation is uncomfortable for Tunisia because olive production is vital to national economy, playing a key role in maintaining social stability and accounts for a large share of employment (30 millions/year). To face this problem, adequate water conservation strategies have been developed since many centuries and most of them have focused on improving the performance of irrigation systems or on adapting crop management systems to reduce the need for irrigation water particularly during summer months. But, at this time, there was no any fundamental or technical research to determine the effective water needs of olive orchards. First studies begin in Tunisia during the seventeenths [64]. Yield-enhancing techniques have been adopted at different scales to ensure the best use of natural water resources, ranging from modern irrigation techniques (drip irrigation and fertigation) to high yielding cultivars grown with high levels of inputs and with densities ranging between 300 and more than 1250 plants/ha. Many research studies show that in environments characterized by alternating wet and dry seasons, adding small amounts of water during the growing season can increase remarkably water productivity [14] [40] [51] [52] [76] [85] [106]. However, this potential of supplemental irrigation must be explored to make better use of the limited resources available, how we can use this fresh water for irrigating olive species which is considered for a long time as a tolerant crop supporting water shortage, vegetating and producing even with very low amounts of water. Nowadays, orchadists, conscious that olive industry is an obvious choice to improve agriculture, take to be informed about the crop water management and how to determine precisely and judiciously water needs for a better use efficiency. To help orchardists, a technical paper has been prepared and widely distributed in 2006 [107], presenting the crop water needs of olive trees depending on soil coverage, age and sites of olive cultivation. In this paper the FAO-PM method was used to compute ET_0 with average climatic data recorded over a twenty-year long period.

Annual course of ETc computed with the PM method shows values ranging between 300 mm for young plantations and up to 800 mm for adult trees following. Highest ETc estimates are those recorded for Gabes, ranging between 335 mm and 894 mm depending on age. Minimum yearly values are recorded for Béja varying from 282 mm and 752 mm. Lysimetric values determined by Nasr (2002) [1] representing the effective need of water are lower than ETc computations (ETc-PM) for Tunis, Nabeul, Sousse and Gabes, which are coastal areas. Annual ratios between the lysimetric values and ETc range between 0.74 and 0.96 for these areas and approximate the unit for both Béja and Sidi Bouzid. At seasonal scale, results showed that ETc increases following season in response to the increasing leaf area, varying for all stations from 15 mm to 61 mm in March, from 87 mm to 254 mm for the coastal areas and from 90 mm to 266 mm for the continental locations from April to November. In most cases and particularly for adult plantations cultivated in Centre and South Tunisia, rainfall amounts didn't meet these water needs. Estimates of seasonal values of (ETc-P), showed negatives values. The deficit of water is present even for young plantations inspite of their low water needs. Rainfall amounts were unable to cover the needs of water for this specie during the period of May-September, i.e. during fruit growth, the most important period that determines the production level. These amounts should be supplied necessary by irrigation. The lowest deficit is that provided for Béja and the highest is observed for Gabes. At annual scale, values of (ETc-P) are positive for Tunis, Nabeul, Sousse, Bizerte and Beja when young olive plantations are considered. For adult trees, values are positive only for Tunis, Bizerte and Béja. This last location is the only case where water needs are covered for olive trees aged 6 to 10 years. So there is a need for irrigating olive plantations aging more than 5 years and especially when olive is cultivated in the western areas. Irrigation is needed during the growing fruit period but also during the other seasons, when shoots grow. It is also necessary during "on" and "off" years. On the other hand the curves relative to rainfall distribution established for the different periods of growing give precious quantitative and qualitative information about water supplies: how irrigation can be made and at which amounts. Indeed, it appears from these results based on the seasonal rainfall frequencies and water needs computed with the PM formulae that irrigation supply is necessary all time for trees aging more than 10 years even for the rainiest locations as Bizerte and Béja where 20% only of the water needs are satisfied by rainfall. For younger plantations, irrigation becomes necessary beginning from the second period of development, *i.e.* April-June for Bizerte, Béja, Nabeul and Tunis. For the other stations, and particularly for Gabes and Sidi-Bouzid, irrigation is necessary for both young and old trees during the early spring period. If the requested amounts are not applied, many problems can arise affecting shoot growth [54], root development, water mineral uptakes [108] [109], fructification [61], olive maturation [74] and production [73] [85] [110], but also the water use efficiency. However, recent studies show that we can reduce water supplies during some periods without affecting production or fruit size, through application of the *concept of deficit irrigation* [50] [55] [56] [111]-[115]. Furthermore, a certain lack of water is appreciated and can improve substantially fructification and oil synthesis, but only if water is applied at specific stages [60] [116].

5. Conclusions

Based on a careful analysis of the history and applications of empirical formula for ET_0 computation, their accuracy have been assessed against the ET_0-PM formula, showing an adequateness of using some of them in Tunisia for different climatic sites. Good relationships were observed between ET_0-PM estimates and those issued from Ivanov, Turc and HS methods, confirming the previous referred conclusion, which indicates that ET_0 can be estimated with a specific formula even when climatic data are lacking. Each region may have an appropriate alternative for ET_0 estimation. This may concern even the remote stations having only minimum and maximum temperature records.

Adequate estimation of ET_0 is of great importance in agricultural and hydrological studies, water resources and watershed management. In particular, it is required for supporting irrigation scheduling, drought management and climate change studies. In this work, irrigation water needs of olive obtained from ET_0-PM computation were determined following sites of olive production, age of the orchard and the growing periods by associating them to the rainfall distribution functions. This allowed us to determine the case studies where irrigation is needed. However, despite a quite good performance of the PM-equation in most applications, and the fact that it is appropriate for a wide range of climates and sites, particularly when it is used for *ETc* estimation and irrigation scheduling purposes, using this method may confront users with the lack of information on radiation and *ETc*. It is thus necessary to adjust them with regard to the local conditions, and particularly in precising the correct values of *ETc*, which should be determined for all growing periods because the crop coefficient concept has not only the problem of it being unique to the reference ET formula use but the change in *ETc* values over time can be a function of days since planting, percent cover soil, humectation, vapour pressure deficit (VPD), solar radiation. Additional research is needed on developing specific crop coefficients that use the PM-equation when calculating ET and a standardized method of calculating the time base for the crop coefficients preferably based on a growing degree day concept. Also, applying irrigation amounts on the basis of ET_0-PM formula needs necessary determination of the effective rainfall amounts which is not easy to measure and many studies use approximate method for its calculation. It remains a big challenge.

Herefore, to have a reliable estimation of ET_0 at a fine spatial resolution over the country, it is important to use accurate methods requiring limited weather data that can be available in a dense network through the country as it is the case for temperature. Indeed, the stations recording the climate variables needed for ET_0-PM estimation are very sparse and in many cases have incomplete records, particularly in the central and southern areas where the Tunisian deserts are situated. That's why to our best knowledge, this work with that of Nasr (2002) [1] is the only one that has estimated ET_0 for the whole country, but we didn't test the possible advantages in using calibrated values for the radiation adjustment coefficient or temperature adjustment for dew point temperature estimation as proposed by Allen (1996) [2]. This calibration is therefore a line to be explored.

Acknowledgements

Authors would like to thank Andolsi Shourouk for her assistance. Computation of ET_0 was supported by this student. Thanks also to Mrs Chaabouni-Hadjtaieb Monia and Pr. Slimani at INAT.

References

[1]   Nasr, Z. (2002) Mesures et estimations de l'évapotranspiration de référence en Tunisie. *Annales de l'INRAT*, **75**, 241-256.

[2]   Allen, R.G. (1996) Assessing Integrity of Weather Data for Reference Evapotranspiration Estimation. *Journal of Irrigation and Drainage Engineering*, **122**, 97-106. http://dx.doi.org/10.1061/(ASCE)0733-9437(1996)122:2(97)

[3]   Hargreaves, G.H. and Samani, Z.A. (1982) Estimating Potential Evapotranspiration. *Journal of the Irrigation and Drainage Division*, **108**, 225-230.

[4]   Hargreaves, G.H. and Samani, Z.A. (1985) Reference Crop Evapotranspiration from Temperature. *Applied Engineering in Agriculture*, **1**, 96-99. http://dx.doi.org/10.13031/2013.26773

[5]   Hargreaves, G.H. and Allen, R.G. (2003) History and Evaluation of Hargreaves Evapotranspiration Equation. *Journal of Irrigation and Drainage Engineering*, **129**, 53-63. http://dx.doi.org/10.1061/(ASCE)0733-9437(2003)129:1(53)

[6]   Rana, G. and Katerji, N. (2000) Measurement and Estimation of Actual Evapotranspiration in the Field under Mediterranean Climate: A Review. *European Journal of Agronomy*, **13**, 125-153. http://dx.doi.org/10.1016/S1161-0301(00)00070-8

[7]   Allen, R.G., Jensen, M.E., Wright, J.L. and Burman, R.D. (1989) Operational Estimates of Reference Evapotranspiration. *Agronomy Journal*, **81**, 650-662. http://dx.doi.org/10.2134/agronj1989.00021962008100040019x

[8]   Allen, R.G., Pereira, L.S., Raes, D. and Smith, M. (1998) Crop Evapotranspiration, Guidelines for Computing Crop Water Requirements. *Irrigation and Drainage Paper* 56, FAO, Rome, 300.

[9]   Habaieb, H. and Masmoudi-Charfi, C. (2003) Calcul des besoins en eau des principales cultures exploitées en Tunisie: Estimation de l'évapotranspiration de référence par différentes formules empiriques. Cas des régions de Tunis, Béja et Bizerte, *Sécheresse*, **14**, 1-9.

[10] Sammis, T.W., Wang, J. and Miller, D.R. (2011) The Transition of the Blaney-Criddle Formula to the Penman-Monteith Equation in the Western US. *Journal of Service Climatology*, **5**, 1-11. www.journalofserviceclimatology.org

[11] Denden, M. and Lemeur, R. (1999) Mesure de la transpiration par le modèle de Penman-Monteith. *Sécheresse*, **10**, 39-44.

[12] Pereira, L.S., Perrier, A., Allen, R.G. and Alves, I. (1999) Evapotranspiration: Review of Concepts and Future Trends. *Journal of Irrigation and Drainage Engineering*, **125**, 45-51. http://dx.doi.org/10.1061/(ASCE)0733-9437(1999)125:2(45)

[13] Villalobos, F.J., Orgaz, F., Testi, L. and Fereres, E. (2000) Measurement and Modeling of Evapotranspiration of Olive (*Olea europaea* L.) Orchards. *European Journal of Agronomy*, **13**, 155-163. http://dx.doi.org/10.1016/S1161-0301(00)00071-X

[14] Fernández, J.E., Palomo, M.J., Díaz-Espejo, A., Clothier, B.E., Green, S.R., Girón, I.R. and Moreno, F. (2001) Heat-Pulse Measurements of Sap Flow in Olives for Automating Irrigation: Tests, Root Flow and Diagnostics of Water Stress. *Agricultural Water Management*, **51**, 99-123. http://dx.doi.org/10.1016/S0378-3774(01)00119-6

[15] Droogers, P. and Allen, R.G. (2002) Estimating Reference Evapotranspiration under Inaccurate Data Conditions. *Irrigation and Drainage Systems*, **16**, 33-45. http://dx.doi.org/10.1023/A:1015508322413

[16] Allen, R.G., Pruitt, W.O., Wright, J.L., Howell, T.A., Ventura, F., Snyder, R., Itenfisu, D., Steduto, P., Berengena, J., Baselga, J., Smith, M., Pereira, L.S., Raes, D., Perrier, A., Alves, I., Walter, I. and Elliott, R. (2006) A Recommendation on Standardized Surface Resistance for Hourly Calculation of Reference ETo by the FAO56 PM-Method. *Agricultural Water Management*, **81**, 1-22. http://dx.doi.org/10.1016/j.agwat.2005.03.007

[17] Gong, L.B, Xu, C.Y., Chen, D.L., Halldin, S. and Chen, Y.D. (2006) Sensitivity of the Penman-Monteith Reference Evapotranspiration to Key Climatic Variables in the Changjiang (Yangtze River) Basin. *Journal of Hydrology*, **329**, 620-629. http://dx.doi.org/10.1016/j.jhydrol.2006.03.027

[18] Orgaz, F., Testi, L., Villalobos, F.J. and Fereres, E. (2006) Water Requirements of Olive Orchards. Determination of Crop Coefficients for Irrigation Scheduling. *Irrigation Science*, **24**, 77-84. http://dx.doi.org/10.1007/s00271-005-0012-x

[19] Cai, J.B., Liu, Y., Lei, T.W. and Pereira, L.S. (2007) Estimating Reference Evapotranspiration with the FAO Penman-Monteith Equation Using Daily Weather Forecast Messages. *Agricultural and Forest Meteorology*, **145**, 22-35. http://dx.doi.org/10.1016/j.agrformet.2007.04.012

[20] Gavilán, P., Berengena, J. and Allen, R.G. (2007) Measuring versus Estimating Net Radiation and Soil Heat Flux: Impact on Penman-Monteith Reference ET Estimates in Semiarid Regions. *Agricultural Water Management*, **89**, 275-286. http://dx.doi.org/10.1016/j.agwat.2007.01.014

[21] López-Moreno, J.I., Hess, T.M. and White, S.M. (2009) Estimation of Reference Evapotranspiration in a Mountainous Mediterranean Site Using the Penman-Monteith Equation with Limited Meteorological Data. *Pirineos*, **164**, 7-31. http://dx.doi.org/10.3989/pirineos.2009.v164.27

[22] Gocic, M. and Trajkovic, S. (2010) Software for Estimating Reference Evapotranspiration Using Limited Weather Data. *Computers and Electronics in Agriculture*, **71**, 158-162. http://dx.doi.org/10.1016/j.compag.2010.01.003

[23] Kra, E.Y. (2010) An Empirical Simplification of the Temperature Penman-Monteith Model for the Tropics. *Journal of Agricultural Science*, **2**, 162-171.

[24] Tabari, H. (2010) Evaluation of Reference Crop Evapotranspiration Equations in Various Climates. *Water Resources Management*, **24**, 2311-2337. http://dx.doi.org/10.1007/s11269-009-9553-8

[25] Allen, R.G., Pereira, L.S., Howell, T.A. and Jensen, M.E. (2011) Evapotranspiration Information Reporting: I. Factors Governing Measurement Accuracy. *Agricultural Water Management*, **98**, 899-920. http://dx.doi.org/10.1016/j.agwat.2010.12.015

[26] Doorenbos, J. and Pruitt, W.O. (1974) Guidelines for Predicting Crop Water Requirements. FAO *Irrigation and Drainage Paper* 24, Rome, 179 p.

[27] Jensen, M.E., Burman, R.D. and Allen, R.G. (1990) Evapotranspiration and Irrigation Water Requirements. ASCE Manuals and Reports on Engineering Practice, 332, American Society of Civil Engineers, New York, 360.

[28] Er-raki, S., Chehbouni, G., Guemouria, N., Ezzahar, J., Duchemin, B., Boulet, G., Hadria, R., Lakhal, A., Chehbouni, A. and Rodriguez, J.C. (2004) Measurement of Evapotranspiration and Development Of Crop Coefficients of Olive (*Olea europaea* L.) Orchards in Semi Arid Region (Marrakech, Morocco). Projet INCO-WADEMED *Actes du Seminaire Modernisation de l'Agriculture Irriguée*, Rabat, du 19 au 23 avril 2004.

[29] Annandale, J.G., Jovanovic, N.Z., Benadé, N., Bchir, A., Boussadia, O., Lemeur, R. and Braham M. (2013) Water Use in Olive Orchards Estimated by Physiologic and Climatic Methods in Tunisia. *European Scientific Journal*, **9**.

[30] Popova, Z., Kercheva, M. and Pereira, L.S. (2006) Validation of the FAO Methodology for Computing ET_0 with Missing Climatic Data. Application to South Bulgaria. *Irrigation and Drainage*, **55**, 201-215. http://dx.doi.org/10.1002/ird.228

[31] Jabloun, M. and Sahli A. (2008) Evaluation of FAO-56 Methodology for Estimating Reference Evapotranspiration Using Limited Climatic Data Application to TUNISIA. *Agricultural Water Management*, **95**, 707-715. http://dx.doi.org/10.1016/j.agwat.2008.01.009

[32] Martinez, C.J. and Thepadia, M. (2010) Estimating Reference Evapotranspiration with Minimum Data in Florida. *Journal of Irrigation and Drainage Engineering*, **136**, 494-501. http://dx.doi.org/10.1061/(ASCE)IR.1943-4774.0000214

[33] Nandagiri, L. and Kovoor, G.M. (2005) Sensitivity of the Food and Agriculture Organization Penman-Monteith Evapotranspiration Estimates to Alternative Procedures for Estimation of Parameters. *Journal of Irrigation and Drainage Engineering*, **131**, 238-248. http://dx.doi.org/10.1061/(ASCE)0733-9437(2005)131:3(238)

[34] Jensen, D.T., Hargreaves, G.H., Temesgen, B. and Allen, R.G. (1997) Computation of ET_0 under Nonideal Conditions. *Journal of Irrigation and Drainage Engineering*, **123**, 394-400. http://dx.doi.org/10.1061/(ASCE)0733-9437(1997)123:5(394)

[35] Thornthwaite, C.W. (1948) An Approach toward a Rational Classification of Climate. *Geographical Review*, **38**, 55-94. http://dx.doi.org/10.2307/210739

[36] Allen, R.G., Smith, M., Perrier, A. and Pereira L.S. (1994) An Update for the Definition of Reference Evapo-transpiration. *ICID Bul.*, **43**, 1-34.

[37] Martí, P. and Zarzo, M. (2012) Multivariate Statistical Monitoring of ET_0: A New Approach for Estimation in Nearby Locations Using Geographical Inputs. *Agricultural and Forest Meteorology*, **152**, 125-134. http://dx.doi.org/10.1016/j.agrformet.2011.08.008

[38] Laroussi, C. and Habaieb, H. (1993) Gestion des ressources en eau en conditions d'aridité, cas de la Tunisie. Etat de

l'agriculture en Méditerranée: Ressources en eau: Développement et gestion dans les pays méditerranéens. Bari: CIHEAM, *Cahiers Options Méditerranéennes*, 92-108.

[39]   Kassas, M. (2005) Aridity, Drought and Desertification. Chapter 7, 95-110. Export Marketing of Gum Arabic from Sudan, World Bank Policy Note March 2007. www.afedonline.org/afedreport/english/book7.pdf http://siteresources.worldbank.org/INTAFRMDTF/Resources/Gum_Arabic_Policy_Note.pdf

[40]   Ben Mechlia, N., Oweis, T., Masmoudi, M., Khatteli, H., Ouessar, M., Sghaier, N., Anane, M. and Sghaier, M. (2009) Assessment of Supplemental Irrigation and Water Harvesting Potential: Methodologies and Case Studies from Tunisia. ICARDA, Aleppo, 36.

[41]   Hamza, M. (2009) La politique de l'eau en Tunisie. Conférence Régionale sur la gouvernance de l'eau, Echanges d'expériences entre l'OCDE et les pays arabes, CITET-Tunis, 8-9 Juillet 2009.

[42]   Ben Ahmed, C., Ben Rouina, B. and Boukhris, M. (2007) Effect of Water Deficit on Olive Trees cv. Chemlali under Field Conditions in Arid Region in Tunisia. *Scientia Horticulturae*, **113**, 267-277. http://dx.doi.org/10.1016/j.scienta.2007.03.020

[43]   Ghrab, M., Gargouri, K. and Ben Mimoun, M. (2008) Long-Term Effect of Dry Conditions and Drought on Fruit Trees Yield in Dryland Areas of Tunisia. *Options Méditerranéennes, Séries A*, **80**, 107-112.

[44]   Ben Rouina, B., Trigui, A., D'andria, R., Boukhriss, M. and Chaieb, M. (2007) Effects of Water Stress and Soil Type on Photosynthesis, Leaf Water Potential and Yield of Olive Trees (*Olea europaea* L. cv Chemlali Sfax). *Australian Journal of Experimental Agriculture*, **47**, 1484-1490. http://dx.doi.org/10.1071/EA05206

[45]   Ben Ahmed, C., Ben Rouina, B., Sensoy, S., Boukhris, M. and Abdallah, F.B. (2009) Saline Water Irrigation Effects on Antioxidant Defense System and Proline Accumulation in Leaves and Roots of Field-Grown Olive. *Journal of Agricultural and Food chemistry*, **57**, 11484-11490. http://dx.doi.org/10.1021/jf901490f

[46]   Bedbabis, S., Ben Rouina, B. and Boukhris, M. (2010) The Effect of Waste Water Irrigation on the Extra Virgin Olive Oil Quality from the Tunisian Cultivar Chemlali. *Scientia Horticulturae*, **125**, 556-561. http://dx.doi.org/10.1016/j.scienta.2010.04.032

[47]   Guerfel, M., Baccouri, B., Boujnah, D. and Zarrouk, M. (2007) Seasonal Changes in Water Relations and Gas Exhange in Leaves of Two Tunisian Olives (*Olea europaea* L.) Cultivars under Water Deficit. *The Journal of Horticultural Science & Biotechnology*, **82**, 721-726.

[48]   Guerfel, M., Baccouri, B., Boujnah, D. and Zarrouk, M. (2007) Evaluation of Morphological and Physiological Traits for Drought Tolerance in 12 Tunisian Olive Varieties (*Olea europaea* L.). *Journal of Agronomy*, **6**, 356-361. http://dx.doi.org/10.3923/ja.2007.356.361

[49]   Boussadia, O., Mechri, B., Benmariem, F., Boussitta, W., Braham, M. and Ben Elhadj, S. (2008) Response to Drought of Two Olive Tree Cultivars (cv Koroneki and Meski). *Scientia Horticulturae*, **116**, 388-393. http://dx.doi.org/10.1016/j.scienta.2008.02.016

[50]   Fernandez, J.E. (2006) Irrigation Management in Olive, Instituto de Recursos Naturales y Agrobiologia de Sevilla (IRNAS), 295-305.

[51]   Chehab, H., Mechri, B., Benmariem, F., Hammami, M., Ben Hadj, S. and Braham, M. (2009) Effect of Different Irrigation Regimes on Carbohydrate Partitioning in Leaves and Wood of Two Table Olive Cultivars (*Olea europaea* L. cv. Meski and Picholine). *Agricultural Water Management*, **96**, 293-298. http://dx.doi.org/10.1016/j.agwat.2008.08.007

[52]   Dabbou, S., Chehab, H., Brahmi F., Esposto S., Elvaggini, R., Tatitcchi, A., Servili, M., Montedoro, G.F. and Hammami, M. (2010) Effect of Three Irrigation Regimes on Arbequina Olive Oil Produced under Tunisian Growing Conditions. *Agricultural Water Management*, **97**, 763-768. http://dx.doi.org/10.1016/j.agwat.2010.01.011

[53]   Masmoudi, M.M. and Ben Mechlia, N. (2003) Deficit Irrigation of Orchards. In: Hamdy, A., Ed., *Regional Action Programme (RAP): Water Resources Management and Water Saving in Irrigated Agriculture (WASIA PROJECT)*, Bari: CIHEAM, 203-216, *Options Méditerranéennes*: Série B, Etudes et Recherches.

[54]   Mezghani-Ayachi, M., Masmoudi-Charfi, C., Gouia, M. and Laabidi, F. (2012) Vegetative and Reproductive Behavior of Some Olive Tree Varieties (*Olea europaea* L.) under Deficit Irrigation Regimes in Semi-Arid Conditions of Central Tunisia. *Scientia Horticulturae*, **146**, 143-152. http://dx.doi.org/10.1016/j.scienta.2012.07.030

[55]   Masmoudi-Charfi, C. and Mezghani-Ayachi, M. (2013) Response of Olive Trees to Deficit Irrigation Regimes: Growth, Yield and Water Relations. Agricultural Research Updates, Vol. 6, Nova Sciences Publishers, New York.

[56]   Fernandez, J.E., Paloma, M.J., Diaz-Espejo, A. and Giron, I.F. (2003) Influence of Partial Soil Wetting on Water Relation Parameters of the Olive Tree. *Agronomie*, **23**, 545-552. http://dx.doi.org/10.1051/agro:2003031

[57]   Ghrab, M., Gargouri, K., Bentaher, H., Chartzoulakis, K., Ayadi, M., Ben Mimoun, M., Masmoudi, M., Ben Mechlia, N. and Psarras, G. (2013) Water Relations and Yield of Olive Tree (cv. Chemlali) in Response to Partial Root-Zone Drying (PRD) Irrigation Technique and Salinity under Arid Climate. *Agricultural Water Management*, **123**, 1-11. http://dx.doi.org/10.1016/j.agwat.2013.03.007

[58] Piedra, P.A., Humanes, G.J., Munoz-Cobo, P. and Martin, S. (1997) Plantations à haute densité. Concepts nécessaires. *Olivae*, **69**.

[59] Larbi, A., Ayadi, M., Ben Dhiab, A. and Msallem, M. (2009) Comparative Study of Tunisian and Foreign Olive Cultivars Sustainability for High Density Planting System. Olivebioteq, Sfax-Tunisia, 177-181.

[60] Rallo, L. (1998) Fructification y Produccion, in El Cultivo del olivo. Junta de Andalucia y Grupo Mundi-Prensa, 107-136.

[61] Rallo, L. and Rapoport, H.F. (2001) Early Growth and Development of the Olive Fruit Mesocarp. *Journal of Horticultural Science and Biotechnology*, **76**, 408-412.

[62] Masmoudi-Charfi, C. (2013) Growth of Young Olive Trees. Special Issue on Plant Growth and Development. *American Journal of Plant Sciences*, **4**, 1316-1344. http://www.scirp.org/journal/ajps

[63] Sanz-Cortès, F., Martinez-Calvo, J., Badenes M.L., Bleiholder, H., Hack, H., Llacer, G. and Meier, U. (2002) Phenological Growth Stages of Olive Trees (*Olea europaea*). *Annals of Applied Biology*, **140**, 151-157. http://dx.doi.org/10.1111/j.1744-7348.2002.tb00167.x

[64] Ben Mechlia, N. and Hamrouni, A. (1978) Alternance et production potentielle chez l'olivier irrigué. *Séminaire International sur l'olivier et autres plantes oléagineuses cultivées en Tunisie*, Mahdia, 3-7 Juillet 1978, 199-208.

[65] Le Bourdelles, J. (1977) Irrigation par goutte à goutte en oléiculture, principe de la méthode, installation et fonctionnement. *Olea*, **24**, 31-49.

[66] Bouaziz, E. (1983) Intensification et irrigation à l'eau saumâtre de l'olivier dans les grandes plaines du Centre Tunisien. *Mémoire de 3ème cycle de l'INAT, Oléiculture-Oléotechnie*, 126 p.

[67] Chehab, H. (2007) Etude écophysiologique, agronomique, de production et relation source-puits chez l'Olivier de table en rapport avec les besoins en eau. Thèse de Doctorat en Sciences Agronomiques, Institut National Agronomique de Tunisie, Tunis.

[68] Masmoudi-Charfi, C., Masmoudi, M.M. and Ben Mechlia, N. (2004) Irrigation de l'olivier: Cas des jeunes plantations intensives. *Revue Ezzaitouna*, **10**, 37-51.

[69] Masmoudi-Charfi, C., Masmoudi, M.M., Mahjoub, I. and Mechlia, N.B. (2007) Water Requirements of Individual Olive Trees in Relation to Canopy and Root Development. *Options Méditerranéennes*, Série B, Studies and Research, Vol. 1. CIHEAM. *Proceedings of the International Conference on Water saving in Mediterranean Agriculture and Future Research Needs*, Valenzano, 14-17 February 2007, 73-80.

[70] Masmoudi-Charfi, C., Ayach-Mezghani, M., Gouia, M., Labidi, F., Lamari, S., Ouled Amor, A. and Bousnina, M. (2010) Water Relations of Olive Trees Cultivated under Deficit Irrigation Regimes. *Scientia Horticulturae*, **125**, 573-578. http://dx.doi.org/10.1016/j.scienta.2010.04.042

[71] Michelakis, N. (1990) Yield Response of Table and Oil Olive Tree Varieties to Different Water Doses under Drip Irrigation. *Acta Horticulturae*, **286**, 271-274.

[72] Michelakis, N. (1995) Effet des disponibilités en eau sur la croissance et le rendement des oliviers. *Olivae*, **56**, 29-39.

[73] Sole Riera, M.A. (1990) The Influence of Auxilary Drip Irrigation with Low Quantities of Water on Olive Trees in Las Garrigas (Cv Arbequina). *Acta Horticulturae*, **286**, 307-310.

[74] Inglese, P., Barone, E. and Gullo, G. (1996) The Effect of Complementary Irrigation on Fruit Growth, Ripening Patter and Soil Characteristics of Olive (*Olea europaea* L.) Cv. Carolea. *Journal of Horticultural Science*, **71**, 257-263.

[75] Fernandez, J.E. and Moreno, F. (1999) Water Use by the Olive Tree. *Journal of Crop Production*, **2**, 101-162. http://dx.doi.org/10.1300/J144v02n02_05

[76] Michelakis, N. (2000) Water Requirements of Olive Tree on the Various Vegetative Stages. *Proceedings of the International Course on Water Management and Irrigation of Olive Orchards*, Cyprus, April 2000, 39-49.

[77] Palomo, M.J., Moreno, F., Fernandez, J.E., Diaz-Espejo, A. and Giron, I.F. (2002) Determining Water Consumptive in Olive Orchards Using the Water Balance Approach. *Agricultural Water Management*, **55**, 15-35. http://dx.doi.org/10.1016/S0378-3774(01)00182-2

[78] Masmoudi-Charfi, C., Masmoudi, M.M., Karray-Abid, J. and Ben Mechlia, N. (2012) The Sap Flow Technique: A Precise Means to Estimate Water Consumption of Young Olive Trees (*Olea europaea* L.). Chapter 2 in Irrigation Management Technologies and Environmentam Impact, Nova Sciences Publishers, New York. https://www.novapublishers.com/catalog/product_info.php.

[79] Cohen, Y. (1991) Determination of Orchard Water Requirement by a Combined Trunk Sap Flow and Meteorology Approach. *Irrigation Science*, **12**, 93-98. http://dx.doi.org/10.1007/BF00190016

[80] Villagra, M.M., Bacchi, O.O.S., Tuon, R.L. and Reichardt, K. (1995) Difficulties of Estimating Evapotranspiration from the Water Balance Equation. *Agricultural and Forest Meteorology*, **72**, 317-325. http://dx.doi.org/10.1016/0168-1923(94)02168-J

[81] Testi, L., Villalobos, F.J. and Orgaz, F. (2004) Evapotranspiration of a Young Irrigated Olive Orchard in Southern Spain. *Agricultural and Forest Meteorology*, **121**, 1-18. http://dx.doi.org/10.1016/j.agrformet.2003.08.005

[82] Todorovic, M., Karic, B. and Pereira, L.S. (2013) Reference Evapotranspiration Estimate with Limited Weather Data across a Range of Mediterranean Climates. *Journal of Hydrology*, **481**, 166-176. http://dx.doi.org/10.1016/j.jhydrol.2012.12.034.

[83] Masmoudi-Charfi, C. (2012) Quantitative Analysis of Soil Water Content in Young Drip Irrigated Olive Orchards. *Advances in Horticultural Sciences*, **26**, 138-147.

[84] Caspari, H.W., Green, S.R. and Edwards, W.R.N. (1993) Transpiration of Well-Watered and Water-Stressed Asian Pear Trees as Determined by Lysimetry, Heat-Pulse, and Estimated by a Penman-Monteith Model. *Agricultural and Forest Meteorology*, **67**, 13-27. http://dx.doi.org/10.1016/0168-1923(93)90047-L

[85] Moriana, A.F., Perez-Lopez, D., Gomez-Rio, A., Salvador, M., Olmedilla, N., Ribas, F. and Fregapane, G. (2006) Irrigation Scheduling for Traditional Low-Density Olive Orchards: Water Relations and Influence on Oil Characteristics.

[86] Razieia, T. and Pereira, L.S. (2013) Reference Estimation of ET_0 with Hargreaves-Samani and FAO-PM Temperature Methods for a Wide Range of Climates in Iran. *Agricultural Water Management*, **121**, 1-18. http://dx.doi.org/10.1016/j.agwat.2012.12.019

[87] Pereira, A.R., Green, E., Villa, N. and Nilson, A. (2005) Penman-Monteith Reference Evapotranspiration Adapted to Estimate Irrigated Tree Transpiration. *Agricultural Water Management*, **83**, 153-161.

[88] Allen, R.G. (2002) Software for Missing Data Error Analysis of Penman-Monteith Reference Evapotranspiration. *Irrigation Science*, **21**, 57-67. http://dx.doi.org/10.1007/s002710100047

[89] Moreno, F., Fernandez, J.E., Clothier, B.E. and Green, S.R. (1996) Transpiration and Root Water Uptake by Olive Trees. *Plant and Soil*, **184**, 85-96. http://dx.doi.org/10.1007/BF00029277

[90] Abid-Karray, J. (2006) Bilan Hydrique d'un système de cultures intercalaires (Olivier—Culture maraîchère) en Tunisie Centrale: Approche expérimentale et essai de modélisation. Thèse de Doctorat, Université de Montpellier II, Montpellier, 172 p.

[91] Granier, A. (1985) Une nouvelle méthode pour la mesure des flux de sève brute dans le tronc des arbres. *Annals of Forest Science*, **42**, 193-200. http://dx.doi.org/10.1051/forest:19850204

[92] Penman, H.L. (1948) Natural Evaporation from Open Water, Bare Soil, and Grass. *Proceedings of the Royal Society*, **A193**, 116-140. http://dx.doi.org/10.1098/rspa.1948.0037

[93] Priestley, C.H.B. and Taylor, R.J. (1972) On the Assessment of Surface Heat Flux and Evaporation Using Large Scale Parameters. *Monthly Weather Review*, **100**, 81-92. http://dx.doi.org/10.1175/1520-0493(1972)100<0081:OTAOSH>2.3.CO;2

[94] Bonachela, S., Orgaz, F., Villalobos, F. and Fereres, J.E. (1999) Measurement and Simulation of Evaporation from Soil in Olive Orchards. *Irrigation Science*, **18**, 205-211. http://dx.doi.org/10.1007/s002710050064

[95] L'olivier, C.O.I. (1997) Encyclopédie Mondiale de l'Olivier. Conseil oléicole international, Madrid, 479 p.

[96] Braham, M. and Boussadia, O. (2013) Annual Report of the National Institute of Olive Tree.

[97] INM (2013) National Institute of Meteorology. www.meteo.tn

[98] Victor, M. and San Diego, P. Potential Evapotranspiration by the Thornthwaite Method. State of University. On-line_Thornthwaite: Potential Evapotranspiration by Thornthwaite Method.

[99] Yunusa, I.A.M., Walker, R.R., Loveys, B.R. and Blackmore, D.H. (2000) Determination of Transpiration in Irrigated Grapevines: Comparison of the Heat-Pulse Technique with Gravimetric and Micrometeorological Methods. *Irrigation Science*, **20**, 1-8. http://dx.doi.org/10.1007/PL00006714

[100] Temesgen, B., Eching, S., Davidoff, B. and Frame, K. (2005) Comparison of Some Reference Evapotranspiration Equations for California. *Journal of Irrigation and Drainage Engineering*, **131**, 73-84. http://dx.doi.org/10.1061/(ASCE)0733-9437(2005)131:1(73)

[101] Alkaeed, O., Flores, C., Jinno, K. and Tsutsumi, A. (2006) Comparison of Several Reference Evapotranspiration Methods for Itoshima Peninsula Area, Fukuoka, Japan. Memoirs of the Faculty of Engineering, Kyushu University, Vol. 66, Vngsas, Fukuoka, 1-14.

[102] Paredes, P. and Rodrigues, G.C. (2010) Necessidades de água para a rega de milho em Portugal Continental considerando condic̣ões de seca. In: Pereira, L.S., Mexia, J.T. and Pires, CA.L., Eds., *Gestão do Risco em Secas*, Métodos, Tecnologias e Desafios, Edic̣ões Colibri e CEER, Lisboa, 301-320.

[103] Trajkovic, S. (2007) Hargreaves versus Penman—Monteith under Humid Conditions. *Journal of Irrigation and Drainage Engineering*, **133**, 38-42. http://dx.doi.org/10.1061/(ASCE)0733-9437(2007)133:1(38)

[104] David, R.M. (2011) The Transition of the Blaney-Criddle Formula to the Penman-Monteith Equation in the Western

United States. *Journal of Service Climatology*, **5**, 1-11.

[105] Masmoudi-Charfi, C., Abid-Karray, J., Gargouri, K., Rhouma, A., Habaieb, H. and Daghari, H. (2012) *Manuel d'Irrigation de l'Olivier, Techniques et Applications*, Institut de l'Olivier, 110 p.

[106] Bonji, G. and Palliotti, A. (1994) Olive in Hand Book of Environmental Physiology of Fruit Crop. Anderson CRC Press Inc., Boca Raton, 165-187.

[107] Masmoudi-Charfi, C. (2006) Irrigation des plantations d'olivier. Document Technique, Institut de l'Olivier, Tunisie.

[108] Palese, A.M., Nuzzo, V., Dichio, B., Celano, G., Romano, M., Xiloyannis, C., Ferreira, M.I. and Jones, H.G. (2000) The Influence of Soil Water Content on Root Density in Young Olive Trees. *Acta Horticulturae*, **537**, 329-336.

[109] Dichio, B., Romano, M., Nuzzo, V. and Xiloyannis, C. (2002) Soil Water Availability and Relationship between Canopy and Roots in Young Olive Trees (Cv. Coratina). *Acta Horticulturae*, **586**, 255-258.

[110] Moriana, A.F., Orgaz, F., Pastor, M. and Fereres, E. (2003) Yield Response of a Mature Olive Orchard to Water Deficits. *Journal of the American Society for Horticultural Science*, **128**, 425-431.

[111] Tognetti, R., D'Andria, R., Lavini, A. and Morelli, G. (2006) The Effect of Deficit Irrigation on Crop Yield and Vegetative Development of *Olea europaea* L. (cvs. Frantoio and Leccino). *European Journal of Agronomy*, **25**, 356-364. http://dx.doi.org/10.1016/j.eja.2006.07.003

[112] Gucci, R., Lodolini, E. and Rapoport, H.F. (2007) Productivity of Olive Trees with Different Water Status and Crop Load. *Journal of Horticultural Science and Biotechnology*, **82**, 648-656.

[113] Melgar, J.C., Mohamed, Y., Navarro, C., Parra, M.A., Benlloch, M. and Fernandez-Escobar, R. (2008) Long-Term Growth and Yield Responses of Olive Trees to Different Irrigation Regimes. *Agricultural Water Management*, **95**, 968-972. http://dx.doi.org/10.1016/j.agwat.2008.03.001

[114] Iniesta, F., Testi, L., Orgaz, F. and Villalobos, F.J. (2009) The Effects of Regulated and Continuous Deficit Irrigation on the Water Use, Growth and Yield of Olive Trees. *European Journal of Agronomy*, **25**, 258-265. http://dx.doi.org/10.1016/j.eja.2008.12.004

[115] Connor, D.J. and Fereres, E. (2005) The Physiology of Adaptation and Yield Expression in Olive. Horticultural Review, Vol. 31, John Wiley & Sons, Inc., Hoboken.

[116] Boulouha, B. (1986) Croissance, Fructification et leur interaction sur la production chez la Picholine Marocaine, *Olea*, **17**, 41-47.

List of Abbreviations

ASCE: American Society of Civil Engineers Standardized Reference ET equation.

BC: Blaney-Criddle formula for ET_0 computation.

c_p: specific heat of the air.

C_{TT}, C_{WT}, C_{HT}, C_{ST}, C_R: functions of temperature, wind, relative humidity, insulation and elevation.

Chr. Harg.: Christiansen-Hargreaves method for ET_0 computation.

e_0: saturation vapor pressure (kPa).

e_a: actual vapor pressure (kPa).

$e_0 - e_a$: vapor pressure deficit of the air (kPa) and e_a is actual vapor pressure (kPa).

$e_0 = 0.5[e_0\{T_{max}\} + e_0\{T_{min}\}]$ and $e_a = [e_{oTmax} RH_{min} + e_{oTmin} RH_{max}]/200$.

ET_0: Reference evapotranspiration (mm/day).

ET_0-**PM or PM-ET_0**: Reference evapotranspiration obtained from Penman-Monteith equation (mm).

ETc: Crop evapotranspiration (mm).

ETc-**PM**: Crop evapotranspiration calculated on the basis of ET_0 Penman-Monteith computation (mm).

ETc-**P**: Climatic water deficit (mm).

ET: Maximum daily, monthly or seasonal evapotranspiration determined under non-stressed conditions (mm).

E: Elevation.

EAG: Eagleman method for ET_0 computation.

F: Rainfall frequency (%).

FAO-PM: Penman-Monteith method of FAO.

G: Soil heat flux density ($MJ \cdot m^{-2} \cdot day^{-1}$), flux of heat into the soil, set equal to zero to represent the condition of an isolated tree.

RH$_{max}$, RH$_{min}$ and RH$_{average}$: Maximum, minimum and average air humidity.

HS: Hargreaves-Samani formula for ET_0 computation.

IA: Index of aridity.

INM: National Institute of Meteorology of Tunisia.

ETc: Crop coefficient.

K$_r$: coefficient introduced to take into account the soil coverage.

LAI: Leaf area index.

N: Maximum sunshine durations (hours/day).

n: Actual sunshine durations (hours/month).

p: daylight hours monthly/annual daylight hours.

i: number of days/month.

P: Rainfall amount (mm)

PM: Penman-Monteith equation.

PMT: Penman Monteith Temperature method.

PO: Penman Original method for ET_0 computation

r$_s$ and r$_a$: bulk surface and aerodynamic resistances.

R$_a$: atmospheric radiation ($Mj/m^2/day$).

R$_n$: net radiation at the crop surface ($MJ \cdot m^{-2} \cdot day^{-1}$), R_n is computed as the algebraic sum of the net short and long short radiations.

ST and SW: Hargreaves-Samani method for ET_0 computation.

T: Tree transpiration (liters of water/day or mm).

T: Mean daily air temperature at 2 m height (°C), u_2 is wind speed at 2 m height ($m \cdot s^{-1}$),

T$_{max}$ and T$_{min}$: maximum and minimum air temperatures (°C),

T_m or $T_{average}$: Mean daily air temperature °C, with T: $°F = (T \times 9/5) + 32$ with $T_{mean} = 0.5 (T_{max} + T_{min})$

ΔT: $T_{max} - T_{min}$.

TW: Thornthwaite method for ET_0 computation.

U$_2$: wind speed measured at 2 m height (m/s).

VPD: vapour pressure deficit (kPa).

Δ: Slope of the saturated vapor pressure curve ($kPa \cdot °C^{-1}$).

γ: Psychrometric constant ($kPa \cdot °C^{-1}$), was set constant and equal to 0.066 $kPa \cdot °C^{-1}$ for all experiments.

Months:

Sept.: September.

Nov.: November.

Dec.: December.

Feb.: February.

2

Comparison between Hydro-Flume and Open Field Head Ditch Irrigation Systems at Kenana Sugar Scheme, Sudan

Daffa Alla M. Abdel[1], Wahab Ali M. Adeeb[2]

[1]Agricultural Engineering Section, Research and Development Department, Kenana Sugar Company, Rabak City, Sudan
[2]Irrigation and Water Management Institute, Gezira University, Wad Medani, Sudan
Email: dafaalla.rabih@kenana.com

Abstract

A study was undertaken in Kenana Sugar Scheme, Sudan during 2009/2010 and 2010/2011 seasons. In this study, the gated pipe (hydro-flume) for furrow irrigation was compared with the conventional open field head ditch irrigation system concerning the volume of irrigation water applied to the field, irrigation efficiencies, the time of cutoff, water and irrigation time saved and the irrigation production efficiency (IPE). To achieve these objectives, two commercial cane fields having the same furrow lengths (2100 m) and slopes were chosen. The study shows that in the open field head ditch irrigation, the irrigation water added was 69.1 mm in the top, 75.7 mm in the middle and 66.1 mm in the end of the furrow. Whereas, the irrigation water added in the gated pipe system was 132.7 mm, 46.1 mm and 101.9 mm, respectively. The present study indicates that the gated-pipe system has a high value of application efficiency (79% - 88%) compared with the open field head ditch (69% - 71%). The percent of deep percolation (PDP) for the gated-pipe system is greater than the PDP obtained under open field head ditch irrigation conditions. Also the percent of runoff (PRO) is higher under the open field head ditch system and the water conveyance efficiency for the open field head ditch is 88%. While the gated pipe needs more advance time but can save 20 to 65 m³ of water/irrigation cycle with better uniformity coefficient (CU) and irrigation production efficiency (IPE) compared with the open field head ditch. From the above mentioned results, it is concluded that under Kenana conditions the gated-pipe system is better than the open field head ditch irrigation system keeping in mind that for more uniform water distribution through irrigated furrows of the long fields of Kenana, increased pressure head at the inlet and/or larger openings of the hydro-flume gates may be necessary.

Keywords

Irrigation, Sugarcane, Gated Pipe, Hydro-Flume, Furrow, Application Efficiency, Percolation

1. Introduction

Surface irrigation is the most widely used irrigation method. This is due to its low capital and maintenance costs, and low energy requirements [1]. Among the many surface irrigation methods, furrow irrigation is one of the most commonly used methods. Furrow irrigation is a method of applying water at a given rate into shallow evenly spaced Canals [2]. In furrow irrigation, the field divided into sectors each of 60 furrows in which irrigation water is applied. The furrows are filled to the desired depth of water and this water is retained until it infiltrates into the soil both vertically and horizontally.

The gated pipe (hydro-flume) is defined as a closed conduct with a circular cross section with water flows inside it (no free surface). The flows result from pressure difference between inlet and outlet and they affected by fluid properties and the flow rate.

The gated pipe furrow irrigation system consists of relatively large diameter pipes of about 0.46 m (18 inches), with gates usually equipped on one side and corresponding to the furrow spacing. The hydro-flume is made of VU and thermally protected low density polyethylene of 700 micron wall thickness for maximum service life time in hot and tropical condition. It is flexible so that no alluvial clings to its wall. The gated pipes as an improvement in furrow irrigation, in which the conventional head ditch and siphon are replaced by an above ground pipe. Gated pipe was introduced to allow more uniform irrigation. Uniformity of flow is determined by setting the gates precisely to deliver equal flow into furrows, the rate of discharge in each furrow was less than with siphon tubes that induced erosion, and less leaching potential. Gated pipe also facilitates the eventual adoption of surge irrigation. Gated-pipes are currently used extensively in sugarcane fields in Upper Egypt. [3] found that using gated-pipe to irrigate long furrow resulted in saving water by 20% and 38% and increasing its use efficiency by 58% and 17% for bean and peas, respectively. [4] stated that varying pipe slope, diameter, number of gates, gate area and mean outflow, affect uniformity of outflows. They added that for the entire typical gated-pipe situation analyzed, maximum flow uniformity is obtained with the pipeline slope uphill in the direction of flow. In Sudan, the gated pipe furrow irrigation system was first introduced for vegetable production in a small scheme in Zaied Elkhair which is located on the eastern bank of the Blue Nile.

In Kenana Scheme, the gated pipe was introduced in year 2003 to allow for more uniform irrigation. The rate of discharge in each furrow was less than with siphon tubes that induced erosion, and less leaching potential. Gated pipe also facilitates the eventual adoption of surge irrigation. The system has high application efficiency when operated properly. In 2001/02, the gated pipe furrow irrigation system was adopted in Kenana Scheme for irrigating sugarcane in large scale. Until January, 2007 about 75% of the total area was serviced by the gated-pipe in place of the open field head ditch system with open canal. Recently (years 2004-2005), the gated-pipe was also adopted in some fields at Sudanese Sugar Company mainly, Asalayia Sugar Scheme.

Irrigation practice in Kenana has subjected to many changes. In year 1981, the water indenting was based on fixed days per cycle, and different sizes of siphon were used in the same field to maintain the cycle regardless of field gradient of furrow length. A irrigation system based on evapotranspiration (10 mm/day) was introduced in 1983. In 1987, an indenting system of irrigation based on the number of operating pumps was adopted. Recently, the individual fields were categorized into three groups according to their length, slops and soil classes as A, B and C system, which are irrigated every 12, 10 and 7 days respectively. The steeper the field, the shorter is the furrow length and the shorter is the irrigation cycle.

Irrigation water is pumped from the White Nile into a main canal and distributed through secondary canals until it reaches an open field head ditch from which it is siphoned through a pipe into the furrows of the field. The Scheme is provided with a drainage net work. In year 2002, the open field head ditch system has been gradually changed to the closed system. The open field head ditch and siphons are replaced by an above ground flexible pipe (Hydro-flume) of 18 inch internal diameter, 100 meter long with adjustable gates spaced at 1.5 meter interval.

1.1. Objectives of the Study

The broad objective of the present study is to study the irrigation production efficiency (IPE) produced through improving on-farm water management. The specific objective is to compare the gated pipe system for furrow irrigation with the open field head ditch system with regards to saving time of irrigation, irrigation production efficiency (IPE) and water distribution uniformity.

1.2. Open Field Head Ditch System

An open field head ditch is an open waterway whose purpose is to carry water from one place to another. Field ditches have smaller dimensions and convey water from the farm entrance to the irrigated fields. Under Kenana conditions, the most commonly used field canal cross-section in irrigation is the trapezoidal cross-section with the following dimensions (**Figure 1** and **Table 1**).

Irrigation water is siphoned out through three or two inches internal diameter siphon tubes from the irrigation ditch. The inflow rate through the siphon tube, which is a function of the pressure head causing flow, was computed using the following formula [5];

$$Q = 0.65 \times 10^{-3} \times a \times \sqrt{2gh} \tag{1}$$

where:

Q = discharge from siphon tube (l/s).

0.65 = coefficient of discharge determined under Kenana condition, [6].

a = area of the cross-section (cm^2).

g = acceleration due to gravity (cm/s^2).

h = pressure head causing flow (cm) which is the difference in elevation between the water surface in the field canal and the center of the outlet under free flow conditions or the water surface above the outlet when the outlet is submerged.

1.3. Gated Pipe Pressure and Flow

As water moves through any pipe, pressure is lost due to turbulence created by the moving water. The amount of pressure lost in a horizontal pipe is related to the velocity head, pipe diameter, roughness and the length of pipe through which the water flows. When velocity increases, the pressure loss increases.

[7] narrates the inflow rate through gated pipe with velocity as stated below;

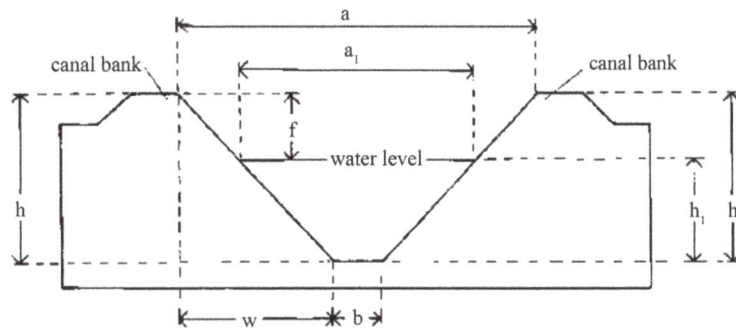

Figure 1. A trapezoidal canal cross-section. Where; The top width of the field canal (a) =2.75 m; The top width of the water level (a_1) = 1.90 m; Height of the field canal (h) = 0.70 m; Water depth in the canal (h_1) = 0.30 m; Bottom width (b) = 1.15 m; Free board (f) = 0.40 m.

Table 1. Dimensions of open field head ditch cross-section.

Parameters	Dimensions
The top width of the open field head ditch	2.75 m
The top width of the water level	1.90 m
Height of the open field head ditch	1.0 m
Water depth in the open field head ditch	0.60 m
Bottom width	1.15 m
Free board	0.40 m

$$Q = CA\sqrt{2gh} \qquad (2)$$

where: Q = discharge from outlet (L/S), C = coefficient of discharge. A = area of the outlet cross-section (cm^2), g = acceleration due to gravity (cm/s^2), h = pressure head causing flow (cm).

[8] investigated the effect of non-uniform outflow distribution along the gated-pipe and he indicated that the open field head ditch method, allow more uniform outflow distribution through siphon tubes along the open field head ditch compared to the gated pipe system, Whereas, the gated pipe allows more outlet water sets than the open field head ditch method. He added that the quantity of water outflow along the gated pipe depend mainly on the alignment and steepness of the gated pipe rather than the number of opening operating at a time or their distance from riser. Because the choice of the gated pipe furrow irrigation system was influenced by the degree of alignment of the pipeline at the field head.

1.4. Loss of Pressure

Water flowing in pipes is always accompanied by a loss of pressure due to friction and the amount of loss is due to smoothness of the inside walls of the pipe, the diameter of the pipe and the velocity of the water flow. The head loss due to friction of a pipe is determined by using [9].

$$h_L = f\frac{L}{D} \times \frac{V^2}{2g} \qquad (3)$$

where: h_L = head loss (m), f = friction factor (dimensionless);
L = pipe length (m), D = pipe diameter (m), V = Average velocity (m/s);
g = Acceleration due gravity.
The friction factor was calculated using [10] as below:

$$f = \frac{0.13}{\mathrm{Re}^{0.172}} \qquad (4)$$

$$\mathrm{Re} = 1.26 \times 10^6 \times Q/D \qquad (5)$$

where: f = friction factor, Re = Reynolds number (more than 10^5 turbulent flow).
Q = Pipe discharge (m^3/hr).
D = Pipe diameter (m).
In straight and long pipe, the loss in head which accounts for the outlets is corrected with a factor (C) suggested by [11] which is a function of the number of the outlets and the friction equation used. The Christiansen factor (C) was expressed as below:

$$C = \left\langle \frac{1}{m+1} \right\rangle + \left\langle \frac{1}{2N} \right\rangle + \left\langle \frac{(m-1)^{0.5}}{6N^2} \right\rangle \qquad (6)$$

where: C = fraction of the head loss under constant outflow conditions, m = 1.85 for Hazen-Williams formula; 1.9 for the Scobey equation and 2.0 for the Darcy-Weisbach equation, N = number of outlets along the pipe.
The Christiansen factor (C) was obtained assuming the following conditions:
■ All the outlets are equally spaced.
■ All the outlets have discharge.
■ No flow at the downstream end of the gated-pipe.
■ The distance between the pipe inlet and the first outlet is equal to the outlet spacing.

2. Materials and Methods

2.1. Site Specification

2.1.1. Location
The study was conducted in the commercial cane fields of Kenana Sugar Estate, (Latitude 13°10'N and longitude 32°40') in heavy clay soils, with 65% clay, 24% silt, 11% sand and a pH of 7.5 - 8.5, during the seasons of 2009/200 and 2010/2011, with the purpose of comparing the gated pipe system for furrow irrigation to the open

field head ditch system with regards to saving time of irrigation, irrigation production efficiency (IPE) and water distribution uniformity. Kenana Scheme is located on the eastern bank of the White Nile River, 300 km south of Khartoum at an altitude of 410 m above mean sea level (msl).

2.1.2. The Soil

Kenana Scheme extends on a clay plain, which dips very gently to the White Nile River plain. The soil is very dark-grayish-brown in color, fine in texture, quite uniform, alkaline in reaction (pH, ranges from 7.5 to 8.5). Non-saline, non-sodic. Kenana soil is characterized by a very low infiltration rate (0.00005 $m^3 \cdot m^{-1} \cdot min^{-1}$) and it contains about 11% sand, 24% silt and 65% clay (mainly montmorillonite). The mean soil water content (0 - 100 cm) at field water capacity was 44% and at wilting point was 20% in volume.

2.1.3. Climate Information

The area of Kenana scheme lies within the tropical dry hot semi-arid climatic zone, with a cool dry season during winter (November-February) followed by a hot dry season (April-June). The rainy season extends from July to October, with peak monthly rainfall in August. The mean annual rainfall is around 340 mm (for the period from 1977 to 2010), while the mean relative humidity 08:00 is around 58%.

2.1.4. Area of the Scheme

Kenana Scheme has a total surface area of about 67,000 hectares (180,000 feddans), only an area of 40,000 hectares (95,000 feddans) is cultivated annually. An area of 6000 hectares is annually left out as semi-fallow.

2.1.5. Irrigation Network

The irrigation network consists of six pumping stations, situated along the main canal, with a total lift of 45 m above the White Nile level. The six pumping stations are connected in series to irrigate the 40,000 hectares. Pumping stations one and two are designed to pump a maximum of 42 $m^3 \cdot s^{-1}$. The other four stations are designed with less capacity because of the diversion of water to primary canals. Irrigation water in the main canal is diverted into the primary canals through gates with flow regulators. The open field head ditches are supplied by the primary canal via off-take pipes. Each open field head ditch irrigates one field, each field is split into five to seven sectors, and each sector consists of 60 furrows with 1.55 m spacing. Furrows run perpendicular to contour lines with lengths ranging from 300 to 2750 m.

2.2. Data Collection

2.2.1. Soil Data

Soil samples for gravimetric moisture determination both for gated pipe irrigation and conventional open field head ditch were collected at furrow top, middle and end for evaluating the soil moisture deficit prior to the irrigation. Soil moisture deficit is a measure of the soil moisture between field capacity and existing moisture content multiplied by the root depth, and it represents the depth of water the irrigation system should supply; which mean the required infiltration depth. At each location along the tested furrow, soil samples were monitored at 20 cm increments to a depth of 100 cm just before and 72 hours after irrigation. Then the gravimetric moisture content (w/w) were calculated and converted to volumetric values (v/v) by multiplying by the dry bulk density [12]. Then the soil moisture change was converted to the infiltrated depth at different sites of the furrow.

2.2.2. Measurement of Hydraulic Parameters

The field comparison measurements started from the third irrigation event, so that the rate of change of infiltration rates and the effect of differences in the intake opportunity times would be minimized. In each tested field a group of four consecutive furrows were tested, using as near constant pressure head as circumstances permitted in the gated pipe and the earthen open field head ditch. The inflow rate, Advance and recession phase, runoff discharge, the time of cutoff and the volume of irrigation water applied to the tested field were used to compare the hydraulic performance.

2.2.3. Measuring Inflow Rate in Gated Pipe

In the present study a graduated glass manometer was placed inside the center of the gates that is inserted into the gated pipe system for measuring the hydraulic head of the gate which equal to distance from the water level in the manometer tube to center of the gated pipe. Stop watch and graduated bucket were also used to calculate the flow rate per unit time (m^3/s).

The distance of each gate from the water source (riser) was calculated by multiplying the gate number by gate spacing which is equal to 1.5 m apart. The amount of irrigation water flowing (Q) and its velocity (V) inside the hydro-flume was calculated using Toricelli equation (Equation (2)). Also the Reynolds numbers (Re) for each tested gated pipe were calculated using Equation (5) [13] and the type of water flow was also determined. Finally, the friction factor was calculated using Equation (4).

2.2.4. Advance and Recession Phase

The most important field data are the advance rates, which can vary throughout the irrigation season. The experimental procedure followed for determination of the advance trajectory was based upon the following steps:

1) Using field stakes and surveying tape, the tested furrows were divided into a number of stations having equal distances between them.

2) As the irrigation water advanced down the furrow, arrival times were recorded at the end of each reach.

When the flow area was equal to 10% of the maximum cross-sectional area, recession times were observed and recorded at each station.

2.2.5. Determination of the Infiltration Parameters (C, k and a)

An average values of 0.0001 m^3/min/m furrow length, 0.446 and 0.002 for the steady-state final infiltration, C, the constants a (as the slope of the regression line) and k (as the y-intercept at time $t = 1$) respectively, as computed by [14] were used in this study. [15] exploits these parameters for the determination of the average infiltrated depth as stated below;

$$Z = kt^a + Ct \qquad (7)$$

where;

Z = infiltrated volume per unit length after an infiltration opportunity time, t.

$$C = \text{the basic intake rate } (m^3 \cdot m^{-1} \cdot min^{-1}) = (Q_{in} - Q_{out})/L \qquad (8)$$

k and a = empirical fitting parameters.

Q_{in} and Q_{out} = the inflow and outflow rates, respectively, in $m^3 \cdot min^{-1}$.

L = furrow length (m).

A comparison of irrigation performance parameters between open field head ditch and gated-pipe:

2.2.6. Water Application Efficiency (Ea)

The Ea was calculated according to [16] as follows;

$$Ea = 100 \times \left\{ \left(Z_{req} \times L \right) / \left(Q_o t_{co} \right) \right\} \qquad (9)$$

where;

Ea = water application efficiency, percent.

Z_{req} = required depth of application.

L = field length (m).

Q_o = discharge/furrow (m^3/m).

t_{co} = cutoff time (minute).

2.2.7. Percent Runoff (PRO)

Losses from the irrigation system via runoff from the field end are indicated in percent runoff;

$$PRO = \text{Volume of runoff/volume of water applied to the field} \qquad (10)$$

2.2.8. Deep Percolation Ratio (DPR)

Losses of water through drainage beyond the root zone, it is defined as:

$$DPR = \text{Volume of deep percolation/volume of water applied to the field} \qquad (11)$$

2.2.9. Christiansen Uniformity (CU)

In the present study the Christiansen Uniformity (CU) as computed using the following formula;

$$CU = 100\left[1.0 - \left(x/Mn\right)\right] \tag{12}$$

where;
   x = sum of the absolute deviation of individual observations from the mean value, M.
   n = number of observations.
   The head loss in the hydro-flume was calculated using Equation (3).

2.2.10. Conveyance Water Loss

A common open field head ditch sector of 200 m was selected for assessing the conveyance water loss. The open field head ditch was 2 m wide. A thin plate rectangular weir was used for monitoring the water flow at the water source. Then the water conveyance efficiency was calculated from the water volumes measured from two sites by the following formula [17]:

$$E_c = V_f / V_t \tag{13}$$

where: E_c is conveyance efficiency, V_f is volume of water delivered to the field and V_t is volume of water delivered from the source.

2.2.11. Water Saving

The water saved is referring to the consumption differences between gated pipe irrigation system and the conventional open field head ditch method.

3. Results and Discussion

3.1. Soil Moisture Content

Table 2 and **Table 3** show the average values of the soil moisture content before and 3 days after irrigation for the two methods of irrigation for depth intervals of 20 cm down to one meter. Results indicated that in fields irrigated by the gated pipe system and due to the continuous wetting effect of irrigation, higher percentage of the volumetric moisture content (three days after irrigation) accumulated at the top layers of the soil. However under conventional furrow irrigation application using open field head ditch, results showed consistent soil water content at different sampling depths. The deep percolation of the soil moisture at infiltration depth below the effective root zone depth at the top of the furrows irrigated by gated pipe was mainly due to the low inflow rate (3 to 4 l/s) and the long irrigation time required to refill the root zone at the field end.

3.2. Distribution of Water Depth

Again **Table 2** and **Tables 3(a)-(c)** present the amount of water added (infiltrated volume) for the soil profile (0 - 100 cm) in the evaluated furrows in different irrigation events (this table is a summary of 8 samples en during the irrigation season). In the open field head ditch furrow irrigated field, the irrigation water added at the three locations was 69.1 mm at the top, 75.7 mm at the middle and 66.1 mm at the end of the furrow. Whereas, the irrigation water added using the gated pipe system at the three locations were 132.7 mm at the top, 46.1 mm at the middle and 101.9 mm at the end of the furrow, respectively.

   Analyses show that (**Table 3(c)**) the mean difference in the volumetric soil moisture content (infiltrated volume) at the top, middle and bottom of the field using open ditch is statistically insgnificant. similar analyses performed for the infiltrated volume using gated pipefurrow irrigation system indicated that the mean difference are highly significat (P $\geq$ 0.5) compared to the open ditch irrigation system.

3.3. Irrigation Parameters for Open Field Head Ditch and Hydro-Flume

Tables 4-7 showed the performance evaluation of the open field head ditch and gated-pipe irrigation systems. These findings indicate that for irrigation number 7 and 11, the gated-pipe system of irrigation has a high value of application efficiency (Ea) (79% and 88%) compared with the Ea of the open field head ditch (69% and 71%). The percent of deep percolation (PDP) for the gated-pipe system is greater than the PDP obtained under open

Table 2. Mean monthly climatic data during 2009/2010 and 2010/2011 seasons in Kenana Sugar Scheme[*].

			Crop season 2009/2010				
Month	Min. temp. (˚C)	Max temp (˚C)	Relative humidity (%) 0800 a.m.	Wind speed (km·day^{-1})	Sunshine (hours)	Radiation M·Jm^{-2} days^{-1}	Rainfall (mm)
Oct-2009	24	34	74	61	9	21.3	0.0
Nov-2009	21	33.6	46	96	10	20.4	0.0
Dec-2009	15.9	30.6	41	107	10	21.8	0.0
Jan-2010	13.6	28	42	144	10	21	0.0
Feb-2010	16.3	33.6	36	121	11	24.2	0.0
March	20.3	36.5	31	129	10	24.4	0.0
April	23.3	39.8	29	114	11	26.5	0.0
May	26.9	40.1	42	85	10	24.7	5.0
June	25.0	35.5	63	130	9	22.8	180.0
July	23.5	31.1	86	96	6	18.4	162.0
August	23.8	30.0	84	79	7	20.1	298.0
September	23.2	29.8	85	64	7	19.8	157.0
Mean	**21.4**	**33.6**	**54.9**	**102.2**	**9.2**	**22.1**	66.8
			Crop season 2010/2011				
Month	Min. temp. (˚C)	Max temp (˚C)	Relative humidity (%) 0800 a.m.	Wind speed (km·day^{-1})	Sunshine (h)	Radiation M·Jm^{-2} days^{-1}	Rainfall (mm)
Oct-2010	23.6	34.1	73	44	10	23	0.0
Nov-2010	23.1	36.1	53	58	10	21.3	0.0
Dec-2010	19.1	33.6	51	104	10	20.4	0.0
Jan-2011	18.3	32.5	48	110	10	21	0.0
Feb-2011	17.4	31.8	38	127	10	22.8	0.0
March	21.1	37.3	32	105	10	24.4	0.0
April	25.3	39.1	41	87	9	23.4	20.2
Ma	25.0	37.9	50	82	9	23.2	8.5
June	24.9	34.6	60	109	9	22.8	47.0
July	24.2	33.3	71	128	8	21.4	163.0
August	23.5	30.0	84	95	6	18.6	86.8
September	23.7	30.7	83	57	7	19.8	102.0
October 2011	23.9	34.9	70	43	10	23.1	12.0
Mean	22.5	34.3	56.8	92.1	9.0	21.9	34.9

[*]Data taken from monthly weather reports of sugarcane research meteorological station at Kenana.

Table 3. (a) Volumetric soil moisture content (Mc) before irrigation((BI) andthree days after irrigation(AI) of the furrow irrigated using open field head ditch; (b) Volumetric soil moisture content (Mc) before irrigation (BI) and three days after irrigation (AI) of the furrow irrigated using gated pipe; (c) Summary of volumetric soil moisture content (Mc) before irrigation (BI) and three days after irrigation (AI) of the furrow irrigated using open ditch and gated pipe.

(a)

Soil depth (cm)	Mc B.I % (v/v)	Mc A.I. % (v/v)	Water added (mm)
Top of the Furrow			
0 - 20	33.0	43.0	20.0
20 - 40	34.1	40.5	12.8
40 - 60	33.6	41.1	15.0
60 - 80	36.8	39.9	6.2
80 - 100	39.1	39.6	1.0
Total	176.6	204.1	55.0
ETp (3 days)			14.1
Mean	35.3	40.8	69.1
Middle of the Furrow			
0 - 20	34.0	42.5	17.0
20 - 40	32.4	41.5	18.2
40 - 60	31.8	37.5	11.4
60 - 80	33.0	39.0	12.0
80 - 100	37.0	38.5	3.0
Total	168.2	199.0	61.6
ETp[*] (3 days)			14.1
Mean	33.6	39.8	75.7
End of the Furrow			
0 - 20	35.0	45.2	20.4
20 - 40	34.6	38.5	7.8
40 - 60	32.0	36.9	9.8
60 - 80	36.0	39.0	6.0
80 - 100	38.0	42.0	8.0
Total	175.6	201.6	52.0
ETp (3 days)			14.1
Mean	35.1	40.3	66.1

ETp[*] = Potential evapotranspiration.

(b)

Soil depth (cm)	Mc B.I % (v/v)	Mc A.I. % (v/v)	Water added (mm)
Top of the Furrow			
0 - 20	32.5	56.4	47.8
20 - 40	35.0	54.2	38.4
40 - 60	36.0	49.0	26.0
60 - 80	40.1	42.5	4.8
80 - 100	42.2	43.0	1.6
Total	185.8	245.1	118.6
ETp (3days)			14.1
Mean	37.2	49.0	132.7
Middle of the Furrow			
0 - 20	30.0	33.0	6.0
20 - 40	32.0	35.0	6.0
40 - 60	33.0	38.0	10.0
60 - 80	34.0	36.0	4.0
80 - 100	32.0	35.0	6.0
Total	161.0	177.0	32.0
ETp (3days)			14.1
Mean	32.2	35.4	46.1
End of the Furrow			
0 - 20	27.0	42.5	31.0
20 - 40	29.2	38.5	18.6
40 - 60	30.4	38.0	15.2
60 - 80	33.0	37.0	8.0
80 - 100	32.5	40.0	15.0
Total	152.1	196.0	87.8
ETp (3days)			14.1
Mean	30.4	39.2	101.9

ETp* = Potential evapotranspiration.

(c)

Location	Mean Water added (mm)	
	Open ditch irrigation	Gated pipe irrigation
Top of the furrow	69.1	132.7
Middle of the furrow	75.7	46.1
Bottom of the furrow	66.1	101.9
Mean	70.3	93.6
CV%	6.99	46.9
SE±	2.84	25.3

Table 4. Evaluation of the open field head ditch irrigation system (irrigation number 7). $N = 0.04$.

K	a	c (m³/min.m)	D_{req} (cm)	Q (lps)	$t_{co(min)}$	V Applied	Slope (%)	Depth (m)
0.002	0.446	0.0001	6.8	6.9	348	144.072	0.04	0.07385

Station	Distance (m)	Advance Time (min)	Recession Time (min)	Infiltration Opport Time (min)	Infiltrated Depth (m)	Excess or Deficit (m)	Abs ($z - z_{mean}$)
1	0	0	330	330	0.1127	0.0447	0.0276
2	200	36	355	319	0.1104	0.0424	0.0253
3	400	61	337	276	0.1012	0.0332	0.0161
4	600	141	350	209	0.0859	0.0179	0.0008
5	800	150	368	218	0.0880	0.0200	0.0030
6	1000	220	380	160	0.0737	0.0057	0.0114
7	1200	262	395	133	0.0664	−0.0016	0.0186
8	1400	348	405	57	0.0421	−0.0259	0.0429

Application efficiency (Ea)	69%
Percent Deep Percolation (PDP)	0%
Percent Run Off (PRO)	31%
Christiansen uniformity (Cu)	79%

Table 5. Evaluation of the open field head ditch irrigation system (irrigation number 11). $N = 0.04$.

K	a	c (m³/min.m)	D_{req} (cm)	Q (lps)	$t_{co(min)}$	V Applied	Slope (%)	Depth (m)
0.002	0.446	0.0001	6.8	6.5	356	138.84	0.04	0.07385

Station	Distance (m)	Advance Time (min)	Recession Time (min)	Infiltration Opport Time (min)	Infiltrated Depth (m)	Excess or Deficit (m)	Abs ($z - z_{mean}$)
1	0	0	341	341	0.1150	0.0470	0.0285
2	200	40	362	322	0.1110	0.0430	0.0245
3	400	63	340	277	0.1014	0.0334	0.0149
4	600	129	355	226	0.0899	0.0219	0.0034
5	800	155	372	217	0.0878	0.0198	0.0013
6	1000	232	391	159	0.0734	0.0054	0.0131
7	1200	271	411	140	0.0684	0.0004	0.0181
8	1400	356	421	65	0.0451	−0.0229	0.0414

Application efficiency (Ea)	71%
Percent Deep Percolation (PDP)	9%
Percent Run Off (PRO)	20%
Christiansen uniformity (Cu)	79%

Table 6. Evaluation of the gated pipe irrigation system (irrigation number 7). $n = 0.04$.

K	a	c (m³/min.m)	D_{req} (cm)	Q (lps)	$t_{co(min)}$	V Applied	Slope (%)	Depth (m)
0.002	0.446	0.0001	6.8	4	460	110.4	0.04	0.05519

Station	Distance (m)	Advance Time (min)	Recession Time (min)	Infiltration Opport Time (min)	Infiltrated Depth (m)	Excess or Deficit (m)	Abs $(z - z_{mean})$
1	0	0	452	452	0.1369	0.0689	0.0366
2	200	48	467	419	0.1305	0.0625	0.0302
3	400	105	482	377	0.1223	0.0543	0.0219
4	600	180	496	316	0.1098	0.0418	0.0094
5	800	252	510	258	0.0972	0.0292	0.0031
6	1000	338	528	190	0.0813	0.0133	0.0190
7	1200	405	543	138	0.0678	−0.0002	0.0325
8	1400	460	560	100	0.0568	−0.0112	0.0435

Application efficiency (Ea)	88%
Percent Deep Percolation (PDP)	7%
Percent Run Off (PRO)	5%
Christiansen uniformity (Cu)	76%

Table 7. Evaluation of the gated pipe irrigation (irrigation number 11). $n = 0.04$.

K	a	c (m³/min·m)	D_{req} (cm)	Q (lps)	$t_{co(min)}$	V Applied	Slope (%)	Depth (m)
0.002	0.446	0.0001	6.8	3.9	525	122.85	0.04	0.05436

Station	Distance (m)	Advance Time (min)	Recession Time (min)	Infiltration Opport Time (min)	Infiltrated Depth (m)	Excess or Deficit (m)	Abs $(z - z_{mean})$
1	0	0	500	500	0.1459	0.0779	0.0393
2	200	60	533	473	0.1409	0.0729	0.0342
3	400	145	551	406	0.1280	0.0600	0.0214
4	600	223	579	356	0.1180	0.0500	0.0114
5	800	285	588	303	0.1070	0.0390	0.0004
6	1000	378	592	214	0.0871	0.0191	0.0195
7	1200	462	610	148	0.0705	0.0025	0.0361
8	1400	525	621	96	0.0555	−0.0125	0.0511

Application efficiency (Ea)	79%
Percent Deep Percolation (PDP)	12%
Percent Run Off (PRO)	9%
Christiansen uniformity (Cu)	70%

field head ditch irrigation (**Tables 4-7**). This is mainly due to the fact that the deep percolation of the soil moisture at infiltration depth below the effective root zone depth at the top of the furrows irrigated by gated pipe was mainly due to the low inflow rate (3 to 4 l/s) and the long irrigation time required to refill the root zone at the field end. The present results also revealed that the percent of runoff (PRO) was higher under the open field head ditch system (20% - 31%) compared with the gated-pipe irrigation system (5% - 9%). Previous results stated that the PRO tends to be directly related to the inflow rate and inversely related to Ea. Results of the comparison between the two system shows that the gated-pipe system is better than the open field head ditch.

The infiltration profile parameters were used to calculate the Christiansen Uniformity Coefficient (CU) using Equation (12). Results showed that the CU for the gated pipe and the open field head ditch methods were equal to 70% - 76% and 79% respectively. The low value of CU was mainly due to the longer contact time which leads to spatial and temporal variations of the soil moisture distribution which is more evident at the top part and along the field irrigated with gated pipe.

3.4. Advance and Recession Phases

Results of the Advance, recession, intake opportunity time and infiltrated volume of the two systems were shown in **Tables 6-9**. It was observed that the intake opportunity time varied with furrow length, irrigation cycle and stage of cane growth, due to the variation of soil moisture condition before irrigation. This variation did not reflect a certain trend in both irrigation methods which implies that the fields were irrigated at random and at different available water depletion levels.

Using the gated pipe irrigation technique, the slow advance rate resulted from low water outflow (3 to 4 L/s) lead to non-uniform intake opportunity time over the field. Therefore using the gated pipe due to irrigation tech-

Table 8. Comparison of the gated pipe system for furrow irrigation (GP) with the open field head ditch (FC) in term of water and time saving.

Irrigation cycle number	Irrigation time (min)		Time saved (min) When using FC	Inflow rate (L/min)		Water added (m³/fed)		Water saved m³ when using gated pipe
	FC	GP		FC	GP	FC	GP	
7		460	112	414	240	277	212	65
8	332	448	116	420	288	268	248	20
11	356	525	169	390	234	267	236	31
14	362	542	180	390	222	272	231	41
Total						1084	927	
Mean						271.00	231.75	

GP = gated pipe system for furrow irrigation; FC = open field head ditch.

Table 9. Comparison of the net, gross and total water applied, cane yield and irrigation production efficiency between the two irrigation systems.

Irrigation method	Net irrigation water added/cycle (m³/fed/cycle)	Gross irrigation water added/cycle (m³/fed/cycle) = Net × Ea	Number of irrigations applied during the season	Total irrigation water applied/ (m³/fed/season)	Cane·yield (ton cane/fed)	Irrigation production efficiency (kg cane/m³ water)
Gated-pipe	231.75	289.69	22	6373.13	50	**7.85**
Open field head ditch	271.00	381.69	18	6870.42	43	**6.26**
SD	23.58	23.43		3.54	3.54	**0.6**
CV%	10.18	8.64		0.05	7.60	**8.47**
SE±	5.03	5.52		2.5	2.50	**0.42**

Data taken from the field management department of Kenana Sugar Scheme.

nique the advance phase should be completed as quickly as possible so that the intake opportunity time over the field will be uniform and then cut the inflow off when enough water has been added to refill the root zone. This can be accomplished with a high, but non-erosive, discharge onto the field. These options are available to solve this problem, partially:

a) Dyke the downstream end to prevent runoff;
b) Reduce the inflow discharge to a rate more closely approximating the cumulative infiltration along the field following the advance phase.

3.5. Water Conveyance Efficiency

The conventional open field head ditch water loss test (Equation (10)) shows that the water lost from 200 m study section for a 180 minute time period was 10 m^3. The volume of water delivered to the field (V_f) was found to be 74 m^3 and the volume of water delivered from the source (V_t) was 84 m^3. Most of the water loss apparently occurred through canal leakage. The water conveyance efficiency for the open field head ditch was equal to 88%.

3.6. Water and Time Saving

Compared with open field head ditch method, the gated pipe system for furrow irrigation can reduce the irrigation quota by 20 to 65 m^3/irrigation cycle depending on the pre-irrigation soil water content. The irrigation time saved when using open field head ditch system was ranging from 112 to 180 minutes per irrigation cycle depending on furrow length and the pre-irrigation soil water content (**Table 8**).

3.7. Irrigation Production Efficiency (IPE)

A comparison of the net, gross and total water applied, cane yield and irrigation production efficiency (IPE) between the two irrigation systems was shown in **Table 9**. Results of the comparison indicate that the difference in the net and total irrigation water applied are highly significant ($P \geq 0.05$). It was found that the field irrigated using the gated pipe system out yielded the field irrigated with open field head ditch but the difference is insignificant ($P \geq 0.05$). Whereas, the obtained value of the irrigation production efficiency (IPE) of the field irrigated with gated-pipe system is statistically significant ($P \geq 0.05$) compared with the IPE of the field irrigated with open field head ditch (**Table 9**).

3.8. Head-Loss through the Gated Pipe

In the present study the calculated friction factor (*f*) values (**Table 10**) were used for the determination of the loss in the total water head of the gated pipe. Results indicated that the magnitude of the loss in the total water head was directly related to the distance from water source (the riser). Higher head produces more water from the gates.

4. Economical Analysis

Following the results of the present study, the total irrigation cost saved for sugarcane production when using cultivar Co 6806 and irrigated when 55% to 60% of the available soil moisture is depleted (treatment D2). was determined as below.

Table 10. Calculation of Reynold number, friction factors and the total head loss.

Distance from riser (m)	Measured inflow rate (m^3/s)	Reynold number (Re)*10^6	Friction factor (*f*) $f = 0.13/Re^{0.172}$	Head loss (m)	Christ. factor (*C*)	Head loss corrected (m)
2	0.002	1.8	0.011	0.031	0.335	0.010
6	0.0036	1.5	0.011	0.090	0.335	0.030
23	0.0062	1.4	0.011	0.284	0.335	0.095
81	0.0045	1.1	0.012	0.667	0.335	0.223
99	0.005	1.8	0.011	1.918	0.335	0.643

5. Conclusions

From the results the following conclusions can be drawn;

The spatial and temporal variations of the soil moisture distribution which is more evident along the field irrigated with gated pipe resulted from prolonged irrigation time. In contrast to the gated pipe system, the open field head ditch furrow irrigation method provides more uniform water distribution along the irrigated furrows.

The present study indicates that the gated-pipe system has a high value of application efficiency (79% - 88%) compared with the open field head ditch (69% - 71%). The percent of deep percolation (PDP) for the gated-pipe system is greater than the PDP obtained under open field head ditch irrigation conditions mainly at the top part of the field due to the longer contact time of irrigation. Also the percent runoff (PRO) is higher under the open field head ditch system and the water conveyance efficiency for the open field head ditch is 88%. While the gated pipe needs more advance time, it can save 20 to 65 m^3 of water/irrigation cycle with slightly lower uniformity coefficient (CU) and higher irrigation production efficiency (IPE) compared with the open field head ditch. In general, the hydro flume irrigation performance parameters are better than the open field head ditch. There are certain parameters, such as CU, which can be improved with more uniform infiltration opportunity time.

6. Recommendations

From the above mentioned results, it is concluded that, under Kenana conditions, the gated-pipe system is better than the open field head ditch irrigation system keeping in mind that for more uniform water distribution through irrigated furrows of the long fields of Kenana, increased pressure head at the inlet and/or larger openings of the hydro-flume gates may be necessary.

References

[1]   Grillo, G., Walker, W.R. and Skogerboe, G.V. (1987) Mathematical Approach for Evaluating the Zero-Inertia Model, Surface Irrigation: Theory and Practice. Utah State University, New Jersey, 328-336.

[2]   FAO (Food and Agriculture Organization) (1993) Small-Scale Irrigation for Arid Zones: Principles and Options. FAO, Rome.

[3]   Raddy and Abdel-Hady, M. (1993) Project Evaluation of Irrigating System on Sandy and Calcareous Soils, Cairo.

[4]   Smith, R.J., Watts, P.J. and Mulder, S.J. (1986) Analysis and Design of Gated Irrigation Pipelines. *Agriculture and Water Management*, **12**, 99-115.

[5]   Michael, A.M. (1978) Irrigation Efficiency Principles of Agricultural Engineering. Vol. 11, Jain Brothers, Jodhpur, 113-178.

[6]   Abdel Wahab, D.M. (1996) Evaluation of Irrigation System under Kenana Condition. Annual Report of Sugarcane Research Department, Kenana Sugar Scheme, Sudan.

[7]   Torricelli, E. (1643) Italian Physicist and Mathematician. http://en.wikipedia.org/wiki/Evangelista_Torricelli

[8]   Abdel Wahab, D. M. (2004) Evaluation of Gated-Pipe Furrow Irrigation System under Kenana Condition. Annual Report of Research & Development.

[9]   Darcy, H. (1845) Recherches Expérimentales Relatives au Mouvement de L'eau Dansles Tuyaux, Mallet-Bachelier, Paris. 268 Pages and Atlas (in French). http://en.wikipedia.org/wiki/Darcy_friction_factor_formulae

[10] Blasius, H. (1913) Blasius Boundary Layer. *Mathematical Physics*, **56**, 1-37. http://naca.larc.nasa.gov/reports/1950/naca-tm-1256

[11] Christiansen, J.E. (1942) Irrigation by Sprinkler. *Agriculture Experimental Station Bulletin*, **37**, 1-124.

[12] Farbrother, H.G. (1973) Water Requirements of Crops in the Gezira. Annual Report of the Gezira Research Station, 139-172.

[13] Reynolds, O. (1883) An Experimental Investigation of the Circumstances Which Determine Whether The Motion of Water Shall Be Direct or Sinuous and of the Law of Resistance in Parallel Channel. *Philosophical Transactions of the Royal Society*, **174**, 935-982.

[14] Abdel Wahab, D.M. (2000) Evaluation, Prediction and Optimization of Long Furrow Irrigation under Kenana Conditions. Ph.D. Dissertation, Irrigation and Water Management Institute, University of Gezira, Wad Medani.

[15] Kostiakov, A.N. (1932) On the Dynamics of the Coefficient of Water Percolation in Soils and on the Necessity for Studying it from a Dynamic Point of View for Purpose of Amelioration. *Transactions of 6th Committee International Society of Soil Science*, Russia, Part A, 17-21.

[16] Hart, W.E., Peri, G. and Skogerboe, G.V. (1979) Irrigation Performance: An Evaluation. *Journal of Irrigation and Drainage Engineering*, **105**, 275-288.

[17] Bos, M.G. (1979) Standards for Irrigation Efficiencies of ICID. *Journal of Irrigation and Drainage Engineering*, **105**, 37-43.

3

Heavy Metals Accumulation in Soil Irrigated with Industrial Effluents of Gadoon Industrial Estate, Pakistan and Its Comparison with Fresh Water Irrigated Soil

Noor Amin[1*], Dawood Ibrar[1], Sultan Alam[2]

[1]Department of Chemistry, Abdul Wali Khan University, Mardan, Pakistan
[2]Department of Chemistry, University of Malakand, Chakdara, Pakistan
Email: *noorulamin_xyz@yahoo.com

Abstract

Wastewater mixed with industrial effluents is used for irrigation in Gadoon Industrial estate and thus contaminating soil. This soil was tested for heavy metal content by using atomic absorption spectrophotometer (perkin elmer 700) and compared with control soil irrigated with tube well water at seven selected spots. Accumulation of the toxic metal was significantly greater in the soil irrigated with industrial effluent than control soil ($p < 0.05$). Manganese (Mn) was the most significant pollutant, accumulated up to 9.95 ppm in the soil irrigated with industrial waste water. It was found that the samples were containing Zn in the range of 1.596 - 6.288, Cu 0.202 - 1.236, Co 0.074 - 0.115, Ni 0.0002 - 0.544, Cr 0.243 - 0.936, Mn 3.667 - 9.955 and Pb 0.488 - 1.259 ppm. No sample was containing the heavy metal above the critical level mentioned in typical and unsafe heavy metal levels in soil.

Keywords

Industrial Effluents, Soil, Tube Well Water, Heavy Metals, Pollutio

1. Introduction

Industry is the backbone of a country for its development and with the growing population, the need for establishing new industries is increasing. Industries on one side manufacture useful products but on the other side

---

*Corresponding author.

time generates waste products, causing various environmental problems. The waste products may be in the form of solid, liquid or gas which lead to the creation of hazards, pollution and losses of energy. The wastes containing different pollutants and heavy metals are discharged into water and soil and ultimately pose a serious threat to human and ecosystem. By-products of different industries like textile, metal, dying chemicals, fertilizers, pesticides, cement, petrochemical, energy and power, leather, sugar processing, construction, steel, engineering and food processing industries are the main contributors to the soil pollution. So the rapid industrialization is accompanied by both direct and indirect adverse effects on environment [1]. Industrial development may result in the generation of industrial effluents. It has been studied that many industries discharge untreated effluents into river and only 10% industries surveyed had primary treatment ranging from oxidation tanks, sedimentation tanks in developing countries [2] [3].

Heavy metal toxicity may results into a number of health problems including a damaged or reduced mental and central nervous function, lower energy levels, and damage to blood composition, lungs, kidneys, liver, and other vital organs. Long-term exposure may result in a slowly progressing physical, muscular, and neurological degenerative process, that mimic Alzheimer's disease, Parkinson's disease, muscular dystrophy, and multiple sclerosis. Allergies are not uncommon and repeated long-term contact with some metals or their compounds may even cause cancer [4] [5].

Heavy metals on the basis of their health importance can be classified into four major groups, as essential, like Cu, Zn, CO, Cr, Mn and Fe, which are micronutrients and are toxic when taken in excess [6] [7], non essential like Ba, Al, Li and Zr, less toxic like Sn and Al, and highly toxic like Hg and Cd. The toxicity limit and recommended or safe intake of some of heavy metals for human health is given in **Table 1**.

In small quantities, certain heavy metals are nutritionally essential for a healthy life. Some of these are referred to as the trace elements (e.g., iron, copper, manganese, and zinc). These elements, or some form of them, are commonly found naturally in foodstuffs, in fruits and vegetables, and in commercially available multivitamin products [8]-[10].

Because of the rapid industrialization, soil pollution by heavy metals is becoming a serious problem. Being an ultimate sink for industrial wastes, almost all industrial wastes are dumped into soil. Heavy metals in wastes find specific adsorption sites in soil where they are retained relatively stronger either on inorganic or organic colloids. [11]-[13]. Research has proven that long term use of sewage effluent for irrigation contaminates soil and crops to such an extent that it becomes toxic to plants and causes deterioration of soil. This contains considerable amount of potentially harmful substances including soluble salts like Fe^{2+}, Cu^{2+}, Zn^{2+}, Mn^{2+}, Ni^{2+} and Pb^{2+}. Additions of these heavy metals are Undesirable [14] [15].

The properties of the soil along with the climate change also changes due to anthropogenic impact. The influence of acid rains on soils and sorption properties of soil has been extensively studied by scientists from various disciplines. In almost all cases, they found that acid rains decrease the ability of binding heavy metals to soil particles. However, for naturally high acidic soils or very weak soils like rusty soils, the effect of acid rains on soils is shown to be much smaller [15].

The present study was conducted with an aim to study the impact of industrial effluents of different industries of Gadoon industrial estate Pakistan, carrying different heavy metals in them and absorbed in soil during irrigation. In this study only seven elements like Cu, Zn, Co, Cr, Pb, Ni and Mn were studied. The results of analysis collected from different locations have been compared and reported.

2. Experimental

Soil samples were collected from seven different sites of the Gadoon industrial estate Gadoon amazai swabi.

Table 1. Toxic limit/recommended/safe intake of heavy metals [12].

Heavy metal	Toxic limit	Recommended intake/Safe intake
Arsenic	3 mg/day for 2 - 3 weeks	15 - 25 micro g/day (adults)
Cadmium	200 micro/kg of fresh weight	15 - 50 micro g/day (adults), 2 - 25 micro g/day children
Lead	>500 micro g/L (Blood)	20 - 280 micro g/day (adults), 10 - 275 micro g/day children
Zinc	150 micro/day	15 micro g/day

The samples identification along with their location is shown in **Table 2**. From each sampling site five soil samples were collected from different location and mixed together. The samples were dried in the sun for four days, grinded and sieved. The samples were then reduced to laboratory samples by using the tabling process. The samples were named as S_1, S_2, S_3, S_4, S_5, S_6 and S_7. The samples were then dried in oven at 110°C for about five hours to remove moisture completely. The chemicals used were Nitric acid, Per-chloric acid, distilled water.

2.1. Nitric-Perchloric Acid Digestion

Nitric-perchoric acid digestion method was performed for sample preparation [7]. One gram of a sample was placed in 250 ml digestion tube and 10 ml of concentrated HNO_3 was added. The mixture was boiled for 30 - 45 minutes to oxidize all easily oxidizable matter. After cooling, 5 ml of 70% $HClO_4$ was added and the mixture was boiled gently till the appearance of dense white fumes. The contents were cooled and 20 ml of distilled water was added, and re boiled to stop the release of any fumes. The solution was cooled again, filtered off through Whatman No. 42 filter paper and transferred to 25 ml volumetric flask. The volume was made up to the mark with distilled water. Blank solution was prepared with the same procedure except the addition of soil sample.

Standards for different elements were prepared from the stock solutions (1000 ppm) using dilution method. For each element different dilutions were made for calibration curve given in **Table 3**.

2.2. Analysis of Trace Metals

Trace metals such as Cu, Co, Fe, Pb, Cr, Mn, Zn and Ni were analyzed in all the samples using atomic absorption spectrophotometer.

2.3. Statistical Analysis

Data obtained during current study was analyzed statistically for mean, standard deviation, ANOVA and Duncan Multiple Range Test (DMRT) by using SPSS for windows, version 16.0 (SPSS Inc., Chicago, IL, USA). Probability less than 0.05 was accepted as significant.

Table 2. Locations at Gadon Amazai from where different samples were collected.

S. No	Sample ID	Sample Location
1	S_1	2 km from Sardar Chemical Industries
2	S_2	2500 m Sardar Chemical Industries and Cherat Paper Sack
3	S_3	1 km Shafi Chemical Industries and Hamza Steel Industries
4	S_4	500 m Shafi Chemical Industries Plot-2
5	S_5	500 m T.W Metal Recycling Industries and Poyal Jadoon Marble Factory
6	S_6	1 km T.W Metal Recycling Industries Plot 11/16
7	S_7	200 m Shafi Enterprises

Table 3. Standards used for different metals.

S/No	Standard Name	Standard Symbol	Concentration (ppm)		
1	Copper	Cu	2	4	8
2	Zinc	Zn	1	2	4
3	Cobalt	Co	7	14	21
4	Nickel	Ni	7	14	28
5	Manganese	Mn	2.5	5	10
6	Chromium	Cr	2	4	8
7	Lead	Pb	12.5	25	50

3. Results and Discussion

The concentration of copper in soil of the study area ranged from 0.202 to 1.236 ppm (**Figure 1**). The highest amount of this metal was present at location 5, situated on a distance of 500 m from T.W Metal Recycling Industries and Poyal Jadoon Marble Factory (**Table 2**). Accumulation of this metal in soil irrigated with industrial effluent was significantly greater than its amount in soil irrigated with tube well water ($p < 0.05$). The Typical background levels for non contaminated soil is 1 - 50 ppm, if the concentration of copper in the soil exceeds 200 ppm it become unsafe for leafy vegetable while the concentration of copper greater than 500ppm become unsafe for garden and children contact [14]. The results show that the study area contains the copper metal in a permissible range.

The concentration of zinc in the studied area varied from 1.596 to 6.288 ppm and its average concentration was 2867 ppm (**Figure 2**). At all locations, the accumulation of Zn was significantly greater in soil irrigated with wastewater than control coil ($p < 0.05$). Zinc is the second most abundant heavy metal found in the study area. The typical background levels of zinc for non contaminated soil is very broad, *i.e.* 9 - 125 ppm, while its concentration above 200 ppm is unsafe for leafy or root vegetables. If the concentration of zinc in soil exceeds 500 ppm, it is considered unsafe for gardens and children contact [15]. The results of the present study show that the study area contains zinc in permissible range.

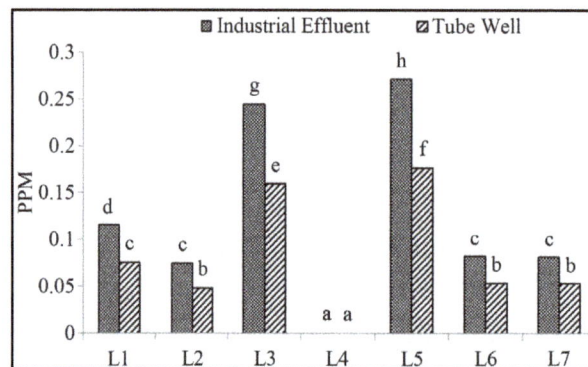

Figure 1. Accumulation of cobalt in soil irrigated with industrial effluent and tube well water at different locations in Gadoon Amazai industrial zone, Pakistan. Bars represents mean value of three replicates and bars labeled with different letters are significantly different from each other (Duncan Multiple Range test; $p < 0.05$).

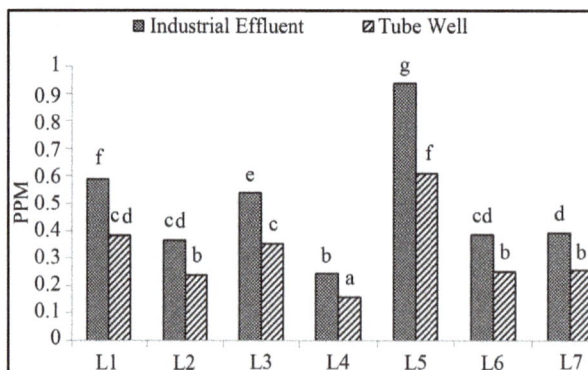

Figure 2. Accumulation of chromium in soil irrigated with industrial effluent and tube well water at different locations in Gadoon Amazai industrial zone, Pakistan. Bars represents mean value of three replicates and bars labeled with different letters are significantly different from each other (Duncan Multiple Range test; $p < 0.05$).

The concentration of cobalt was in the range of 0.001 to 0.271 ppm at different locations in the study area and its average concentration was 0.124 ppm (**Figure 3**). The highest concentration of this metal was also in the soil irrigated with contaminated water at location 5 ($p < 0.05$). Concentration of this metal was well below the toxic level in the study area and minimum among all metals detected over there.

From **Figure 4**, it is obvious, that the concentration of Nickel in the studied area was below the detection limit in soil from L1 and L2. Amount of this metal was greatest (0.38 - 0.544 ppm) in wastewater irrigated soil of location 5 and 6, which was significantly greater than the concentration of Ni in other locations ($p < 0.05$). Although the abundance of Ni was significantly greater in soil supplied with wastewater than soil irrigated with tube well water (**Figure 5**), nevertheless, it was well below the critical levels of this metal [15]. Nickle is the sixth abundant heavy metal found in the study area. The typical background levels of nickel for non contaminated soil is very broad, *i.e.* 0.5 - 50 ppm, while its concentration above 200 ppm is unsafe for leafy or root vegetables. If the concentration of nickel in soil exceeds 500 ppm, it is considered unsafe for gardens and children contact [15] [16].

Chromium was the fifth abundant heavy metal in all the locations of the study area (**Figure 6**). Its concentration varied from 0.243 in sample-4 to 0.936 ppm in sample-5 and its average concentration was found to be 0.493 ppm. Soil of location 5 had the greatest concentration of this metal ($p < 0.05$), irrespective of the water source used for irrigation.

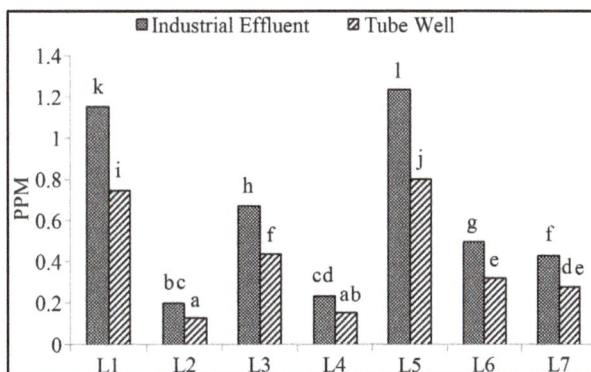

Figure 3. Accumulation of copper in soil irrigated with industrial effluent and tube well water at different locations in Gadoon Amazai industrial zone, Pakistan. Bars represents mean value of three replicates and bars labeled with different letters are significantly different from each other (Duncan Multiple Range test; $p < 0.05$).

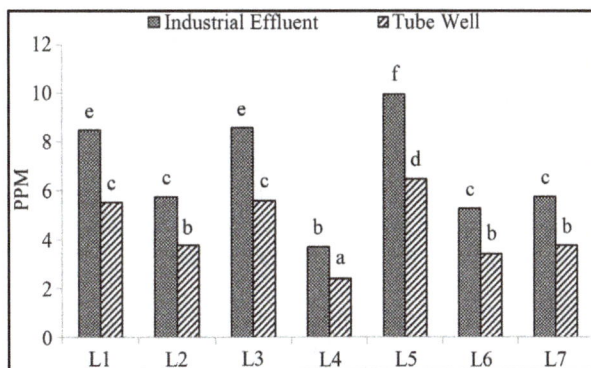

Figure 4. Accumulation of manganese in soil irrigated with industrial effluent and tube well water at different locations in Gadoon Amazai industrial zone, Pakistan. Bars represents mean value of three replicates and bars labeled with different letters are significantly different from each other (Duncan Multiple Range test; $p < 0.05$).

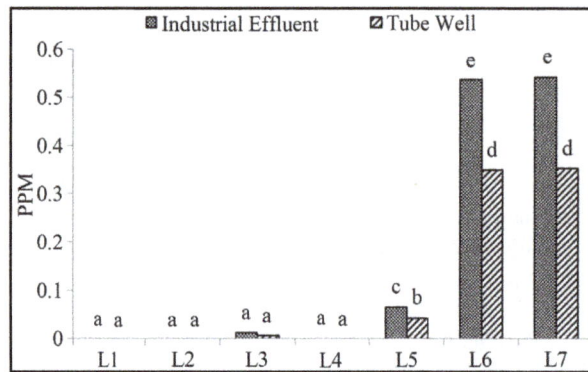

Figure 5. Accumulation of nickel in soil irrigated with industrial effluent and tube well water at different locations in Gadoon Amazai industrial zone, Pakistan. Bars represents mean value of three replicates and bars labeled with different letters are significantly different from each other (Duncan Multiple Range test; $p < 0.05$).

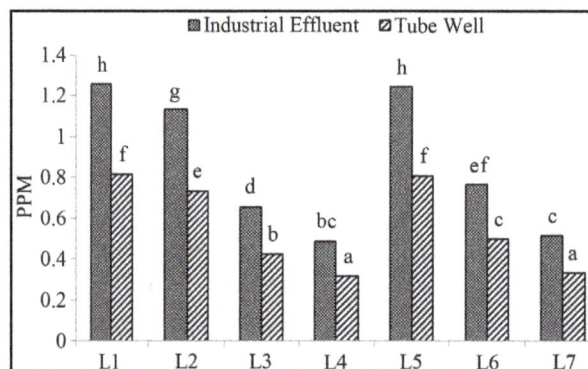

Figure 6. Accumulation of lead in soil irrigated with industrial effluent and tube well water at different locations in Gadoon Amazai industrial zone, Pakistan. Bars represents mean value of three replicates and bars labeled with different letters are significantly different from each other (Duncan Multiple Range test; $p < 0.05$).

The most abundant heavy metal found in the study area was manganese which varied from 3.667 in sample-4 to 9.955 ppm in sample-5 (**Figure 7**).

Lead in the study area varied from 0.488 in sample 4 to 1.259 ppm in sample (**Figure 8**). Lead was the third most abundant heavy metal found in our study area. The Typical background level of lead for non contaminated soil is 10 - 70 ppm. When the concentration of lead goes above 500 ppm, it become unsafe for leafy or root vegetables. If the concentration of zinc in soil exceeds 1000 ppm, it is then considered unsafe for gardens and children contact [17] [18]. The result of the present study shows that the study area contains zinc in permissible range. Overall the concentration of heavy elements in the study area are given in the order Mn > Zn > Pb > Cu > Cr > Ni > Co.

The result showed a high level of zinc and manganese and low level of cobalt and nickel in soil samples. The variations of the metal contents observed in these soil samples depends on the physical and chemical nature of the soil and absorption capacity of each metal in the soil which is altered by the innumerable environmental factors and nature of soil (Reference). Location 5 contained greater amount of heavy metals, which was elevated to even greatest by irrigation with industrial effluent. Although at all locations, the metals were well below their critical values; however, continuous monitoring of the soil may help to predict any increase in future.

Soils normally contain low background levels of heavy metals. Excessive levels of heavy metals can be hazardous to man, animals and plants. Heavy metals regulated by the EPA are arsenic, cadmium, copper, lead, nickel, selenium, and Zinc. The limits of some heavy metals for non contaminated soil and plants and garden are given in **Table 4**. Unsafe levels of heavy metals affect the soil texture, organic matter and pH.

4. Conclusion

The present study showed that heavy metals found in soil irrigated by the effluents of Gadoon industrial area are in the order of Mn > Zn > Pb > Cu > Cr > Ni > Co and their concentration is below the typical background

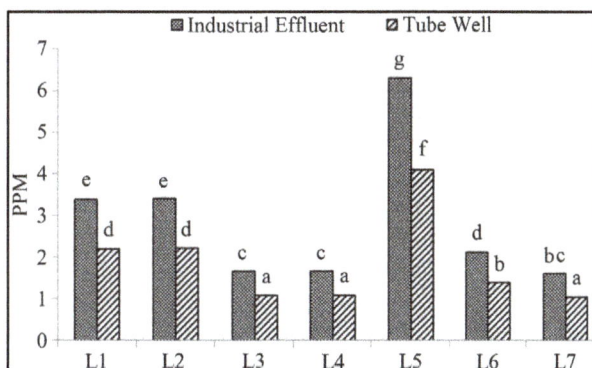

Figure 7. Accumulation of zinc in soil irrigated with industrial effluent and tube well water at different locations in Gadoon Amazai industrial zone, Pakistan. Bars represents mean value of three replicates and bars labeled with different letters are significantly different from each other (Duncan Multiple Range test; $p < 0.05$).

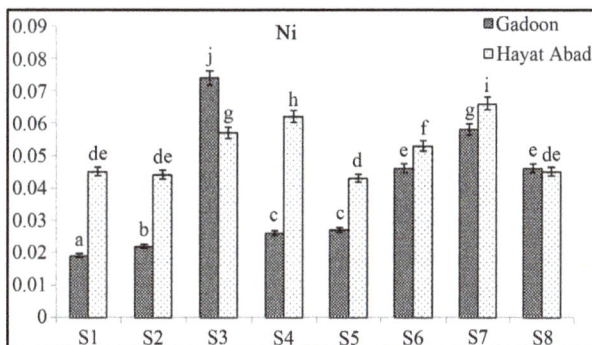

Figure 8. Mean concentration of Ni in the discharge of various industries found in the industrial area of Gadoon Amazai, Sawabi and Hayat Abad, Peshawar. Bars labeled with different alphabets are significantly different from each other (Duncan Multiple Range Test; $p < 0.05$).

Table 4. Typical and unsafe heavy metal soil levels [13].

Heavy metal	Typical background levels for non contaminated soil (ppm)	Unsafe for leafy or root vegetables (ppm)	Unsafe for gardens and children contact (ppm)
Cadmium	0.1 - 1.0	>10	50
Copper	1 - 50	>200	500
Lead	10 - 70	>500	1000
Nickel	0.5 - 50	>200	500
Zinc	9 - 125	>200	500

level for non-contaminated soil as given in **Table 4**. This shows that plants and vegetables grown in this soil should have no adverse effects on animals and human beings. However, it may be emphasized that prolong use of industrial effluent for irrigation may lead to accumulation of heavy metals to toxic level in the soil.

References

[1]   Amin, N., Hussain, S., Alamzeb, A. and Begum, S. (2013) Accumulation of Heavy Metals in Edible Parts of Vegetables Irrigated with Waste Water and Their Daily Intake to Adults and Children, District Mardan, Pakistan. *Food Chemistry*, **136**, 1515-1532. http://dx.doi.org/10.1016/j.foodchem.2012.09.058

[2]   Amin, N., Shah, M.T. and Ali, K. (2009) Investigation of Raw Material for the Manufacturing of Sulphate-Resisting Cement in Darukhula Nizampur, NWFP, Pakistan. *Magazine of Concrete Research*, **61**, 779-785. http://dx.doi.org/10.1680/macr.2008.61.10.779

[3]   Ali, K., Amin, N. and Shah, M.T. (2008) Chemical Study of Limestone and Clay for Cement Manufacturing in Darukhula, Nizampur District, Nowshera, North West Frontier Province (N.W.F.P.), Pakistan. *Chinese Journal of Geochemistry*, **27**, 242-248. http://dx.doi.org/10.1007/s11631-008-0242-8

[4]   Wang, L.X., Guo, Z.H., Xiao, X.Y., Chen, T.B., Liao, X.Y. and Song, J. (2008) Variations and Trends of the Freezing and Thawing Index along the Qinghai-Xizang Railway for 1966-2004. *Journal of Geographical Sciences*, **18**, 3-16. http://dx.doi.org/10.1007/s11442-008-0003-y

[5]   Rattan, R.K., Datta, S.P., Chhonkar, P.K., Suribabu, K. and Singh. A.K. (2005) Long-Term Impact of Irrigation with Sewage Effluents on Heavy Metal Content in Soils, Crops and Groundwater—A Case Study. *Agriculture, Ecosystems & Environment*, **109**, 310-322. http://dx.doi.org/10.1016/j.agee.2005.02.025

[6]   Amin, N. (2010) Chemical Activation of Bagasse Ash in Cementitious System and Its Impact on Strength Development. *Journal of the Chemical Society of Pakistan*, 32, 481-484.

[7]   Sharma, R.K., Agarwal, M. and Marshall, F. (2007) Heavy Metals Contamination of Soil and Vegetables in Suburban Areas of Varanasi, India. *Ecotoxicology and Environmental Safety*, **66**, 258-266.

[8]   Sridhara, N., Chary, C.T. and Samuel, S.R.D. (2008) *Asian Journal of Ecotoxicology*, **69**, 43.

[9]   Zhou, S.L., Lu, C.F. and Wan, H.Y. (2005) *Journal of Henan Normal University* (*Natural Science*), **33**, 12.

[10]   Xie, Z.M., Li, J., Chen, J.J. and Wu, W.H. (2006) *Asian Journal of Ecotoxicology*, **1**, 2.

[11]   Zhang, C.S. (2006) Using Multivariate Analyses and GIS to Identify Pollutants and Their Spatial Patterns in Urban Soils in Galway, Ireland. *Environmental Pollution*, **142**, 501-511. http://dx.doi.org/10.1016/j.envpol.2005.10.028

[12]   US EPA National Recommended Water Quality Criteria. Federal Register, Care (1998).

[13]   Cui, Y.J., Zhu, Y.G., Zhai, R.H., Chen, D.Y., Huang, Y.Z. and Qiu, Y. (2004) *Environment International*, **3**, 25.

[14]   Ding, A.F., Pan, G.X. and Li, L.Q. (2006) Distribution of PAHs in Particle-Size Fractions of Selected Paddy Soils from Tai Lake Region, China and Its Environmental Significance. *Acta Sci. Circum.*, **26**, 293-299.

[15]   Melloul, O.L., Hassani, A. and Bouhoum, K. (2002) *International Journal of Environmental Health*, **23**, 21.

[16]   Fleck, J.A., Grigal, D.F. and Nater, E.A. (1999) Mercury Uptake by Trees: An Observational Experiment. *Water Air and Soil Pollution*, **115**, 513-523. http://dx.doi.org/10.1023/A:1005194608598

[17]   Khan, S., Cao, Q., Zheng, Y.M., Huang, Y.Z. and Zhu, Y.G. (2008) Health Risks of Heavy Metals in Contaminated Soils and Food Crops Irrigated with Wastewater in Beijing, China. *Environmental Pollution*, **152**, 686-692. http://dx.doi.org/10.1016/j.envpol.2007.06.056

[18]   Lupton, G.P., Kao, G.F., Johnson, F.B., Graham, J.H. and Helwig, E.B. (1985) Cutaneous Mercury Granuloma. *Journal of American Academy of Dermatology*, **12**, 296-303. http://dx.doi.org/10.1016/S0190-9622(85)80039-6

4

Spectral Crop Coefficient Approach for Estimating Daily Crop Water Use

Nithya Rajan[1*], Stephan J. Maas[2]

[1]Texas A & M AgriLife Research and Extension Center, Vernon, Texas, USA
[2]Texas Tech University and Texas A & M AgriLife Research Center, Lubbock, Texas, USA
Email: [*]nrajan@ag.tamu.edu

Abstract

While the amount of water used by a crop can be measured using lysimeters or eddy covariance systems, it is more common to estimate this quantity based on weather data and crop-related factors. Among these approaches, the standard crop coefficient method has gained widespread use. A limitation of the standard crop coefficient approach is that it applies to "standard conditions" that are invariant from field to field. In this article, we describe a method for estimating daily crop water use (CWU) that is specific to individual fields. This method, the "spectral crop coefficient" approach, utilizes a crop coefficient numerically equivalent to the crop ground cover observed in a field using remote sensing. This "spectral crop coefficient" K_{sp} is multiplied by potential evapotranspiration determined from standard weather observations to estimate CWU. We present results from a study involving three farmers' fields in the Texas High Plains in which CWU estimated using the K_{sp} approach is compared to observed values obtained from eddy covariance measurements. Statistical analysis of the results suggests that the K_{sp} approach can produce reasonably accurate estimates of daily CWU under a variety of irrigation strategies from fully irrigated to dryland. These results suggest that the K_{sp} approach could be effectively used in applications such as operational irrigation scheduling, where its field-specific nature could minimize over-irrigation and support water conservation.

Keywords

Crop Coefficient, Evapotranspiration, Water Use, Ground Cover, Remote Sensing

1. Introduction

Information on the amount of water used by a growing crop is useful in a number of applications. These include

---

[*]Corresponding author.

irrigation scheduling, selecting water-efficient crops and cropping systems, and assessing the impact of crops on regional water resources. Crop water use can be measured in the field using devices such as weighing lysimeters and eddy covariance systems. Weighing lysimeters are immovable and thus can only collect data on water use from the fields in which they have been installed. Eddy covariance systems can be moved from field to field to make measurements, but their substantial cost often limits the number that is deployed in a project. While accurate, these devices are not appropriate for collecting water use data from a large number of fields within a region.

An alternative approach to measuring crop water use is to estimate it. Equations that take into account the energy balance and aerodynamic characteristics of the plant canopy, such as the Penman-Monteith Equation, have been used to estimate the evapotranspiration of crops and other vegetation [1]. A simplification of this approach commonly used in irrigation scheduling involves splitting the computations into two steps. In the first step, potential evapotranspiration (PET) for a hypothetical reference crop is calculated from weather data using a modified form of the Penman-Monteith Equation. In the second step, this "reference evapotranspiration" (ET_0) is multiplied by an empirically determined factor (the "crop coefficient" K_c) to produce a value of "crop evapotranspiration" (ET_c) presumably appropriate for the crop under investigation. This "crop coefficient approach" has been standardized for application to a wide range of crops around the world [2].

Standard crop coefficients are designed to estimate ET_c under "standard conditions" which represent "the upper envelope of crop ET where no limitations are placed on crop growth or ET due to water shortage, crop density, or disease, weed, insect, or salinity pressures" [2]. Thus, the standard crop coefficient approach can indicate how much water a crop would be using if it were growing under "standard" (non-limited) conditions. It can't tell you how much water the crop in a particular field is actually using, particularly if the growth of the crop is limited by management practices such as deficit irrigation. In such situations, the standard crop coefficient approach will likely result in over-estimation of the amount of water used by the crop.

Standard crop coefficients are often determined from field studies involving measurements of ET made using weighing lysimeters. They are determined as the ratio of measured ET to calculated ET_0. Thus, crop coefficients are not physical quantities and cannot be directly measured in the field. This feature makes it difficult to adjust the values of crop coefficients to account for non-standard growing conditions. A few attempts have been made to use remote sensing observations to evaluate or adjust the values of crop coefficients [3]-[9]. In these cases, empirical relationships were developed between K_c and remote sensing data, usually in the form of a vegetation index like NDVI. Such relationships are likely to be site-specific.

Over the past decade, we have conducted research to develop a method for estimating crop water use that does not suffer from the limitations of the standard crop coefficient approach. The result is the "spectral crop coefficient" (K_{sp}) approach [10]. Characteristics of the spectral crop coefficient are as follows: 1) it is specific to individual fields (*i.e.*, it does not represent "standard conditions" but rather the actual conditions occurring in a field); 2) it represents a physical quantity that can be measured in a field; and 3) it can be used to estimate the actual amount of water taken up and used by the growing crop (*i.e.*, crop water use rather than ET). In this article, we will provide an overview of the K_{sp} approach and present results from recent studies demonstrating its effectiveness in estimating crop water use under a variety of conditions.

2. Materials and Methods

2.1. Spectral Crop Coefficient Approach

Crop water use (CWU) can be defined as the water from irrigation, precipitation, and/or soil moisture reserves actually used by the growing crop. It is the water taken up by the plant roots and either used in photosynthesis (PS) or lost through transpiration (TR). So,

$$CWU = PS + TR \tag{1}$$

In general, TR is much greater than PS, so

$$CWU \approx TR \tag{2}$$

CWU is not the same as ET, since ET includes soil evaporation.

In the spectral crop coefficient approach, daily CWU can be estimated using the following relationship,

$$CWU = K_{sp} \cdot PET_{fc} \cdot F_{stress} \tag{3}$$

where PET_{fc} is the PET of a well-watered (non-stressed) crop with complete ground cover, and F_{stress} is a stress factor with a value ranging from 1 (no water stress) to 0 (fully stressed, non-transpiring plants) [10]. K_{sp} is numerically equivalent to the ground cover (GC) of the crop, so it will have a value ranging from 1 (complete ground cover) to 0 (bare soil). Under non-stress conditions ($F_{stress} = 1$), Equation (2) essentially states that CWU for a field is equal to the PET of the crop calculated as if it had 100% GC, multiplied by the fraction of GC actually present in the field. F_{stress} is introduced into the equation to reduce CWU under conditions where stomatal closure significantly reduces transpiration. Previous studies have suggested that, for crops that are acclimated to their field environment, the value of F_{stress} should be approximately 1 [10].

PET_{fc} in $kg \cdot m^{-2} \cdot s^{-1}$ is calculated using a form of the Penman-Monteith Equation,

$$PET_{fc} = \left(\frac{1}{\lambda}\right) \frac{\Delta R_{n,fc} + \rho_a c_p (e_s - e_a)/r_{a,fc}}{\Delta + \gamma (1 + r_{cp}/r_{a,fc})} \tag{4}$$

in which $R_{n,fc}$ is the net radiation in $MJ \cdot m^{-2} \cdot s^{-1}$, $(e_s - e_a)$ is the vapor pressure deficit of the air in kPa, ρ_a is the air density in $kg \cdot m^{-3}$, c_p is the specific heat of air at constant pressure in $MJ \cdot kg^{-1} \cdot {}^{\circ}C^{-1}$, Δ is the slope of the saturation vapor pressure curve in $kPa \cdot {}^{\circ}C^{-1}$, γ is the psychrometric constant in $kPa \cdot {}^{\circ}C^{-1}$, λ is the latent heat of vaporization in $MJ \cdot kg^{-1}$, and r_{cp} and $r_{a,fc}$ are the canopy and aerodynamic resistances, respectively, in $s \cdot m^{-1}$ [10]. In this expression, the values of $R_{n,fc}$ and $r_{a,fc}$ are representative of a crop with complete ground cover, and r_{cp} represents the canopy resistance of a well-watered, unstressed crop. Soil heat flux has been omitted from this expression since it should be small under full canopy conditions. The weather observations required to evaluate Equation (4) are the same as those needed to calculate ET_0 in the standard crop coefficient approach.

Since K_{sp} is equivalent to crop GC, it can be measured for a field. For programs involving many fields within an agricultural region, measuring GC using standard photographic or light-interception techniques may be time-consuming and labor-intensive. In these cases, GC can be estimated with reasonable accuracy using multispectral imagery from satellite or airborne remote sensing systems [11] [12]. Values of GC on days between remote sensing acquisitions can be estimated through interpolation. Thus, the spectral crop coefficient approach is well-suited for application on a regional scale using operational sources of remote sensing imagery and weather data from existing observing networks.

2.2. Study Sites

Performance of the K_{sp} approach under different environmental conditions was demonstrated using data from three commercial fields in the Southern High Plains of Texas. The three fields used in this study (hereafter referred to as Field 1, Field 2, and Field 3) were part of the Texas Alliance for Water Conservation (TAWC) Demonstration Project, a large project conducted in this region to promote conservation of regional water resources. All three fields were planted to cotton (*Gossypium hirsutum* L.) and managed in a manner consistent with current farming practices in the region. All three fields were rectangular in shape. The study was conducted in Fields 1 and 3 in 2008 and in Field 2 in 2010. Field 1 contained 14.5 ha and was located approximately 3.25 km southwest of Lockney, TX, USA (34°7'30"N; 101°26'31.2" W). It was irrigated using subsurface drip irrigation and had crop rows oriented north-south. Field 1 received approximately 600 mm of irrigation during the growing season, and the irrigation was terminated on 8 September 2008 (Day 252). Field 2 was located approximately 1 km south of Lockney and contained 18.0 ha. It was irrigated using subsurface drip irrigation and had crop rows oriented east-west. Field 2 received approximately 100 mm of irrigation during the growing season, and the irrigation was terminated on 28 August 2010 (Day 240). Field 3 was located approximately 3.25 km north of Lockney and contained 20.6 ha. It had crop rows oriented east-west and was not irrigated ("dryland"). Field 1 was planted on 5 May 2008 (Day 126), Field 2 on 15 May 2010 (Day 135), and Field 3 on 21 May 2008 (Day 142). The soil in all three fields is a Pullman clay loam (fine, mixed, thermic, Torrertic Paleustoll) with 0% - 1% slope [13]. The climate of the study region is semi-arid with an average annual precipitation of approximately 460 mm [14].

2.3. Eddy Covariance Data

Daily observed values of CWU for comparison with corresponding estimates calculated using the Ksp approach were obtained from surface energy balance measurements made using an eddy covariance (EC) system operated at each field. In 2008, the EC system was mounted on a trailer that could be placed at the edge of the field (Field 1 and Field 3) to make measurements when the wind was blowing across the field in the direction of the trailer. The trailer had an erectable mast holding the sonic anemometer, infrared gas analyzer, and temperature/humidity probe. These sensors were mounted on the mast according to recommendations provided by the manufacturer [15]. These sensors were maintained at a height of 2 m above the height of the plant canopy. Sensors used to measure soil heat flux were buried in the field according to recommendations provided by the manufacturer [15]. A net radiometer was mounted on a tripod in the field approximately 2 m above the plant canopy. Data from all sensors were measured and recorded at a 20-Hz sampling rate using a CR3000 data logger (Campbell Scientific, Logan, UT). The raw high frequency data from the EC system were processed using a set of procedures described by Rajan *et al.* [10]. Thirty-minute average values of sensible heat flux, latent heat flux (ET), soil heat flux, and net radiation were calculated from the processed data.

The trailer supporting the EC system was situated along the north edge of the field it was measuring, approximately halfway between the east and west borders of the field. Data from the EC system were excluded from analysis if the wind direction was less than 110° or more than 250° (measured clockwise from north) in order to exclude situations where the fetch was not dominated by the field environment. In 2010, the EC system was mounted on a tripod located near the center of the field (Field 2). In this case, data did not need to be excluded based on wind direction.

EC data were used in this study only for those days when the soil surface in the field was observed to be dry. This was intended to minimize the contribution of soil evaporation to the overall latent heat flux measured by the EC system so that the measured ET approximated CWU as previously defined. Use of subsurface drip irrigation in Fields 1 and 2 resulted in the soil surface being dry except immediately following rainfall events. Since Field 3 was not irrigated, its soil surface was also dry except immediately following rainfall events.

In 2008, the EC system was stationed at Field 1 from 7 August (Day 220) through 4 October (Day 278), resulting in 14 days for which the wind direction and soil dryness criteria were met. During that year, a second EC system was stationed at Field 3 from 31 August (Day 244) through 3 October (Day 277), resulting in 8 days for which the wind direction and soil dryness criteria were met. In 2010, the EC system was stationed at Field 2 from 20 August (Day 232) through 21 September (Day 264), resulting in 32 days for which the soil dryness criterion was met. As these numbers suggest, placing the EC system in the middle of the field in 2010 resulted in significantly fewer days being excluded from analysis.

Energy balance closure was examined by summing the corresponding 30-minute average values of sensible heat flux, latent heat flux, soil heat flux, and net radiation. The resulting sum was usually less than zero, indicating that all the energy in the system was not accounted for by the EC measurements. This situation is common for EC measurements of agricultural and natural ecosystems [16]-[18]. In this study, it was assumed that the deficit in energy preventing closure resulted from underestimating the sensible and latent heat flux components. In these situations, the sensible and latent heat fluxes were adjusted by partitioning the residual in the energy balance between these two components according to the measured Bowen Ratio [19].

Daily CWU was obtained by summing the 30-minute average values of latent heat flux measured over the course of a day, divided by the latent heat of vaporization. Values of latent heat flux measured during the night were near zero and were excluded from the daily calculation.

2.4. Satellite Data

Satellite data for estimating crop GC (= K_{sp}) were obtained from Landsat imagery. In 2008, Landsat-5 Thematic Mapper (TM) imagery containing the study site was acquired on 5 dates spanning the period when EC systems were stationed at Fields 1 and 3: 6 August (Day 219), 22 August (Day 235), 7 September (Day 251), 23 September (Day 267), and 9 October (Day 283). These acquisitions were supplemented by Landsat-7 Enhanced Thematic Mapper (ETM+) imagery acquired on 2 additional dates: 15 September (Day 259) and 1 October (Day 275). In 2010, Landsat-5 TM imagery containing the study site was acquired on 3 dates spanning the period when the EC system was stationed at Field 2: 12 August (Day 224), 28 August (Day 240), and 29 September (Day 272). These acquisitions were supplemented by Landsat-7 ETM+ imagery acquired on one additional date: 5 September (Day 252).

Each image, located according to the Landsat World Reference System (WRS-2) along Path 30 at Row 36, was obtained from the U.S. Geological Survey (USGS) EarthExplorer website [20]. Pixel size in the imagery was specified as 30 m, and Systematic Correction (L1G) was applied by USGS to the image data. In Systematic

Correction, the image is rotated, aligned, and georeferenced to a user-defined map projection (WGS84), and is radiometrically corrected based on characteristics of the sensor [21].

Data extracted from the Landsat imagery were used to estimate crop GC for the study fields on each of the image acquisition dates using the procedure described by Maas and Rajan [11]. Image data analysis was carried out using ENVI image processing software (ITT, Boulder, CO). A scatterplot of each image (excluding portions containing clouds, cloud shadows, and water bodies) was constructed by plotting pixel digital count (DC) values in Band 4 (NIR spectral band) versus corresponding DC values in Band 3 (red spectral band). The bare soil line and the point corresponding to 100% GC was then identified by visual inspection of each scatterplot, allowing calculation of the value of the Perpendicular Vegetation Index (PVI) for 100% GC for each image. The boundaries of the study fields were delineated in each Landsat image and average DC values in the red and NIR spectral bands were calculated for the pixels within the boundaries. These average values were used along with the appropriate equation of the bare soil line to calculate the value of PVI corresponding to each field. GC for each field on each image acquisition date was then calculated by dividing the PVI value corresponding to the field by the appropriate value of PVI corresponding to 100% GC. For each field, values of crop GC for days between satellite image acquisition dates were determined using linear interpolation.

2.5. Weather Data and PET

Since our intent was to demonstrate the application of the K_{sp} approach under operational conditions, we chose to use commonly available weather observations to evaluate PET_{fc} in Equation 3 rather than the detailed environmental measurements available from the EC systems stationed at the study fields. Daily weather data for the study period were obtained from the West Texas Mesonet (WTM), an operational network of observing stations scattered across the region including the study site [22]. Specifically, data were obtained from the WTM station at Plainview, Texas, located approximately 30 km to the west of the study site. Data included average daily air temperature (T_a), average daily dew point temperature (T_d), average daily wind speed (u), and total daily solar irradiance (Q_s). Values of T_a, T_d, and u were measured at 2 m above the surface. Values of T_a and T_d were used to evaluate the vapor pressure deficit in Equation (4). Values of u were used to evaluate the aerodynamic resistance in Equation 4 using the following equation,

$$r_{a,fc} = \frac{\left[\ln\left(\dfrac{z-d}{z_0}\right)\right]^2}{k^2 u} \tag{5}$$

where z is the measurement height, d the zero-plane displacement, z_0 the surface roughness, and k the Von Karman constant (0.41) [23]. In evaluating Equation 5, z-d was assumed to be 2 m. The value of z_0 for a cotton canopy with full cover was assumed to be 4 cm based on the results of previous studies [10]. Canopy resistance in Equation 4 was assumed to be 50 s/m, a value considered appropriate for non-stressed crop canopies with complete ground cover [24]. Net radiation in Equation 4 was evaluated from Q_s and T_a using procedures described by Allen *et al.* [2]. In calculating $R_{n,fc}$, the albedo of the surface was assumed to be 0.23, which is appropriate for vegetation with 100% GC [2].

2.6. Statistical Analysis

Estimates of CWU were calculated using Equation (3) for days with EC measurements. In these calculations, the value of F_{stress} was set equal to 1 based on the assumption that the crop was acclimated to its environment. A paired t-test was used to determine if the average of the estimated CWU values was significantly different from the average of the observed values of CWU from the EC measurements [25]. In this calculation, the values from all three fields were combined to increase the sample size.

Estimated values of CWU were plotted versus corresponding observed values of CWU from the EC measurements. The distribution of points was fit using simple linear regression analysis. Data from all three fields were combined in this analysis to maximize the range of CWU values present in the data set and thereby make the regression more robust. Student's t tests were used to determine if the slope of the regression was significantly different from 1, and if the intercept of the regression was significantly different from zero [25]. Were that not the case, then one could conclude that the regression was not significantly different from the 1:1 line, and that the K_{sp} approach did a reasonably good job of estimating CWU for the study.

As a test of whether the assumption that $F_{stress} = 1$ was valid, Equation 3 was re-arranged and the value of F_{stress} was calculated as follows,

$$F_{stress} = CWU_{obs} \big/ \left(K_{sp} \cdot PET_{fc} \right) \qquad (6)$$

in which CWU_{obs} is the observed value of CWU from the EC measurements. A Student's t test was used to determine if the average value of F_{stress} from this analysis was significantly different from 1 [25]. If this average value were not significantly different from 1, then one could conclude that the crops in this study generally were acclimated to their surrounding environments. To further test this hypothesis, Student's t tests were used to determine if the slope of the regression through these points was significantly different from zero, and if the intercept of the regression was significantly different from 1 [25]. Were that not the case, then one could conclude that the regression was not significantly different from a line with a constant y-value (F_{stress}) of 1.

The difference was determined for each pair of estimated and observed CWU values and used to calculate the Average Absolute Error (AAE) according to the equation,

$$AAE = \frac{\sum_{i=1}^{n} \left| CWU_{est} - CWU_{obs} \right|}{n} \qquad (7)$$

where n is the number of observations. Data from all three study fields were combined for this analysis. AAE can be considered as a measure of the overall accuracy of the estimation approach.

3. Results and Discussion

Figure 1 shows GC values estimated from Landsat image data spanning the periods during which EC measurements were made at the three study fields. The differences in GC values among the three fields can largely be attributed to the type of irrigation each received. Field 1 received 600 mm of irrigation, an amount that would normally result in maximum growth of the crop (*i.e.*, Field 1 was "fully irrigated"). GC for this field remained high during the period of EC measurements. The irrigation for this field was terminated on Day 252, and effects of this termination on crop GC occurred largely after the last Landsat overpass date (Day 259). Field 2 was "deficit irrigated" using only a modest amount of irrigation (100 mm) such that the growth of the crop was below the maximum level but still capable of producing acceptable yields. Deficit irrigation is a common practice among cotton farmers in this region. Compared to Field 1, the farmer began reducing irrigation earlier for Field

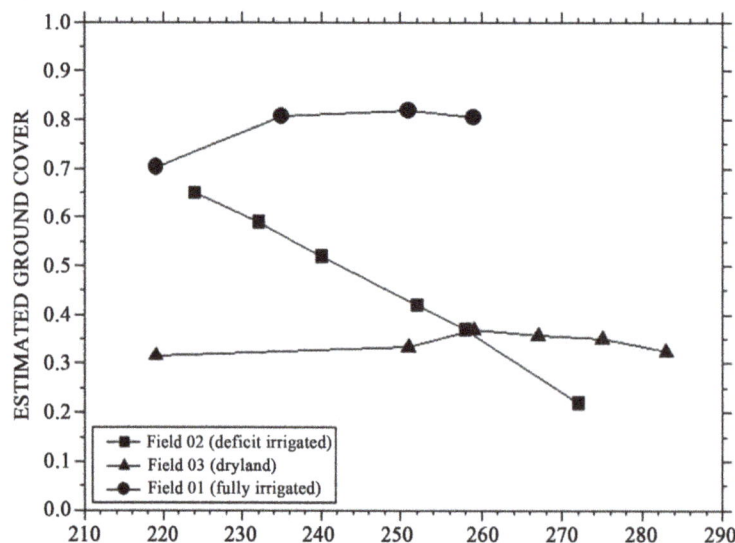

Figure 1. Ground cover values (solid symbols) estimated from Landsat image data spanning the periods during which eddy covariance measurements were made at the three study fields. Straight line segments connecting symbols were used to determine interpolated values of GC for days between Landsat overpass dates.

2, and terminated irrigation altogether on an earlier date (Day 240). The result was a steady decline in crop GC for Field 2 over the period of EC measurements. Field 3 was a dryland field receiving no irrigation. Its GC remained relatively constant over the period of EC measurements, with a small increase in GC between Day 251 and Day 259 resulting from a rain event. These results show that both the magnitude and temporal dynamics of crop GC can vary markedly among fields in an agricultural region, emphasizing that methods for estimating CWU need to be field-specific.

An example of CWU estimates produced by the K_{sp} method is presented in **Figure 2**. This figure shows daily CWU estimates for Field 2, along with their corresponding observed CWU values measured using EC. Both the observed and estimated CWU values exhibit a steady decrease over time as a result of the decrease in crop GC over the study period (see **Figure 1**). As shown in **Figure 2**, the least-squares linear regression fit to the observed values is very close to the corresponding least-squares linear regression fit to the estimated values over the study period. This suggests that the K_{sp} method responded appropriately to the crop and environmental factors driving CWU in this field. The average absolute error (AAE) between estimated and observed CWU values was 0.69 mm/day. Results for Fields 1 and 3 are qualitatively similar to those for Field 2.

Table 1 presents average daily estimated and observed CWU (mm/day) for each of the three fields in the study. The table also shows the average daily estimated and observed CWU for the combined data from the three fields. The paired t-test performed on the combined data resulted in $t = -0.0297$ with 56 df. This value was less than the corresponding critical value ($t_{\alpha=0.05,56df} = 2.004$), which suggests that there was no significant difference between the average estimated and observed values of CWU in this study.

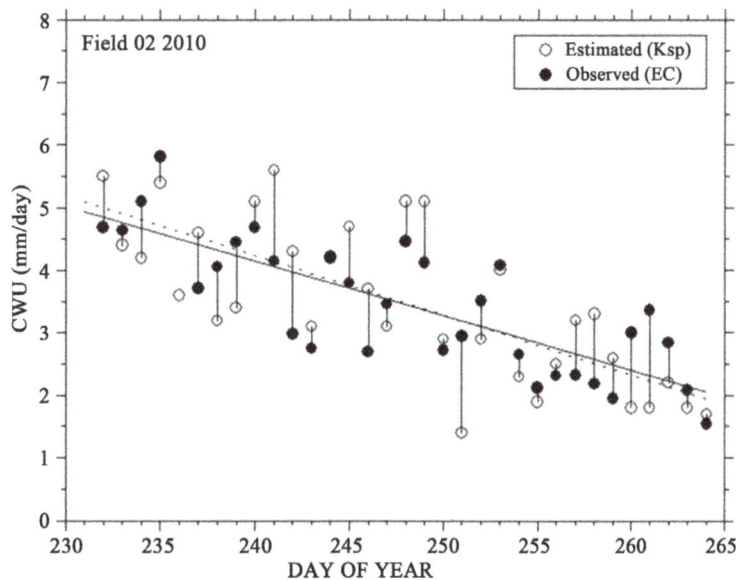

Figure 2. Estimated daily values of CWU (open circles) plotted along with their corresponding observed values of CWU (solid circles) for Field 2 in the study. Vertical lines connect the pairs of estimated and observed values. The solid line passing through the set of points is the least-squares fit to the observed values, while the dashed line is the least-squares fit to the estimated values.

Table 1. Average daily CWU (mm/day) for each of the three fields and for the combination of the data from the three fields.

	detamitsE	devresbO
dleiF 01	5.68	5.47
Field 02	3.47	3.43
dleiF 03	2.15	1.90
All 3 Fields	3.80	3.72

Figure 3 shows observed values of CWU plotted versus their corresponding values of estimated CWU for the three fields in the study. The diagonal solid line in the figure represents the 1:1 line. Points in the figure tend to cluster along the 1:1 line over the range of CWU values. The dashed line represents the simple linear regression fit to the points in the figure. This regression line has a slope of 0.926 and a y-intercept of 0.278, and explains approximately 80% of the total variance in the data. The t-test performed to determine if the regression slope was significantly different from 1 resulted in $t = -1.112$ with 54 df. This value was less than the corresponding critical value ($t_{\alpha=0.05,54df} = 2.005$), which suggests that there was no significant difference between the slope of the regression and 1. The t-test performed to determine if the regression intercept was significantly different from zero resulted in $t = 1.033$ with 54 df. This value was less than the corresponding critical value ($t_{\alpha=0.05,54df} = 2.005$), which suggests that there was no significant difference between the y-intercept of the regression and zero. Overall, these results suggest that there was no significant difference between the regression line and the 1:1 line in this study.

Figure 4 shows values of F_{stress} calculated using Equation (6) plotted versus their corresponding values of CWU_{obs}. There is considerable scatter in the points over the range of observed CWU, but they tend to cluster around a value of 1 with an average value of 1.028 for the combined data set. The t-test performed to determine if this average value was significantly different from 1 resulted in t = 0.725 with 56 df. This value was less than the corresponding critical value ($t_{\alpha=0.05,56df} = 2.004$), which suggests that there was no significant difference between the average value of F_{stress} and 1 in this study.

The dashed line in **Figure 4** represents the simple linear regression fit to the points in the figure. This regression line has a slope of −0.0392 and a y-intercept of 1.174. The t-test performed to determine if the regression slope was significantly different from 0 resulted in $t = -1.650$ with 54 df. This value was less than the corresponding critical value ($t_{\alpha=0.05,54df} = 2.005$), which suggests that there was no significant difference between the slope of the regression and zero. The t-test performed to determine if the regression intercept was significantly different from 1 resulted in $t = 1.806$ with 54 df. This value was less than the corresponding critical value ($t_{\alpha=0.05,54df} = 2.005$), which suggests that there was no significant difference between the y-intercept of the regression and 1. Overall, these results suggest that there was no significant difference between the regression line

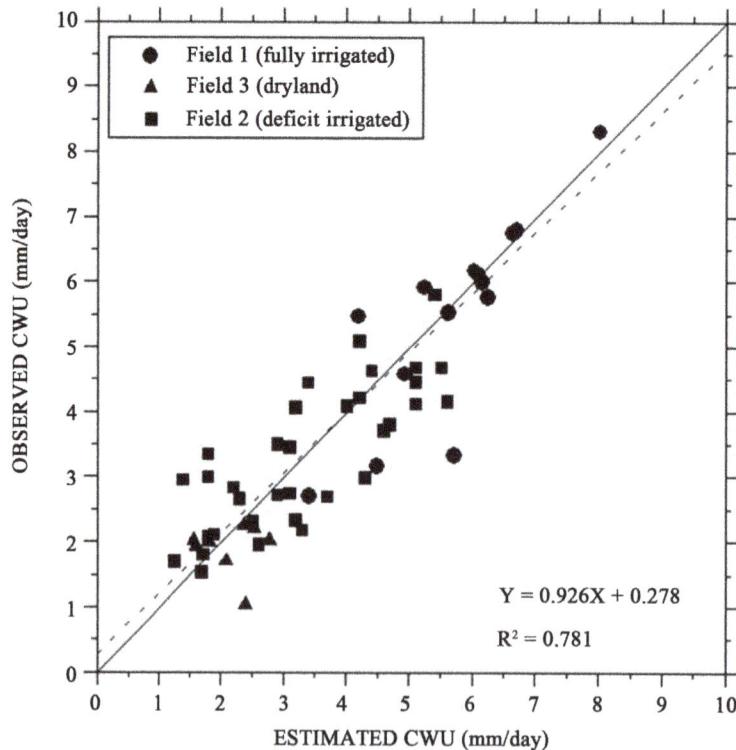

Figure 3. Estimated versus observed CWU for the three cotton fields in the study. The solid diagonal line is the 1:1 line, while the dashed line is the least-squares regression fit to the data.

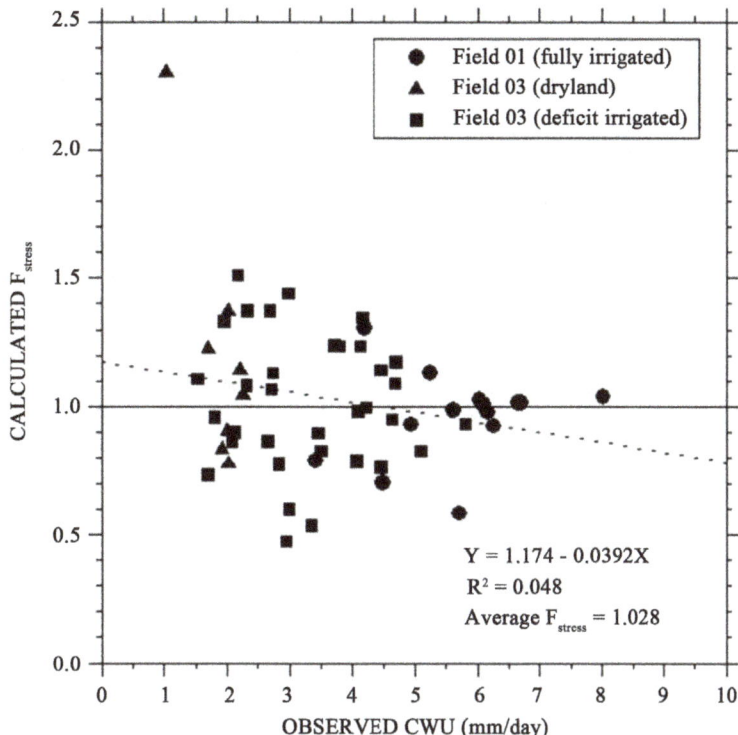

Figure 4. Calculated values of the stress factor F_{stress} plotted versus their corresponding values of observed CWU for the three cotton fields in the study. The solid horizontal line represents a value of F_{stress} equal to 1. The dashed line represents the least-squares regression fit to the data.

and a line with a constant y-value (F_{stress}) of 1. Note that, if the extreme point in **Figure 4** (x = 1.042, y = 2.302) is removed from the analysis, the results (in terms of the conclusions drawn from the statistical tests on the slope and y-intercept of the regression and the average F_{stress} value) remain the same.

There was considerable scatter among the individual values plotted in **Figure 3** and **Figure 4**. Since the data used to determine estimated and observed values of CWU came from a number of diverse sources (satellite imagery, weather network observations, and in-field EC measurements), the opportunity existed for random errors (which might compound or cancel) to be introduced in multiple stages of the calculations. Still, the statistical analysis of the results suggests that the K_{sp} approach performed reasonably well in estimating daily CWU for the three fields. The K_{sp} approach produced results that were field-specific, and it appeared to perform reasonably well for all three types of irrigation (fully irrigated, deficit irrigated, and not irrigated) used in this study. The AAE for CWU estimates for the combined data from the three fields was 0.62 mm/day suggesting that, on average, estimates of CWU produced by the approach should have sub-mm accuracy.

The results presented in **Figure 3** and **Figure 4** suggest that differences in CWU among the types of irrigation were predominantly the result of differences in crop GC rather than differences in stress induced by stomatal closure. During the study period, average values of GC deduced from the satellite data for the crops in Fields 1 (fully irrigated), Field 2 (deficit irrigated), and Field 3 (dryland) were 0.82, 0.48 and 0.35, respectively. Thus, the crops in this study appeared to be acclimated to their surrounding environments and were controlling their seasonal transpirational losses through manipulation of their leaf area in order to maintain their stomata in an open and functional condition. A physiological mechanism for this behavior has been described [26]. As plants deplete the water in the soil profile, they begin to experience reduced leaf growth and increased leaf senescence at a value of available soil water of around 50% of field capacity. For these plants, stomatal closure and reduced leaf transpiration is delayed until levels of available soil water fall to around 30% of field capacity. Thus, these plants have the opportunity to conserve soil water by reducing the surface area of the leaf canopy while maintaining photosynthesis in the remaining leaves.

Since leaves grow and senesce at a relatively slow rate as compared to stomatal function, this mechanism for acclimatization would be most effective when changes in environmental conditions (such as soil water) are gradual, as generally occurs in field environments. Abrupt changes, as when the irrigation to well-watered crops is terminated, can result in periods of stomatal closure while the plants adjust their canopy size through accelerated leaf senescence [10].

The K_{sp} approach is well-suited for applications such as irrigation scheduling. Its use of operational weather observations (the same as used in the standard crop coefficient approach) and readily available satellite image data make it easy to apply on a regional scale, while the evaluation of the crop coefficient from actual satellite-derived field observations allows its CWU estimates to be field-specific. Use of field-specific CWU estimates should help minimize over-irrigation of fields and thus contribute to water conservation.

4. Conclusion

Results of this study suggest that the spectral crop coefficient approach is effective in estimating the CWU of agricultural fields under a variety of irrigation conditions ranging from fully irrigated to dryland. In this approach, CWU can be evaluated from standard weather observations from operational weather observing networks and readily available medium-resolution multispectral satellite imagery. The use of satellite data in the method allows the estimates of CWU to be specific to individual fields, as opposed to the standard crop coefficient approach in which estimates of crop ET assume "standard conditions" that are invariant from field to field. Under field conditions where the crop is acclimated to its environment, water use is controlled more by crop ground cover than by stomatal closure. On average, CWU estimates produced by the K_{sp} approach might be expected to have sub-mm accuracy. The K_{sp} approach is well-suited for operational applications such as irrigation scheduling, where its use could contribute to water conservation by minimizing over-irrigation.

Acknowledgements

The authors wish to thank the Texas Alliance for Water Conservation (TAWC) Demonstration Project, funded through the Texas Water Development Board, for the resources to conduct this study.

References

[1] Burman, R.D. (2003) Evapotranspiration Formulas. In: Stewart, B.A. and Howell, T.A., Eds., *Encyclopedia of Water Science*, Marcel Dekker, New York, 253-257.

[2] Allen, R.G., Pereira, L.S., Raes, D. and Smith, M. (1998) Crop Evaporation: Guidelines for Computing Crop Water Requirements. Irrigation and Drainage Paper 56, Food and Agriculture Organization of the United Nations, Rome.

[3] Heilman, J.L., W.E. Heilman, and D.G. Moore (1982) Evaluating the Crop Coefficient Using Spectral Reflectance. *Agronomy Journal*, **74**, 967-971. http://dx.doi.org/10.2134/agronj1982.00021962007400060010x

[4] Bausch, W.C. and Neale, C.M.U. (1987) Crop Coefficients Derived from Reflected Canopy Radiation: A Concept. *Transactions of ASAE*, **30**, 703-709. http://dx.doi.org/10.13031/2013.30463

[5] Bausch, W.C. and Neale, C.M.U. (1989) Spectral Inputs Improve Corn Crop Coefficients and Irrigation Scheduling. *Transactions of ASAE*, **32**, 1901-1908. http://dx.doi.org/10.13031/2013.31241

[6] Neale, C.M.U., Bausch, W.C. and Heerman, D.F. (1989) Development of Reflectance Based Crop Coefficients for Corn. *Transactions of ASAE*, **28**, 773-780.

[7] Bausch, W.C. (1995) Remote-sensing of Crop Coefficients for Improving the Irrigation Scheduling of Corn. *Agricultural Water Management*, **27**, 55-68. http://dx.doi.org/10.1016/0378-3774(95)01125-3

[8] Hunsaker, D.J., Pinter Jr., P.J., Barnes, E.M. and Kimball, B.A. (2003) Estimating Cotton Evapotranspiration Crop Coefficients with a Multispectral Vegetation Index. *Irrigation Science*, **22**, 95-104. http://dx.doi.org/10.1007/s00271-003-0074-6

[9] Hunsaker, D.J., Pinter Jr., P.J. and Kimball, B.A. (2005) Wheat Basal Crop Coefficients Determined by Normalized Difference Vegetation Index. *Irrigation Science*, **24**, 1-14. http://dx.doi.org/10.1007/s00271-005-0001-0

[10] Rajan, N., Maas, S.J. and Kathilankal, J. (2010) Estimating Crop Water Use of Cotton in the Texas High Plains. *Agronomy Journal*, **102**, 1641-1651. http://dx.doi.org/10.2134/agronj2010.0076

[11] Maas, S.J. and Rajan, N. (2008) Estimating Ground Cover of Field Crops Using Medium-Resolution Multispectral Satellite Imagery. *Agronomy Journal*, **100**, 320-327. http://dx.doi.org/10.2134/agrojnl2007.0140

[12] Rajan, N. and Maas, S.J. (2009) Mapping Crop Ground Cover Using Airborne Multispectral Digital Imagery. *Preci-*

sion Agriculture, **10**, 304-318. http://dx.doi.org/10.1007/s11119-009-9116-2

[13] NRCS (1978) Soil Survey of Floyd County, Texas. Natural Resource Conservation Service, US Department of Agriculture, Washington DC.

[14] Larkin, T.J. and Bowmar, G.W. (1983) Climatic Atlas of Texas. Publication LP-192, Texas Department of Water Resources, Austin.

[15] Campbell Scientific (2006) Open-Path Eddy Covariance System Operator's Manual (Revised 9/06). Campbell Scientific, Inc., Logan.

[16] Foken, T. (2008) The Energy Balance Closure Problem: An Overview. *Ecological Applications*, **18**, 1351-1367. http://dx.doi.org/10.1890/06-0922.1

[17] Wilson, K., Goldstein, A., Falge, E., Aubinet, M., Baldocchi, D., Berbigier, P., Bernhofer, C., Ceulmans, R., Dolman, H., Field, C., Grelle, A., Ibrom, A., Law, B.E., Kowalsky, A., Meyers, T., Moncrieff, J., Monson, R., Oechel, W., Tenhunen, J., Valentini, R. and Verma, S. (2002) Energy Balance Closure at FLUXNET Sites. *Agricultural and Forest Meteorology*, **113**, 223-243. http://dx.doi.org/10.1016/S0168-1923(02)00109-0

[18] Mauder, M., Liebenthal, C., Göckede, M., Leps, J., Beyrich, F. and Foken, T. (2006) Processing and Quality Control of Flux Data during LITFASS-2003. *Boundary-Layer Meteorology*, **121**, 67-88. http://dx.doi.org/10.1007/s10546-006-9094-0

[19] Twine, T.E., Kustas, W.P., Norman, J.M., Cook, D.R., Houser, P.R., Meyers, T.P., Prueger, J.P., Starks, P.J. and Wesley, M.L. (2000) Correcting Eddy-Covariance Flux Underestimates over a Grassland. *Agricultural and Forest Meteorology*, **103**, 279-300. http://dx.doi.org/10.1016/S0168-1923(00)00123-4

[20] http://edcsns17.cr.usgs.gov/EarthExplorer/

[21] Chander, G. and Markham, B. (2003) Revised Landsat-5 TM Radiometric Calibration Procedures and Postcalibration Dynamic Ranges. *IEEE Transactions on Geoscience and Remote Sensing*, **41**, 2674-2677. http://dx.doi.org/10.1109/TGRS.2003.818464

[22] http://www.mesonet.ttu.edu/

[23] Monteith, J.L. and Unsworth, M. (1990) Principles of Environmental Physics. Edward Asner Publishers, London.

[24] Allen, R.G., Pruitt, W.O., Wright, J.L., Howell, T.A., Ventura, F., Snyder, R., Itenfisu, D., Steduto, P., Berengena, J., Yrisarry, J.B., Smith, M., Pereira, L.B., Raes, D., Perrier, A., Alves, I., Walter, I. and Elliott, R. (2006) A Recommendation on Standardized Surface Resistance for Hourly Calculation of Reference ET_0 by the FAO56 Penman-Monteith Method. *Agricultural Water Management*, **81**, 1-22. http://dx.doi.org/10.1016/j.agwat.2005.03.007

[25] Ostle, B. and Mensing, R.W. (1982) Statistics in Research. Iowa State University Press, Ames.

[26] Rosenthal, W.D., Arkin, G.F., Shouse, P.J. and Jordan, W.R. (1987) Water Deficit Effects on Transpiration and Leaf Growth. *Agronomy Journal*, **79**, 1019-1026. http://dx.doi.org/10.2134/agronj1987.00021962007900060014x

5

Evaluation of Plant Densities and Various Irrigation Regimes of Sorghum (*Sorghum bicolor* L.) under Low Water Supply

Ali A. Alderfasi[1*], Mostafa M. Selim[1], Bushra A. Alhammad[2]

[1]Department of Plant Production, Faculty of Food and Agriculture Science, King Saud University, Riyadh, Saudi Arabia
[2]Faculty of Sciences and Humanities, Salman Bin Abdulaziz University, Alkharj, Saudi Arabia
Email: [*]aderfasi@gmail.com

Abstract

Drought stress, during growth season along with plant density, is an important problem that needs attention. In order to investigate the influence of both factors in increasing the water use efficiency, field experiments were laid out in split-plot design at Agriculture Research Station, Collage of Food and Agriculture Sciences, King Saud University, to investigate the effects of irrigation intervals *viz.*, irrigation every (6, 9 and 12 days) under different plant densities *i.e.*, (6, 8 and 10 plants/m²) on growth, yield and yield component parameters as well as grain quality of sorghum local variety (Gizani). Results revealed that almost all growth, yield and yield component parameters were significantly influenced by both factors as well as their interaction. Chemical composition of seeds, leaf proline content and WUE were also considered. Severe drought stress condition caused gradual decrease in most of the growth characters as compared to watered treatment and reflected in decreasing yield and yield component characters. Increasing plant densities led to raise biomass production and seed yield per unit area and not able to compensated the low number and weight of grains per panicle. Contrary, low plant density, under adequate irrigation conditions, can be compensated by a high number of grains per panicle and high weight of the grain. Maximum seed yield per hectare was recorded by the interactional effects of most watered treatments (irrigation every 6 days) and plant density of 10 plants per square meter.

Keywords

Sorghum, Plant Density, Plant Population, Irrigation, Water Use Efficiency, Drought Stress

---

[*]Corresponding author.

1. Introduction

Human demand for food increases with increasing world population. It is expected that human population will increase to over 8 billion by the year 2020 and this will worsen the current scenario of food security. It is predicted that at least 10 million people will be hungry and malnourished in the world by the end of this century [1] Thus, in order to feed the world's population agricultural production must be doubled, produced in more environmentally sustainable ways and more profitable crops with shorter growing period should be used [2].

The picture is more aggravate under arid and semiarid regions, with the fact that over a quarter of land area on the earth is considered as arid and semiarid regions [3]. In these regions, water deficit is the main factor that limits crops performance. Limitation of water source, irregular annual rainfall during growth season and lack of sources management cause severe decreasing in crops yield [4]. Therefore, drought stress during growth season is an important problem that needs to attention [5]. Sorghum (*Sorghum bicolor* (L.) Moench) is known to be relatively more tolerant to drought conditions than other crops [6]. The resistance of sorghum to harsh climatic conditions ranked it to the fifth important cereal crop in the world [7] [8]. Therefore, it is fair to assume that sorghum has a great potential for arid areas. Sorghum plays an important role as a staple food grain in many developing countries [9]. Drought stress occurrence and intensity affect characteristics like plant height, plant dry weight, specific leaf area, seed yield and yield components [10]. Thus, the effect of water stress on plants under water deficit conditions in the soil could be determined from smaller sizes of leaf or low heights in plants or decreased wet or dry weight as well as lower yield and yield components [11]. The severity of loss depends on many factors *viz.*, timing, length and severity of the drought period.

Plant density is the second important factor in crop yield. The use of proper planting density of sorghum crop plays an important role in the efficient use of water irrigation applied and reaches to optimum yield performance. The relationship between plant density and yield of cereals has been studied extensively, but conflicting reports have led to a renewed interest in the effects of high plant densities on yield of cereals. [12] [13] reported that increasing plant density up to 166 and 333 thousand plants per hectare for tall and short types of sorghum respectively decreased plant height, stem diameter, number of green leaves and leaf area per plant, while grain yield for both types increased by increasing plant density. Some researchers were recorded the highest dry matter production at high plant densities [14] [15]. While others expressed that a greater number of grains per panicle and higher weight of grains compensated the lower plant densities [16]. [17] reported that plant density affectd the post-flowering source/sink ratio through its effects on plant leaf area, the amount of light intercepted per plant and kernel number per plant. Recent research reports of [18] showed a 4% increase in yield for corn grown in 76-cm row spacing compared with 56-cm row spacing. [19] found that, the proper plant density is the best way for improving sorghum production under arid condition. Although there are a huge number of research reports, there is not accurate information of the appropriate plant density and irrigation schedules that can follow by farmers for maximum yield performance. Therefore, the present investigation was focused on evaluation of sorghum responses to plant densities and irrigation schedules on growth, grain yield and yield component characters as well as grain chemical constituents of sorghum under drought stress conditions.

2. Material and Methods

2.1. Experimental Design, Field Site Description and Plant Material

Factorial experiments based on split-plot design with four replications were done in two successive summer seasons of 2011 and 2012 to evaluate growth, yield and yield components parameters of local sorghum variety (Gizani) under three irrigation intervals *viz.*, every (6, 9 and 12 days) were assigned in main plots and three plant densities *i.e.*, (6, 8 and 10 plants/m^2 achieve by plants spaced 30, 25 and 20 cm) were randomly distributed in sub-plots at Agriculture Research Station, Collage of Food and Agriculture Sciences, King Saud University, Deerab, South Riyadh region, Saudi Arabia (24.42°N latitude and 46.44°E longitudes, Altitude 600 m). The region is under arid climate conditions, with high temperatures and truncated rainfall during the summer and low temperatures and little rainfall during the winter season. Maximum and minimum mean temperature and relative humidity during the two growing seasons are presented in (**Table 1**).

2.2. Soil and Irrigation Water Characters

Prior to the field experiment, the field soil site was sampled 0 - 60 cm depth from five sites for physical and chemical analyses according to the methods described by [20] and [21], results are shown in (**Table 2** and **Table**

3). Chemical properties of the irrigation water used were also analyzed according to the methods described by American Public Health Association, [22], results are presented in (**Table 4**). Total number of irrigation and amount of water used for each treatment over the growing seasons are mentioned in (**Table 5**).

2.3. Field Preparation and Sowing

Seed bed was prepared before sowing as recommended; field was ploughed three cross harrowing's with tractor followed by a thorough harrowing to break the clods. Phosphorus fertilizer was applied at the rate of 70 kg. P_2O_5/ha. as the form of superphosphate (16% P_2O_5), whereas potassium as the form of potassium sulphate (42% K_2O), by the rate of 100 kg K_2O was applied broadcasting during soil preparation, the field was properly leveled

Table 1. Monthly maximum, minimum, mean temperature and relative humidity during 2011 and 2012 seasons.

Month	Temperature (°C)						Relative humidity (%)	
	Maximum		Minimum		Mean			
	2011	2012	2011	2012	2011	2012	2011	2012
May	40.2	41.5	22.9	23.4	31.6	32.5	10.6	10.4
June	42.7	44.4	24.8	25.9	33.8	35.2	10.2	10.3
July	44.9	45.7	26.7	27.3	35.8	36.5	12.7	10.4
August	44.9	45.9	27.8	26.9	36.4	36.4	10.9	10.8
September	41.8	42.7	24.4	23.3	33.1	33.3	11.2	11.4
October	37.2	38.2	19.4	18.5	28.3	33.3	11.3	11.6

Table 2. Physical analyzes of the experimental soil site during the two growing seasons.

Properties season	Saturation %	pH soil paste (1:5)	EC (dS/m)	O.M %	CaCO₃ %	Field capacity (%)	Wilting point (%)	Sand (%)	Silt (%)	Clay (%)	Soil texture (%)
First season	29.70	7.86	3.88	0.46	29.42	16.30	7.67	57.92	27.20	14.88	Sandy loam
Second season	28.12	7.81	3.91	0.47	29.63	16.42	7.71	57.82	27.25	14.90	Sandy loam

Table 3. Chemical analyzes of the experimental soil site during the two growing seasons.

Properties season	Available macro and micro nutrients (ppm)						
	N	P	K	Fe	Mn	Zn	Cu
First season	35.40	14.80	243.50	3.27	2.44	6.07	0.70
Second season	35.80	12.76	251.42	3.24	2.61	6.13	0.74

Table 4. Chemical analyzes of the irrigation water during the two growing seasons.

Properties season	pH	EC (dS/m)	O.M %	Soluble cations (meq./L)				Soluble anions (meq/L)			Total NPK (ppm)		
				Ca⁺	Mg⁺	Na⁺	K⁺	HCO₃⁻	Cl⁻	SO₃⁻	N	P	K
First season	7.10	1.45	0.02	6.30	1.75	7.35	0.44	2.40	4.85	9.14	10.50	9.23	17.00
Second season	7.17	1.73	0.02	5.50	1.87	7.65	0.46	2.60	4.80	8.56	11.01	9.42	17.12

Table 5. Total number of irrigation and amount of water used for each treatment over the growing season in both seasons.

Water regime treatments	Mean water apply (m³/ha)		Number of irrigations over growing season	
	First season	Second season	First season	Second season
Weekly irrigation	10,000	10,050	20	21
Irrigation every 9 days	7000	7000	14	14
Irrigation every 12 days	5000	5000	10	10

and divided into plots each one (3 × 3.5 m) included six ridges, three meters long and 50 cm apart, total experimental unit area was 10.5 m^2 earmarked with raised bunds all around to minimize the movement of water. Pipe type was laid to facilitate irrigation to plots individually (irrigation network).Sowing was done on 14th and 16th April in the first and second seasons, respectively. Plants were thinned according to plant densities 20 days after sowing. Nitrogen fertilizer was applied by the recommended dose of N (100 Kg N/ha.) two times, once after thinning and the second 45 days later. Water irrigation applied according to the experimental treatments by using flowed irrigation system, through line pipe provide with meter gages for measuring water applied over the growing season.

2.4. Treatments Details

Nine treatments were investigated in the study in a split-plot design arrangement in randomized complete block design with irrigation treatments (every 6, 9 and 12 days) as the main plot, and plant density (6, 8 and 10 plants/m^2) as the sub-plot in three replications.

2.5. Measurement of Plant Parameters

During growth stages number of days to 50% of flowering was determined. After 80 days from sowing a plant sample of 10 plants was taken for studying some growth parameters *viz.*, plant height (cm), stem diameter (cm), number of green leaves per plant and leaf area (cm^2) of the second upper leaf using the following formula: Leaf area (cm^2) = leaf length (cm) × maximum leaf width (cm) × 0.747.

2.5.1. Harvesting, Measuring Yield and Yield Component Characters

At harvest, two central rows of each sub-plot were hand pulled and completely air dried and threshed, then seed yield per hectare as well as biological yield per hectare were determined. Sub sample of ten plants was taken for determining yield component characters viz., head Length, cm; head weight, gm; grain weight per head gm; total plant weight kg, 1000 seed and grain yield per unit area.

2.5.2. Calculation Yield Parameters
1) *Irrigation water-use efficiency*
Irrigation water use efficiency (IWUE) kg/m^3 is defined as the ratio of the crop yield (final economic yield) to irrigation water applied over the growing season, including rainfall [23] as follows: water use efficiency (WUE) (kg/m^3) = economical yield in kg divided by seasonal water used.
2) *Harvest and crop index*
Harvest index (HI) and crop index (CI) were also calculated using the formula suggested by [24] as follows:
HI = Grain yield (kg·ha^{-1})/Biological yield (kg·ha^{-1}) × 100.
Biological yield = Grain yield + straw yield, CI = Grain yield (kg·ha^{-1})/Straw yield (kg·ha^{-1}) × 100.

2.5.3. Chemical Analyses
1) *Proline content*
Leaf proline content was determined in fully expanded uppermost leaves at full flowering stage using the method of [25].
2) *Chemical composition of grains*
Crude Protein content was estimated by determining nitrogen content in absolutely dry seeds using micro-kjeldahl method according to the procedures described by [26] and then nitrogen percentage was multiplied by 6.25 and protein yield was calculated.

Total carbohydrates percentage were determined using spectrophotometer and total soluble sugar using soxhelt apparatus by the method described by [27].

2.6. Statistical Analysis

The experimental design was laid out in split-plot design with 4 replications according to the methods described by [28]. Main plots were occupied by three water intervals, whereas plant densities were arranged in sub-plots. The data collected during the experiment processes was subjected to analysis of variance according to split-design and upon obtaining significant differences, Least Significant Differences (LSD) at 0.05% of probability was calculated and then differences among treatments means were compared.

3. Results and Discussion

3.1. Effect on Growth Characters

Analysis of variance of the data for growth characters presented in (**Table 6**) showed that severe drought stress condition caused gradual decrease in most of the growth characters except stem diameter and number of green leaves per plant *viz.*, time required to reach to 50% flowering, plant height and leaf area per plant. [29] reported that the time of flowering and maturity for the three mungbean cultivars was shortened under stress condition compared to well-watered conditions. Data obtained also reveal that, the most watered treatment, irrigation every 6 days exceeded the other treatments and recorded the highest values of each of growth character in both seasons, whereas the minimum values were obtained in the less irrigated treatment, irrigation every 9 followed by 12 days (**Table 6**). The contribution of water stress in most of growth characters may be due to the availability of more resources (nutrient + water) generative parts of the plant for receiving to maximum growth rate. These results confirm the finding of [11] who concluded that the effect of water stress on plant growth under water deficit conditions in the soil could be determined from smaller sizes of leaf or low plant heights or decreased fresh or dry weight and later lower yield and yield component characters. In the same concern, [10] reported that drought stress affected on growth characteristics like plant height, plant dry weight and specific leaf area.

The data presented in the same table also evident that increasing plant competition for water and nutrients, which are induced at high plant densities, resulted in decreased the time required to reach to 50% of flowering, plant height, stem diameter, number of green per plant, and leaf area in both seasons. Similar results were obtained by [13], he found that increasing plant density decreased growth characters of sorghum. [17] reported that plant density affects the source/sink ratio through its effects on plant leaf area. Furthermore, the same trend of

Table 6. Means of growth characters of sorghum under different irrigation intervals and plant densities as well as their interaction in 2011 and 2012 seasons.

Irrigation intervals (A)	Plant density, m² (B)	2011 season					2012 season				
		50% Flowering	Plant height cm	Stem diameter, cm	No. of leaves/ plant	Leaf area cm²	50% flowering	Plant height cm	Stem diameter cm	No. of leaves/ plant	Leaf area cm²
6	6	77.0	269.5	1.4	9.7	591.7	79.0	276.5	1.4	10.0	615.4
	8	74.0	255.5	1.4	9.8	582.5	76.0	261.0	1.3	9.9	584.5
	10	72.0	237.9	1.2	8.8	563.4	73.0	243.5	1.3	8.6	584.2
General mean		74.3	254.3	1.3	9.4	579.2	76.0	260.3		9.5	594.7
9	6	73.0	230.7	1.2	8.7	553.3	73.0	236.5	1.1	8.8	578.2
	8	72.0	222.5	1.1	8.6	546.8	73.0	225.0	1.0	8.9	554.6
	10	70.0	218.8	1.0	8.8	537.4	68.0	219.7	1.1	8.9	536.2
General mean		71.7	224.0	1.1	8.7	545.8	71.3	227.0		8.9	556.3
12	6	70.0	226.5	1.0	8.0	542.6	70.0	223.6	0.9	7.9	534.8
	8	72.0	215.8	1.0	7.8	522.7	73.0	220.5	0.9	7.8	522.7
	10	68.0	207.5	0.9	7.7	498.6	68.0	210.4	0.9	7.6	492.7
General mean		70.0	216.6	0.97	7.8	521.3	70.3	218.2		7.8	516.7
General mean of factor (B)	6	73.3	242.2	1.2	8.8	562.5	74.0	252.8	1.1	8.9	576.1
	8	72.7	231.3	1.2	8.7	550.7	74.0	235.5	1.1	8.9	553.9
	10	70.0	221.4	1.0	8.4	533.1	69.7	224.5	1.1	8.4	537.7
LSD	A	1.4	10.16	ns	1.2	28.4	1.0	12.4	ns	0.82	30.7
	B	0.5	6.9	ns	ns	10.9	2.3	9.2	ns	ns	14.9
	A × B	1.2	5.7	ns	1.0	4.0	1.8	7.6	ns	0.74	18.4

decreasing plant growth with increasing plant densities was also observed by [19]. The interaction of plant densities and variable water supplies are presented in (**Table 6**), clearly obvious that, a general the highest density suffered a higher water stress than the other densities. Since the higher number of plants per m^2 extracted the soil stored water and dissolving nutrients more quickly than the lower densities and then suffering a faster water stress. Data clearly indicate that, sowing sorghum by high plant densities 8 and 10 plants per m^2 and irrigation every 9 and 12 days registered low values of growth characters compared to the low plant density and most watered treatment. These results confirm the important role of both factors in the efficient use of inputs used especially at low input conditions. These results are in line with those reported by [9]-[11].

3.2. Effect on Yield Component Characters

Yield component characters is a complex trait affected by many factors *i.e.*, nutrients absorption; photosynthesis; management agronomic practices and the mutual effects of genetic constituencies in various environments [30]. Regarding to the effect of various water regime, the data presented in (**Table 7(a)**) showed that, water stress is seriously damaged almost all the yield component characters as a result of previously effect on growth characters. In general gradual decrease in panicle length, panicle weight, and grain weight per panicle, 1000 grain weight and number of grains per panicle were observed as the time of applied water irrigation increased in both seasons. The maximum mean values of these parameters over the both seasons were obtained with the most watered treatments which irrigated every 6 days with values of 20.7, 101.59, 86.9, 34.15 and 252.12, respectively. Whereas, the minimum mean values were 17.67, 83.05, 56.79, 30.0 and 185.75, respectively recorded in the less irrigated treatments, irrigation every 12 days in the both seasons (**Table 7(a)**). These results are similar to those found by [31] and [32], they reported that an obviously decrease in most of yield component characters of cereal under water deficit conditions. Compatible results were also reported by [33] who found that a water deficit during the flowering stage in sorghum reduced the number of grains per panicle that is partly compensated by an increase in grain weight.

Plant density in this study was categorized in class b (**Table 7**), results indicated that panicle length, and panicle weight, grain weight per panicle, 1000 grain weight and number of grains per panicle were decreased as plant density increased. These results confirmed the fact that the increasing number of plants per m^2, need high amount of nutrients and soil stored water to cover all plants requirements. Consequently large number of plants per unit area suffering a faster water stress, than the other densities. Thus, a low plant density, under adequate irrigation conditions can be compensated the expected low yield by a high panicle weight, large number of grains per panicle and high weight of the grain per panicle [34] and [35]. Comparable results in the review were obtained by [16] [17] and [19]. Regarding levels of plant density, the experiments of [36] and [37] indicated that sorghum has the capacity of compensating some yield component characters under certain limiting conditions, which contribute in an important way to the final yield of the crop. The interaction of water intervals × plant densities had also a significant effect on panicle length, and panicle weight, grain weight per panicle, 1000 grain weight and number of grains per panicle in both seasons. The maximum value of each parameter was registered when plants were sown by plant density 6 plants per m^2 and received irrigation water every 6 days, followed by sowing plants by 8 plants per m^2 at the same irrigation level. Results in the same table, also clear that increasing plant densities and time of applied water irrigation induced a decreased in all yield component characters of sorghum. The highest decrement was observed at high plant density (10 plants/m^2 and received irrigation water every 12 days). Such results are probably early, since the effect was detected in most of growth characters, which may be due to low availability of water and nutrient and consequently less photosynthesis because of high inter-specific competition and high rate of respiration as a result of enhanced mutual shading [37]. These results confirm the important effect of water stress and plant density as well as their interactions on the yield components of sorghum.

3.3. Effect on Grain Yield and Yield Parameters

A general and gradual decrease in grain yield production, biological yield, harvest index and crop index was observed as the time of applied irrigation water in the three sorghum plant densities increased. Thus, the maximum values of all of these parameters were obtained in the most watered treatments, generally in treatment irrigated every 6 days and the minimum values in the less irrigated treatments, which irrigation was located every 9 and 12 days in the same sorghum densities (**Table 7(a)**). The effect of water stress on plant growth as well as

Table 7. (a) Means of the yield and yield components of sorghum under different irrigation intervals and plant densities as well as their interaction in 2011 and 2012 seasons; (b) Means of the yield and yield components parameters of sorghum under different irrigation intervals and plant densities as well as their interaction in 2011 and 2012 seasons.

(a)

Irrigation intervals	Plant density M²	2011 season						2012 Season					
		Panicle length cm	Panicle weight gm	Grain weight/panicle gm	1000 grain weight gm	No. Of grains per panicle	W U E Kg/M³/Ha	Panicle length cm	Panicle weight gm	Grain weight/panicle, gm	1000 Grain weight, gm	No. Of Grains per panicle	WUE Kg/M³/Ha
6	6	22.2	120.9	103.7	35.5	310.4	0.336	23.3	117.9	109.3	36.7	305.4	0.356
	8	20.3	102.8	80.6	33.2	240.9	0.344	20.8	91.1	93.6	34.8	250.8	0.365
	10	18.9	87.9	66.7	31.5	204.7	0.391	18.7	88.9	67.5	33.2	200.5	0.378
General mean		20.47	103.87	83.67	33.40	252.0	0.357	20.93	99.3	90.13	34.9	252.23	0.366
9	6	20.0	89.8	68.5	33.4	209.8	0.437	20.3	91.8	78.8	34.5	222.4	0.451
	8	17.8	92.4	65.4	31.2	200.7	0.449	18.2	96.1	74.7	32.4	210.9	0.463
	10	17.0	82.7	61.5	30.1	200.0	0.471	17.5	86.4	64.5	31.7	204.0	0.477
General mean		18.27	88.30	65.13	31.57	203.5	0.452	18.67	91.43	72.67	32.87	212.43	0.464
12	6	18.7	86.5	59.3	30.2	190.8	0.520	17.9	87.6	55.4	31.4	177.0	0.544
	8	18.0	82.4	56.2	29.4	191.6	0.520	17.2	84.3	58.7	30.2	190.8	0.568
	10	17.2	76.8	54.6	28.5	188.7	0.568	17.0	80.7	56.5	30.3	180.6	0.588
General mean		17.97	81.90	56.70	29.37	190.37	0.536	17.37	84.2	56.87	30.63	182.80	0.567
General mean of factor (B)	6	20.3	99.07	77.17	33.03	237.0	0.431	20.5	99.10	81.17	34.20	234.93	0.450
	8	18.7	92.53	67.40	31.27	211.1	0.438	18.7	90.58	75.67	32.47	217.50	0.465
	10	17.7	82.47	60.93	30.03	197.8	0.477	17.7	85.33	62.83	31.73	195.03	0.481
LSD	A	1.0	4.8	8.3	1.3	18.2	---	1.1	5.2	14.2	1.8	20.4	---
	B	0.82	6.2	5.4	0.94	10.3	---	0.89	3.6	11.6	0.66	14.8	---
	A × B	0.94	7.5	1.8	1.2	12.4	----	0.70	4.2	8.5	1.2	10.7	---

(b)

Irrigation intervals	Plant density m²	2011 season				2012 season			
		Grain yield (g/m²)	Biological yield (g/m²)	HI %	CI %	Grain yield, (g/m²)	Biological yield (g/m²)	HI %	CI %
6	6	336	1132.3	0.30	0.42	356	1166.4	0.30	0.44
	8	344	1143.9	0.30	0.43	364	1210.5	0.30	0.43
	10	391	1338.0	0.29	0.41	378	1293.0	0.29	0.41
General mean		357.0	1204.7	0.30	0.42	366.0	1223.3	0.30	0.43
9	6	306	891.6	0.52	0.52	316	899.2	0.45	0.54
	8	314	800.0	0.39	0.65	324	857.4	0.38	0.61
	10	330.0	905.5	0.36	0.57	334	976.5	0.34	0.52
General mean		316.7	865.7	0.40	0.58	324.7	844.4	0.39	0.56
12	6	260	504.3	0.51	1.06	272	577.3	0.47	0.89
	8	260	514.9	0.51	1.02	284	566.9	0.50	1.00
	10	284	624.4	0.45	0.83	274	603.4	0.45	0.83
General mean		268.0	547.9	0.49	0.96	283.3	582.5	0.47	0.91
General mean of factor (B)	6	300.7	769.4	0.42	0.64	314.7	814.3	0.41	0.62
	8	306.0	819.6	0.40	0.60	324.0	878.3	0.39	0.68
	10	335.0	956.0	0.37	0.54	335.3	957.6	0.36	0.59
LSD	A	22.4	45.20	---	---	32.8	87.9	---	---
	B	18.3	40.89	---	---	8.7	42.8	---	---
	A × B	21.0	12.60	---	---	9.6	62.4	---	---

final yield and yield component characters of sorghum were investigated by some researchers, they concluded that, identification the optimum time of water supplies can reach to the optimal yield performance. Compatible results were obtained by [10] and [11]. The analysis of the yield component parameters presented in (**Table 7(a)**) indicated that the reduction in panicle length, and panicle weight, grain weight per panicle, 1000 grain weight and number of grains per panicle is associated with increasing in plant densities. Contrary grain yield per hectare is partly compensated by an increase in plant density (**Table 7(a)**). Thus, the highest plant density registered the highest grain yield 335.0 g/m^2 as compared to low plant density 300.7 g/m^2. These results are similar to those found by [12] [18] and [19], however, other authors did not found yield compensation processes in other cereals for different plant populations [38]. Significant differences were also found in the interactions between the different irrigation treatments and plant densities, since the higher number of plants per m^2 (10 plants per m^2) and application irrigation water every 6 days were followed. The data of the two years presented in (**Table 7(b)**) indicated that a high plant density under deficit irrigation conditions did not present any advantage in the sorghum yield. Such effect may be attributed to the fact that, plants grown under more moisture content in root zone, photosynthesis rate and photosynthetic substrates translocation is expected to be high compared to plants under stress condition and consequently seed yield of these plants are also expected to have less.

3.4. Effect on Chemical Composition of Leaves and Grains

Proline content reduced under moderate water stress, but increased under severe water stress (**Table 8**). In agreement with our results, [39] reported an increase in proline content of several plant species in water stress conditions. They further suggested that the accumulation of proline may play an important role in drought adaptation in the tested species. It is also clear from (**Table 8**) that, increasing plant density increased leaf proline content. Thus, the highest number of plants per m^2 (10 plants/m^2) registered the highest values of leaf proline content 3.27 and 3.36 (μmol/g FW) in the first and second seasons, respectively. Furthermore, the interactions of both factors recorded significant differences, sowing sorghum by the highest number of plants per m^2 under low

Table 8. Grain quality parameters of Sorghum under different irrigation intervals and plant densities as well as their interaction in 2011 and 2012 seasons.

Irrigation intervals	Plant density m^2	2011 season			2012 season		
		Leaves proline content (μmol/g FW)	Crude protein %	Total carbohydrate %	Leaves proline content (μmol/g FW)	Crude protein %	Total carbohydrate %
6	6	2.38	10.26	84.55	2.72	12.46	82.65
	8	2.78	9.12	83.52	2.82	10.25	82.12
	10	3.00	8.22	83.22	2.89	9.72	81.42
General mean		2.72	9.20	83.76	2.81	10.81	82.06
9	6	3.10	9.82	83.00	3.57	10.22	81.21
	8	3.24	8.62	82.69	3.30	9..32	81.00
	10	3.28	8.10	82.42	3.34	8.00	80.45
General mean		3.21	8.85	82.70	3.40	9.18	80.89
12	6	3.00	8.24	82.32	3.38	9.44	80.12
	8	3.33	7.65	82.00	3.42	8.12	80.00
	10	3.52	7.23	82.00	3.86	8.00	80.00
General mean		3.28	7.71	82.11	3.55	8.52	80.04
General mean of factor (B)	6	2.83	9.44	83.29	3.22	10.71	81.33
	8	3.12	8.46	82.74	3.18	9.23	81.04
	10	3.27	7.85	82.55	3.36	8.57	80.62
LSD	A	0.36	----	----	0.54	----	----
	B	0.28	----	----	0.12	----	----
	A × B	0.21	----	----	0.34	----	----

moisture content in root zone recorded the highest values of leaf proline content. Such effect may be attributed to high competition between plants for water and dissolving nutrients and low photosynthetic substrates translocation. From the data in the same (**Table 8**), it could be recognized that the highest crude protein and carbohydrate content were recorded under the watered irrigation treatment (irrigation every 6 days), after that a gradual decrease was noticed as water deficit was increased. In addition, carbohydrate content seems to be more stable with slight changes compared to crude protein in both seasons.

The effect of different plant densities on leaves proline content and grains chemical composition, data obtained reveal that increasing number of plants per square meter decreased either leaves proline content or grains chemical composition, whereas carbohydrate content did not gave a stable or defined trend on both seasons. The interaction effect in (**Table 8**), clearly show that increasing plant densities under high level of water deficit, mostly increased leaves proline content and in the same times decreased grains chemical composition.

4. Conclusions

Decreasing water supply by increasing the time between irrigations produced a significant reduction of the aerial sorghum characters *viz.*, stem diameter, number of green leaves per plant, time required to reach to 50% flowering, plant height, leaf area per plant and all of these parameters put in a reduction in the grain yield. The most important parameter among yield component parameters under investigation is harvest index, which is increased as water stress increased. Such results indicated that grain yield was more sensitive to water stress than the aerial dry matter production. For the three plant densities, the relationship between grain yield and harvest index was similar under the three water irrigation applied.

The results indicated that important compensation processes occurred between the different sorghum yields components. Thus, the lesser number of plants per m^2 of the lower sorghum densities were compensated with a greater production of tillers, a greater number of grains per panicle and a higher weight of these grains. The result of this study suggests that sowing sorghum using less density could be an alternative crop if water supply is limiting under water deficit conditions.

Acknowledgements

The authors would like to extend their sincere appreciation to the Deanship of Scientific Research at King Saud University for financially supporting this work.

References

[1]   FAO (2003) (Food and Agriculture Organization), FAO Yearbook: Production. Vol. 55, FAO, Rome, 164-166.

[2]   Borlaug, N.E. and Dowswell, C.R. (2005) Feeding a World of Ten Billion People: A 21st Century Challenge. In: Tuberosa, R., Phillips, R.L. and Gale, M., Eds., *Proceedings of the International Congress in the Wake of the Double Helix: From the Green Revolution to the Gene Revolution*, Bologna, 27-31 May 2003, 3-23.

[3]   Komeili, H.R., Rashed-Mohassel, M.H., Ghodsi, M. and Zare-FeizAbadi, A. (2008) Evaluation of Modern Wheat Genotypes in Drought Resistance Condition. *Agricultural Researches*, **4**, 301-312.

[4]   Zhang, S.O. and Outlaw Jr., W.H. (2001) Abscisic Acid Introduced into Transpiration Stream Accumulates in the Guard Cell Apoplast and Causes Stomatal Closure. *Plant, Cell & Environment*, **24**, 1045-1054.
       http://dx.doi.org/10.1046/j.1365-3040.2001.00755.x

[5]   Banon, S., Fernandez, J.A., Franco, J.A., Torrecilas, A., Alarcon, J.J. and Sanchez-Blanco, M.J. (2004) Effects of Water Stress and Night Temperature Preconditioning on Water Relation and Anatomical Change of *Lotus creticus* Plants. *Science Horticulture*, **101**, 333-342. http://dx.doi.org/10.1016/j.scienta.2003.11.007

[6]   Igartua, E., Gracia, M.P. and Lasa, J.M. (1994) Characterization and Genetic Control of Germination—Emergence Responses of Grain Sorghum to Salinity. *Euphytica*, **76**, 185-193. http://dx.doi.org/10.1007/BF00022163

[7]   Kole, C. (2001) Wild Crop Relatives: Genomic and Breeding Resource Cereals. Institute of Natural Research.

[8]   Agrama, H.A. and Tuinstra, M.R. (2003) Phylogenetic Diversity and Relationships among Sorghum Accessions Using SSRs and PAPDs. *African Journal of Biotechnology*, **2**, 334-340. http://dx.doi.org/10.5897/AJB2003.000-1069

[9]   Buah, S.S.J. and Mwinkaara, S. (2009) Response of Sorghum to Nitrogen Fertilizer and Plant Density in the Guinasavana Zone. *Agronomy Journal*, **8**, 124-130. http://dx.doi.org/10.3923/ja.2009.124.130

[10]   Hosseinian Maleki, S. and Mirshekari, B. (2011) Irrigation Period in Three Rapeseed Cultivars Influences Crop Phe-

nology and Yield. *Journal of Food, Agriculture and Environment*, **9**, 446-448.

[11] Gohari, A.A. (2012) Effect of Soil Water on Plant Height and Root Depth and Some Agronomic Traits in Common Bean (*Phaseolus vulgaris*) under Biological Phosphorous Fertilizer and Irrigation Management. *International Research Journal of Applied and Basic Sciences*, **3**, 848-853.

[12] Ma, B.L., Dwyer, L.M. and Costa, C. (2003) Row Spacing and Fertilizer Nitrogen Effects on Plant Growth and Grain Yield of Maize. *Canadian Journal of Plant Science*, **83**, 241-247.

[13] Selim, M.M. (1995) Evaluation of Some Grain Sorghum Genotypes Grown under Different Plant Densities and Levels of Nitrogen Fertilization. *Egyptian Journal of Agronomy*, **20**, 83-97.

[14] Fischer, K.S. and Wilson, G.L. (1975) Studies of Grain Production in *Sorghum bicolor* (L. Moench).V. Effect of Planting Density on Growth and Yield. *Australian Journal of Agricultural Research*, **26**, 31-41. http://dx.doi.org/10.1071/AR9750031

[15] Ferraris, R. and Charles-Edwards, D.A. (1986) A Comparative Analysis of the Growth of Sweet and Forage Sorghum Crop. I. Dry Matter Production, Phenology and Morphology. *Australian Journal of Agricultural Research*, **37**, 495-512.

[16] Berenguer, M.J. and Faci, J.M. (2001) *Sorghum* (*Sorghum bicolor* L. *Moench*) Yield Compensation Processes under Different Plant Densities and Variable Water Supply. *European Journal of Agronomy*, **15**, 43-55. http://dx.doi.org/10.1016/S1161-0301(01)00095-8

[17] Borrás, L., Maddonni, G.A. and Otego, M.E. (2003) Leaf Senescence in Maize Hybrids: Plant Population, Row Spacing and Kernel Set Effects. *Field Crops Research*, **82**, 13-26. http://dx.doi.org/10.1016/S0378-4290(03)00002-9

[18] Charles, A.S. and Charles, S.W. (2006) Corn Response to Nitrogen Rate, Row Spacing, and Plant Density in Eastern Nebraska. *Agronomy Journal*, **94**, 529-535.

[19] Zand, N., Shakiba, M.-R., Moghaddam-Vahed, M. and Dabbagh-Mohammadai-nasab, A. (2014) Response of *Sorghum* to Nitrogen Fertilizer at Different Plant Densities. *International Journal of Farming and Allied Sciences*, **3**, 71-74.

[20] Cottenie, A., Verlo, M., Kjekens, L. and Camerlynch, R. (1982) Chemical Analysis of Plant and Soil. Laboratory of Analytical Agrochemistry. State University, Gent, Belgium, Article No. 42, 80-284.

[21] But, R. (2004) Soil Survey Laboratory Manual Report No. 42 USDA. National Resources Conservation Service, Washington DC.

[22] American Public Health Association (APHA) (1992) Standard Methods for Examination of Water and Wastewater. 18th Edition, APHA, AWWA, WPCF, NY, Washington DC.

[23] Bos, M.G. (1985) Summary of ICID Definition of Irrigation Efficiency. *ICID Bulletin*, **34**, 28-31.

[24] Donald, C.M. and Hamblin, J. (1976) The Biological Yield and Harvest Index of Cereals as Agronomic and Plant Breeding Criteria. *Advances in Agronomy*, **28**, 361-405. http://dx.doi.org/10.1016/S0065-2113(08)60559-3

[25] Bates, L.S., Waldren, E.P. and Teare, I.D. (1973) Rapid Determination of Free Proline for Water Stress Studies. *Plant and Soil*, **39**, 205-207. http://dx.doi.org/10.1007/BF00018060

[26] AOAC (2000) Official Methods of Analysis. 25th Edition, Association of Official Analysis Chemists, Washington DC.

[27] Dubois, M., Gilles, K.A., Hamilton, J., Roberts, R. and Smith, F. (1956) Colorimetric Method for Determination of Sugar and Related Substances. *Analytical Chemistry*, **28**, 350-356. http://dx.doi.org/10.1021/ac60111a017

[28] Gomez, K.A. and Gomez, A. (1984) Statistical Procedure for Agricultural Research—Hand Book. John Wiley & Sons, New York.

[29] Sadeghipour, O. (2009) The Influence of Water Stress on Biomass and Harvest Index in Three Mung Bean Cultivars. *Asian Journal of Plant Sciences*, **8**, 245-249. http://dx.doi.org/10.3923/ajps.2009.245.249

[30] Shiri, M., Momeni, H. and Geranmayeh, B. (2013) The Survey of the Morphological and Physiological Basis of Maize Grain Yield under Drought Stress Condition through Path Analysis. *Technical Journal of Engineering and Applied Sciences*, **3**, 3647-3651.

[31] Krieg, D.R. and Lascano, R.J. (1990) Sorghum. In: Stewart, B.A. and Nielsen, D.R., Eds., *Irrigation of Agricultural Crops*, American Society of Agronomy, Madison, 719-740.

[32] Tyagi, A.P., Mor, B.R. and Singh, D.P. (1998) Path Analyses in Upland Cotton (*G. hirsutum* L.). *The Indian Journal of Agricultural Science*, **22**, 137-142.

[33] Hussein, M.M. and Alva, A.K. (2014) Growth, Yield and Water Use Efficiency of Forage *Sorghum* as Affected by NPK Fertilizer and Deficit Irrigation. *American Journal of Plant Sciences*, **5**, 2134-2140. http://dx.doi.org/10.4236/ajps.2014.513225

[34] Ismail, A.M.A. and Ali, A.H. (1996) Effect of Nitrogen Rates and Plant Densities on Some Morphological Characters

and Yield of Grain *sorghum*. *Arab Gulf Journal of Scientific Research*, **4**, 49-58.

[35] Swamya, B.P.M., Upadhyaya, H.D., Goudara, P.V.K., Kullaiswamya, B.Y. and Singh, S. (1997) Phenotypic Variation for Agronomic Characteristics in a Groundnut Core Collection for Asia. *Field Crops Research*, **84**, 359-370. http://dx.doi.org/10.1016/S0378-4290(03)00102-3

[36] Abuzar, M.R., Sadozai, G.U., Baloch, M.S., Baloch, A.A., Shah, I.H., Javaid, T. and Hussain, N. (2011) Effect of Plant Population Densities on Yield of Maize. *Journal of Animal and Plant Sciences*, **21**, 962-965.

[37] Zamir, M.S.I., Ahmad, A.H., Javeed, H.M.R. and Latif, T. (2011) Growth and Yield Behaviour of Two Maize Hybrids (*Zea mays* L.) towards Different Plant Spacing. *Cercetari Agronomice in Moldova*, **44**, 33-40. http://dx.doi.org/10.2478/v10298-012-0030-9

[38] Guberac, V., Martincic, J., Maric, S., Bede, M., Jurisic, M. and Rozman, V. (2000) Grain Yield Components of Winter Wheat New Cultivars in Correlation with Sowing Rate. *Cereal Research Communication*, **28**, 307-314.

[39] Wang, S., Wan, C., Wang, Y., Chen, H., Zhou, Z., Fu, H. and Sosebee, R.E. (2004) The Characteristics of Na^+, K^+ and Free Proline Distribution in Several Drought-Resistant Plants of the Alxa Desert, China. *Journal of Arid Environments*, **56**, 525-539. http://dx.doi.org/10.1016/S0140-1963(03)00063-6

Waterlogging in the New Reclaimed Areas Northeast El Fayoum, Western Desert, Egypt, Reasons and Solutions*

El Sayed Ali El Abd[1], Maged Mostafa El Osta[2]

[1]Geology Department, Desert Research Centre, Cairo, Egypt
[2]Geology Department, Faculty of Science, Damanhour University, Damanhour, Egypt
Email: drmagedelosta.edu.alex@hotmail.com

Abstract

The waterlogging in the new reclaimed areas has become a major concern in the area Northeast El Fayoum, Western Desert, Egypt. It is not only endangering the structures and properties but also causing major environmental problem affecting the health of the area, habitats, and the biotic of the land community, as well as the deteriorating of Egypt's Pharaonic monuments (El Lahun and Hawarah pyramids). Both the daily seepage from excess irrigation water and the presence of impervious clay or limestone beds at shallow depths may represent the main contributor of groundwater rising in the shallow aquifer. This paper investigates the interplay of the hydrogeological characteristics, soil properties and recent land reclamation projects on the distribution of waterlogging and salinization within the study area. The field observations show that new reclaimed areas have been recently cultivated in distant areas from the old agricultural land. These new cultivations have developed widespread waterlogging, soil salinization and deterioration of Egypt's Pharaonic monuments as a result of rising groundwater related problems. In this paper, the data used come from database of drillings for eleven observation wells distributed inside the whole area to measure periodic water levels. The soil litho-units are mainly composed of coarse sand, sandy clay, silt and fractured limestone underlined by impervious clay or limestone, thus limiting the downward percolation of excess irrigation water and therefore develops waterlogging. The drainage networks and suitable irrigation methods have to be considered when planning for a new cultivation in dry land to better control waterlogging and salinization hazard. It is highly recommended in this research that newly small and deep cut drainage canals network should be constructed and connected to the master drainage canal to dewater the excess irrigation water

---

*Waterlogging in the new reclaimed areas Northeast El Fayoum, Western Desert, Egypt, reasons and solutions.

and to prevent the waterlogging in the concerned area.

Keywords

Observation Wells, El Lahun and Hawarah Pyramids, Soils, Waterlogging, Drainage Canals Network

1. Introduction

In Egypt, the new reclaimed areas are being affected by soil salinity and waterlogging. In El Fayoum depression, Western Desert, Egypt, waterlogging problem is stated to be serious in the areas lying in the lower reaches of the concerned area [1] [2], reported how soil salinity and waterlogging problems have developed worldwide, and the speed with which they are advancing at present. Plants that are waterlogged are very susceptible to salinity, especially in their early growth stages [3]. Waterlogging and salinity problems pose a serious threat to the world's productive agricultural land. Disturbance of the natural balance by introducing irrigation causes a rising water table, where natural drainage sinks cannot cope with the increase in ground water recharge [4]. Recharge to deep aquifers is closely linked to the incidence of waterlogging [5] and to the development of land salinization. The major artificial causes of waterlogging in the command areas are seepage from water conveyance systems [6], breakages of regulatory structure, silting and weed growth in canals [7]. Lack of surface and sub-surface drainage, poor maintenance of drainage system, over irrigation and growing water intensive crops are some of major causes of poor realization of benefits from the irrigation systems [8] [9] indicated that due to the accumulation of organic matter, soil color is generally darker in poorly drained areas than well drained soils. Thus, a proper assessment of these waterlogged areas is a prerequisite for finding a solution to the problem. In general, for mapping of waterlogged areas, conventional technique such as ground survey observation wells are uses. There is growing concern about the decline in soil fertility, changes in water table depth, deterioration in the quality of irrigation water and rising salinity in the new reclaimed areas of Egypt.

El Fayoum depression is one of the most important agriculture lands of Egypt due to the good soil cover and the high crop productivity. The historical places as Hawwara and Lahun pyramids helped El Fayoum to be one of the wonderful tourism places in Egypt. It is exposed to several problems as waterlogging and soil salinization due to the human activities in the new reclaimed area located in the desert land in the periphery of the old cultivated land. These problems reduce the crop productivity in the old cultivated land. Tourism activities also leave its negative effect on the total income of El Fayoum governorate.

Hence, in the present study, a systematic attempt has been made for rapid, reliable assessment and delineation of the surface and sub-surface waterlogged areas in all the irrigation command areas of the Northeast El Fayoum, Western Desert, Egypt. As well as, a trial is made through this search to give an explanation for the geological and hydrological factors that leading to the evolution of these problems and accordingly suggesting suitable solutions to minimize their harmful effects.

2. Study Area Description

The study area, *i.e.* Northeast El Fayoum is located southwest of Cairo by about 100 Km and northwest of BeniSuef by about 30 Km, and situated between 29°10'N and 29°20'N latitudes and 30°50'E and 31°00'E longitudes (**Figure 1**). It is now considered as the bread basket for the excess irrigation water from the new reclamation areas. The overall climate of this area is characterized by hot, long and dry summer and warm, short winter with scarce precipitation. Also, great temperature differences between summer and winter and between day and night characterizes this belt. The temperature ranges between 46.7°C and 35.1°C in summer months and from 20°C to 21.5°C during winter months. The average annual rainfall is around 10.3 mm which is relatively low. The total annual evaporation intensity reaches 2296 mm/year, and the annual mean of relative humidity is 51.6 %.

The geomorphology, stratigraphy and structure of El Fayoum area have attracted the attention of many researchers including [10]-[17], as well as the field observations of the present authors. Geomorphologically, the study area has been subdivided into the following units (**Figure 2**):

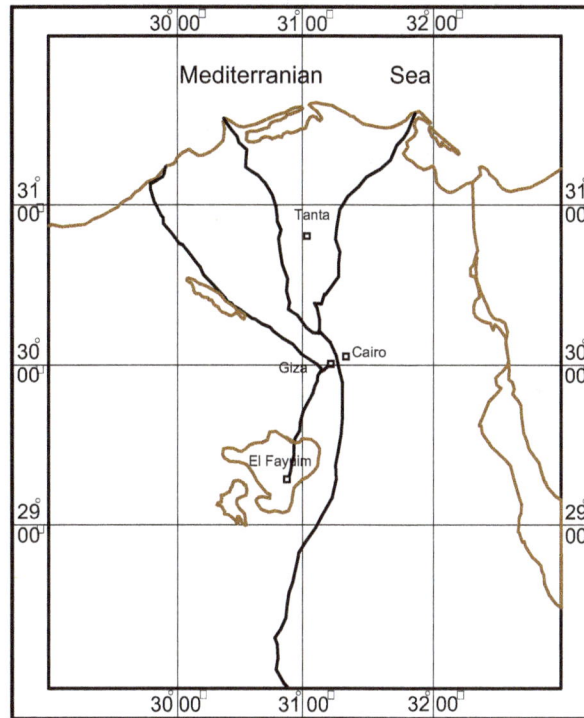

Figure 1. Location map of the study area.

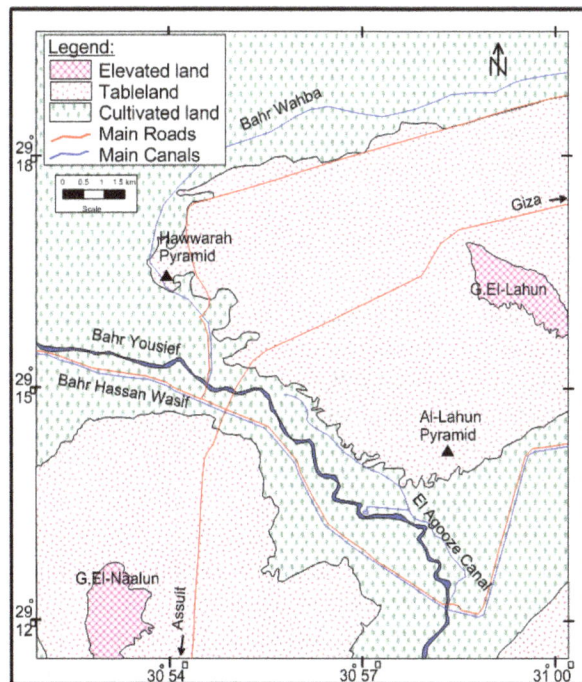

Figure 2. Geomorphologic map of the study area.

2.1. Tableland

Tableland Surrounds El Fayoum depression from the east and separating it from the Nile valley. Its elevation ranges between +30 m and +100 m above sea level and slopes with different degrees towards the Nile Valley and El Fayoum depression (**Figure 3**). It is mainly composed of Middle Eocene fractured limestone and dis-

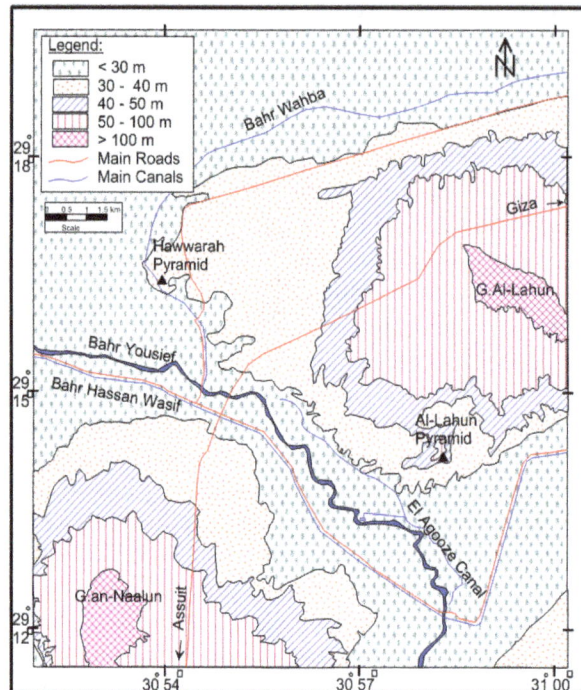

Figure 3. Orographic map of the study area.

sected by short drainage lines that are directed towards the Nile Valley and El Fayoum depression.

2.2. Elevated Land

Elevated land, is the high land overlooking the tableland and includes Gebel El Lahun (+144 m) and Gebel El Naalun (+157 m). It is also as the tableland mainly composed of Middle Eocene fractured limestone and dissected by short drainage lines that are directed towards the Nile Valley and El Fayoum depression in the direction of the slope.

2.3. Cultivated Land

Cultivated Land is the eastern part of El Fayoum depression which characterized by good soil. Its surface is covered by lacustrine deposits with ground elevation less than 30 m above sea level (**Figure 3**). The outlet of El Fayoum depression to the Nile Valley is a part of this unit which represented by a narrow path Known as "Hawaret El Maktaa" between Gebel El Lahun and Gebel El Naalun. Through that path runs Bahr Youssef the main canal carrying the Nile water to the net of irrigation canals dissecting the El Fayoum depression floor. Reclamation projects are now under execution in the eastern peripheries of this land using Nile water in irrigation and create waterlogging problems in the neighboring old lands.

Geologically, El Fayoum and vicinities are occupied by sedimentary rocks belonging to Tertiary and Quaternary Eras (**Figure 4**). The surface exposure has a thickness of about 848 m [1]. Quaternary deposits are distinguished into, Holocene and Pleistocene deposits as follows:

The Holocene deposits are distinguished into; Aeolian deposits, young lacustrine deposits and young Nilotic deposits. Aeolian deposits are composed of loose quartz sand. Young lacustrine deposits are well defined in El Fayoum depression, associated with the lake development and composed of fine sand and clay with thin relics of gypsum and carbonate materials forming the agriculture soil. Young Nilotic deposits have a variable thickness (1 m to 12 m) as a result of the seasonal Nile floods and composed of silt and fine sand dominated by quartz grains and heavy minerals.

The Pleistocene deposits are differentiated into Old lacustrine and Old Nilotic deposits. Old lacustrine deposits (45 m thick) are present within the depression area in the form of terraces at levels +43 m, +30 m and +25 m indicating fresh water lake feed by the Nile during the Pleistocene times. They consist mainly of clay, fine to

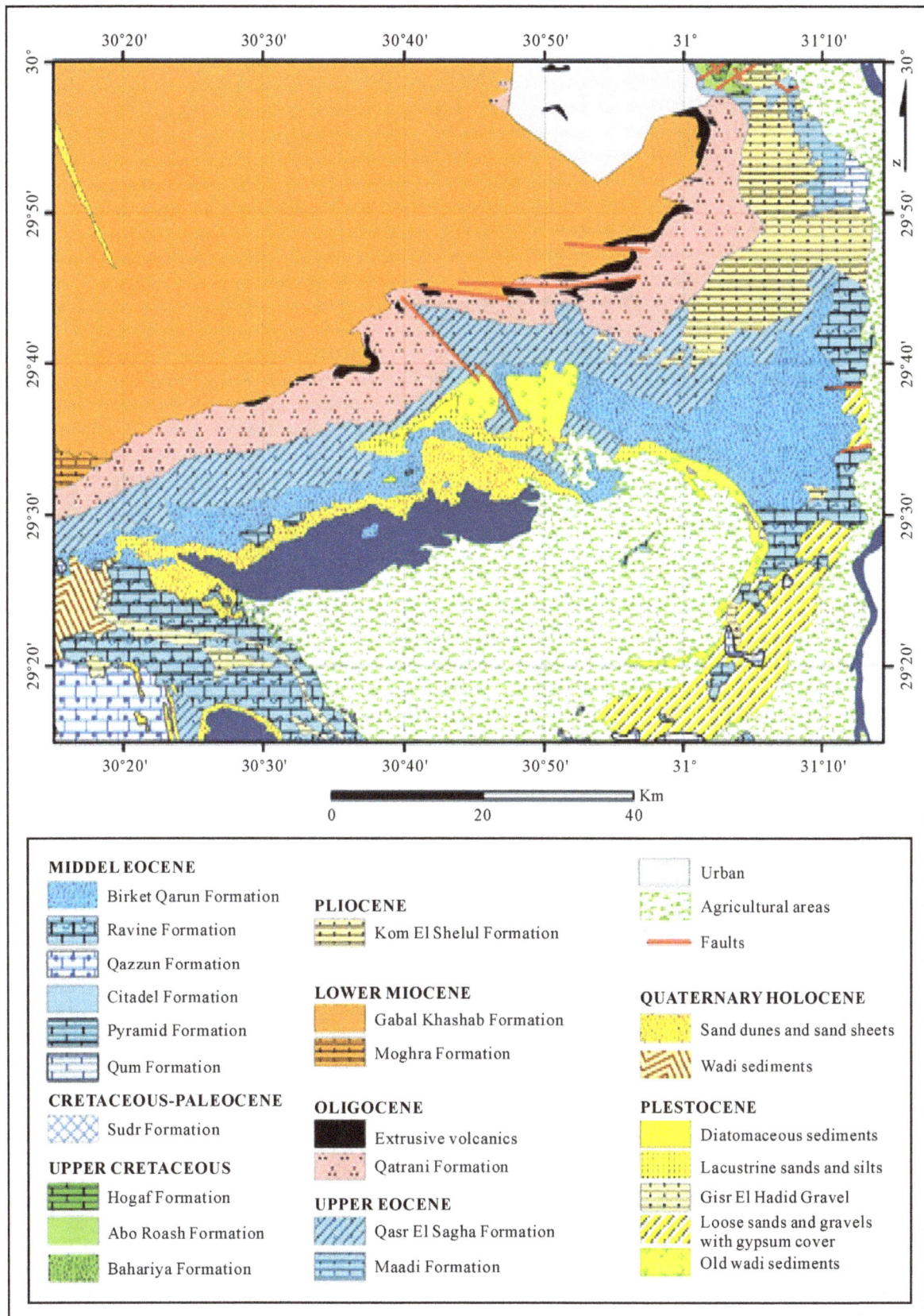

Figure 4. Geologic map of the study area [21].

medium coarse sand with considerable amounts of calcareous and gypsiferous materials. They show high content of mica, crystalline gypsum, iron, manganese oxides, sand debris, calcareous fragments and organic and fossil remains. Old Nilotic deposits (190 m thick) are recognized at the Nile-Fayoum divide where they are differentiated into terraces at levels +89 m, +112 m, +134 m and +167 m. They are composed of sand and gravels forming a good groundwater aquifer.

Tertiary deposits are differentiated into Pliocene Miocene, Oligocene and Eocene rocks as follows:

The Pliocene deposits (90 m thick) are defined in the Nile-Fayoum divide forming of fossiliferous sandstone in some places overlain by Quaternary sand and gravels [11].

The Miocene rocks (20 m thick) are exposed at Gebel Qatrani to the north of El Fayoum depression overlying basalt exposures. They are composed of a series of alternating beds of sand and gravels with silicified wood remains.

The Oligocene rocks (275 m thick) are encountered underneath the Miocene rocks at Gebel Qatrani area and capped by basalt sheet. They are composed of sand and sandstone with shale and marl interbeds. These rocks are rich in silicified wood and land animals (crocodiles, tortoises and turtles).

The Eocene rocks (390 m thick) have a wide distribution in El Fayoum-Wadi El Rayan area [10]. They are distinguished into Upper and Middle Eocene rocks. Upper Eocene rocks are formed of a series of escarpments overlooking El Fayoum depression from the eastern side. The rock succession is composed mainly of sand (69%), shale (19%) and limestone (12%) with high fossil content. These facies and faunal content indicate shallow marine environment. The Upper Eocene rocks were distinguished into two formations, the upper Qasr El Sagha Formation and the lower BirketQarun Formation, both are dominated at the northern extremities of the cultivated land fringing Qarun lake and the hill mass of GaretGehannam on the west side of El Fayoum depression. The thickness of both formations is about 180 m and 50 m respectively. On the other hand, Middle Eocene rocks constitute the oldest exposed rocks in El Fayoum area and its vicinities. They are represented mainly by limestone and marl with shale intercalations. They are discriminated into two formations, Raveine Formation and Wadi El Rayan Formation. The former appears below the alluvial deposits in the deep water channels and consists of gypseous shale, marl, limestone and sand. The other Formation is exposed at Wadi El Rayan where its outcrop composed of hard white limestone full of nummulitegizahensis, argillaceous sand and sandy shale.

Structurally, El-Faiyum depression is a structurally-controlled tectonic basin, marked by northeast-striking faults along its northern margin [18]. Qarun Lake is located in the northern part of El-Faiyum depression, at a right-step between two strands of the NE striking fault system. The E-W structures probably related to the relative motion between south Europe and north Egypt and closure of the Neotethys [19]. The continuation of these faults in the present area is only detected from magnetic and seismic studies.

The area of study is a new reclaimed area located in the eastern peripheries of the old cultivated land in El Fayoum depression. Nile water is the source of its irrigation by lifting. Bahr Wahba bonded the study area from north and northwest and Bahr yousief and El Agooze canal from south and southwest respectively. It dominated by two famous Egypt's Pharaonic monument pyramids; Hawarah and Al-Lahun (**Figure 2**). These pyramids are exposed to different deterioration processes (aging) caused by internal and external stresses due to the mineral composition of the building materials, climate factors and groundwater rising. All the existing elements of the pyramids are constructed from mud bricks which consist of quartz, kaolinite, calcite, montmorillonite, microcline and gypsum, as well as Eocene limestone which consists essentially of calcite and small amounts of halite [20]. There are two types of cements: clay cement consisting of quartz, orthoclase, calcite and illite, jointing the wall's structure of the pyramid and gypsum cement consisting of calcite and traces of gypsum and quartz jointing the stones. Wasp nests from the pyramid consist of fine grains of quartz and calcite joined by wasp saliva, while the salts are halite and gypsum. The actual state of the building materials of the pyramid is poor; the mud bricks are more friable and the limestone is weakened and highly porous due to exposure to deteriorating factors. The area of study is covered by fractured white Fossiliferous Limestone interbeded by impervious layer (clay and marl beds) and elevated than the surrounding old cultivated land.

Due to the traditional irrigation methods many dangerous lakes was formed and waterlogging appeared in the area causing serious damage in the new and old lands. Field observation shows that the land surface slopes towards these localities where, ground surface elevation reaches less than +30 m (**Figure 3**). These areas receive both return flow of irrigation water and groundwater from the surrounding parts where depths to groundwater reaches less than one meter and groundwater levels coincide with the land surface under these localities (**Figure 5**). As well as, the moisture is drawn up into the stones of ancient buildings through capillary action and the

Figure 5. Field Photos show the developed waterlogging in the study area.

water table rises inside the Hawarah and Al-Lahun pyramids. Other factors such as the lack of advanced drainage system and the presence of impermeable shale and clay beds at shallow depths raised groundwater to and near the surface at some localities forming waterlogging and surface water ponding (**Figure 5**), this phenomenon destroyed the majority of the new reclaimed areas and the Egypt's greatest monuments could be vulnerable with collapse if action is not taken.

3. Methods and Materials

To study waterlogging problem in the concerned area, the following were preformed:

1) Survey of El Fayum toposheets at 1:50,000 scale was used for preparation of the base maps and for drilling observation wells.

2) Field work was conducted two times in the depression for this search from 2008 to 2010.

3) Using the interpreted Landsat ETM+ images and their photo-mosaics, the rock units have been checked and photo-graphed.

4) Eleven observation wells were constructed to measure periodic water levels (**Figure 6**).

5) Rock sample collection during construction of observation wells to study the type of subsurface rocks (**Figure 7** and **Table 1**).

6) The aquifers properties and groundwater flow were examined using their hydraulic parameters and groundwater levels of the observation wells and 6 lakes during 2008 and 2010 to construct water level map and determine water flow direction.

7) To assess the waterlogged areas induced by rise in the ground water level near to the surface in the study area, ground water table data pertaining to pre- and post-monsoon seasons of the year 2008-2010 were collected

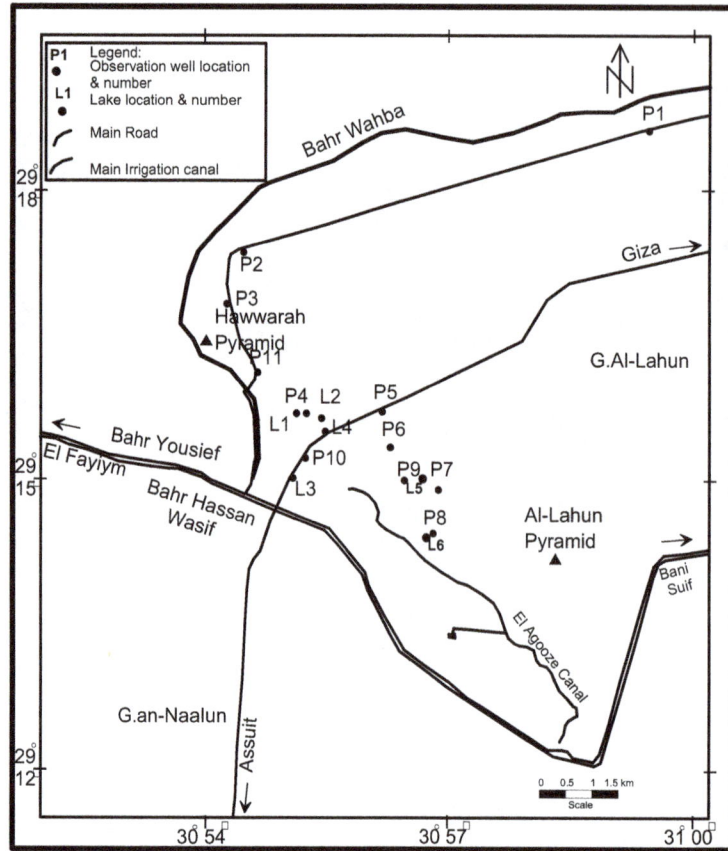

Figure 6. Location of the observation wells and lakes in the study area.

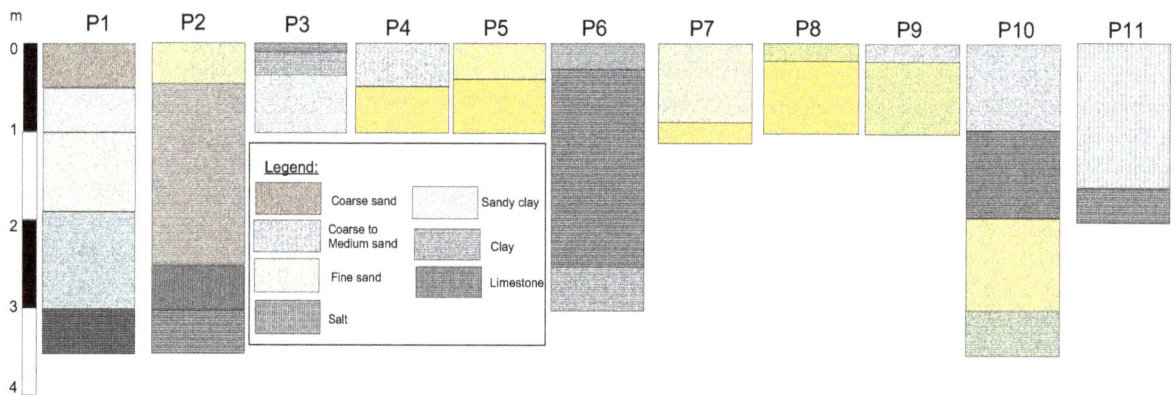

Figure 7. Lithologic logs of the observation wells in the study area.

for all the observation wells monitored by the author's which are spread all over the study area (**Figure 9** and **Figure 10**).

8) Water samples were collected to study the hydrochemicall characteristics of the subsoil water.

4. Results and Discussion

Hydrogeologically, the various litho-units of the shallow aquifer in the study area can be grouped by coarse sand, sandy clay, silt and fractured limestone underlined by impervious clay (P1, P2, P3, P6, P9, P10 and P11) or limestone (P4, P5, P7 and P8) which prevent downward of irrigation water (**Figure 7**).

The observations on groundwater fluctuation (**Figure 8**) from January (2008) to January (2010) demonstrate a

Table 1. Date of observation wells and lakes in the study area.

Observation Well & lake No.	Total depth (m)	Casing (m)	Ground elevation (m)	Depth to water 1/2008	Water level 1/2008	Depth to water 1/2010	Water level 1/2010
P1	3.5	0.18	25.82	1.5	24.32	1.45	24.37
P2	3.4	0.31	29.224	0.97	28.254	0.95	28.274
P3	1	0.13	27.339	0.28	27.059	0.26	27.079
P4	0.8	0.25	25.359	0.4	24.959	0.38	24.979
P5	1	0.3	32.25	0.55	31.7	0.5	31.75
P6	3	0.39	34.861	0.83	34.031	0.80	34.061
P7	1.1	0.35	39.197	0.55	38.647	0.53	38.667
P8	1	0.15	34.015	0.25	33.765	0.21	33.805
P9	1	0.15	34.488	0.27	34.218	0.22	34.268
P10	3.5	0.16	25.375	0.45	24.925	0.37	25.005
P11	2	0.22	27.152	0.56	26.592	0.48	26.672
L1			24.734	0	24.734	+0.05	24.784
L2			24.743	0	24.743	+0.09	24.833
L3			25.573	0	25.573	+0.06	25.633
L4			27.445	0	27.445	+0.05	27.495
L5			35.609	0	35.609	+0.08	35.689
L6			33.765	0	33.765	+0.04	33.805

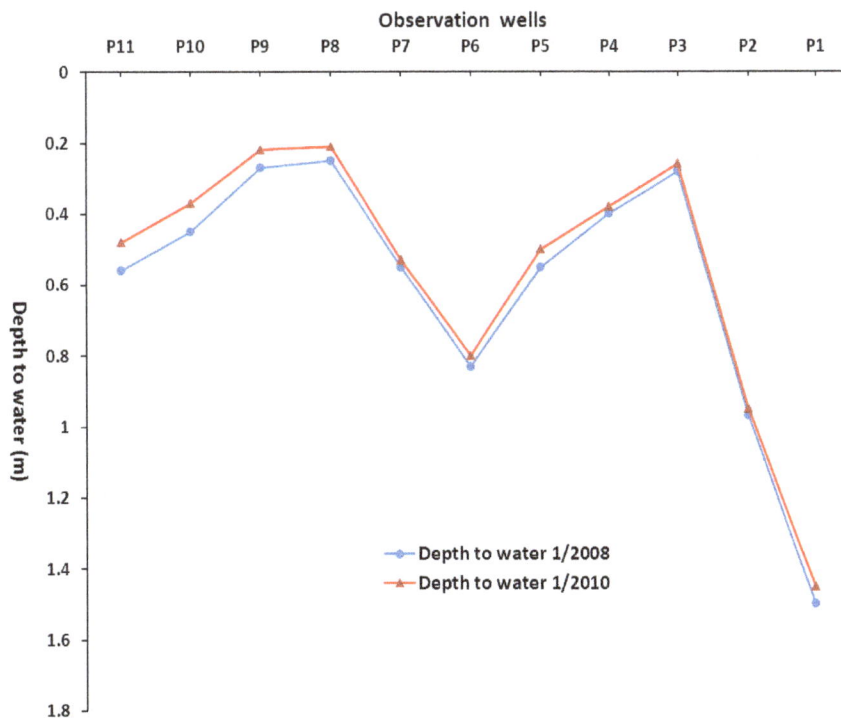

Figure 8. Depth to water fluctuation in the study area.

relatively increase in water table by about 0.02 m (the lowest at P3) to 0.08 m (the highest at P11). The increase of groundwater levels occurred in practically all piezometers with different magnitudes with an annual average of 0.05 m. On the other hand, the water tables of the command area for the year 2008-2010 are delineated and the integrated maps are shown in **Figure 9** and **Figure 10**. Comparing water levels of 1/2008 (**Figure 9**) and water level of 1/2010 (**Figure 10**), cleared that all the observation wells and the lakes have increase in the water level from 0.09 m (L2) to 0.02 m (p3). As well as, the subsoil water moves from the high level land located in the middle of the concerned area towards the low land (old cultivated land) located in the periphery of the new reclaimed land and the two ancient pyramids (Hawarah and Al-Lahun) causing serious waterlogging and damage of the old cultivated land. Groundwater threatens more than the great monument's stability. If water wicks up into the statue, it will leach salts from the limestone and deposit them on the surface of the stones. Eventually, the salts will accelerate the flaking erosion and the fall of clay plaster and cement that has long afflicted the Hawarah and Al-Lahun pyramids as well as the salt crystallization in the entrance (**Figure 11**). The deteriorating mechanism of salt solution depends on two sources of water: humidity from condensation and groundwater from capillary rise In order to protect the antiquities from further accelerated deterioration, lowering of groundwater to safe levels was needed.

The hydrogeochemical characteristics of the sub-soil water in the study area are discussed through the chemical analysis of eleven sub-soil water samples (beside irrigation canals water samples) collected during January 2008 (**Table 2**). The chemical analyses deal with the determination of major cations (Na^+, K^+, Ca^{2+} and Mg^{2+}) and anions (CO_3^-, HCO_3^-, SO_4^{2-} and Cl^-). According to Chebotarev classification for salinity [22], the subsoil water of the study area can be classified from brackish water to high saline water. The salt content of the water samples ranges from 32,816 mg/l (P8) to 4224.4 mg/l (P11). Furthermore, the salinity increases generally towards the northwest of the study area (**Figure 12**). Hyper-saline water is restricted to high land area reflecting the effect of gypsiferous shale beside the infiltration of high saline excess irrigation water. Saline water is restricted to low land in the periphery of the old cultivated land (P1, P2 and P11) reflecting the effect of fresh

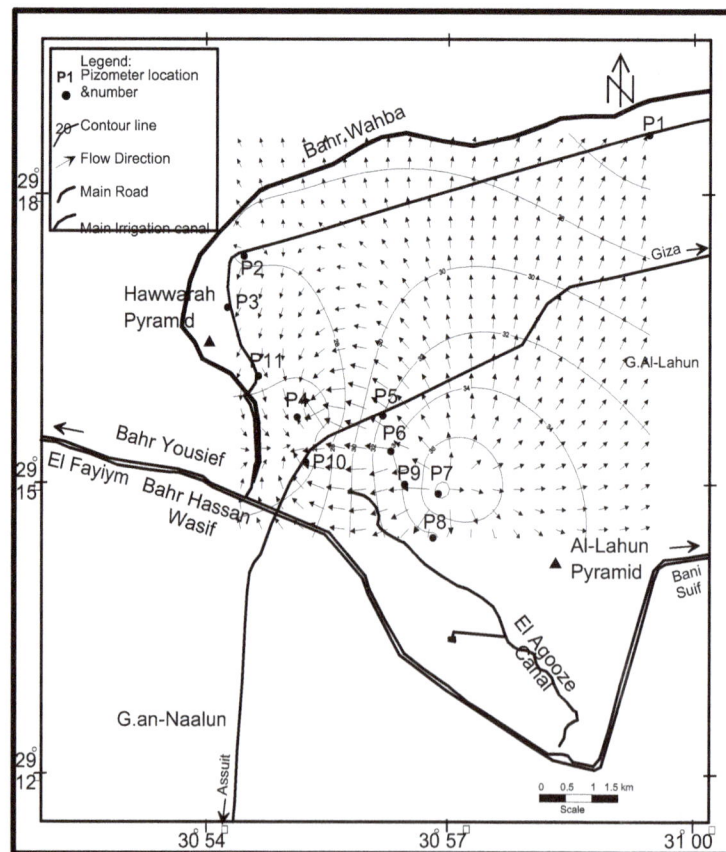

Figure 9. Water level contour map in the study area (2008).

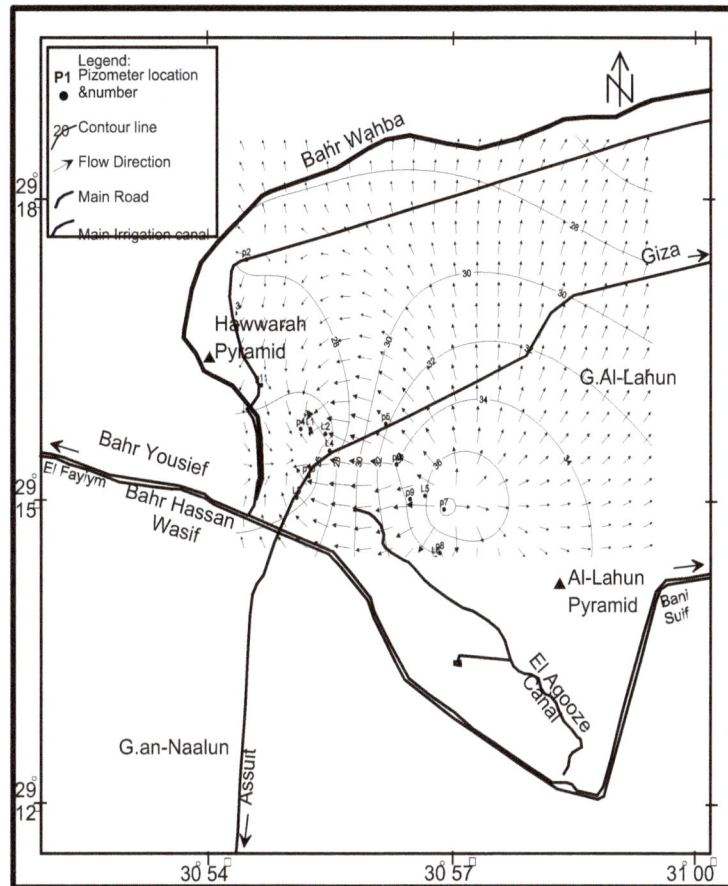

Figure 10. Water level contour map in the study area (2010).

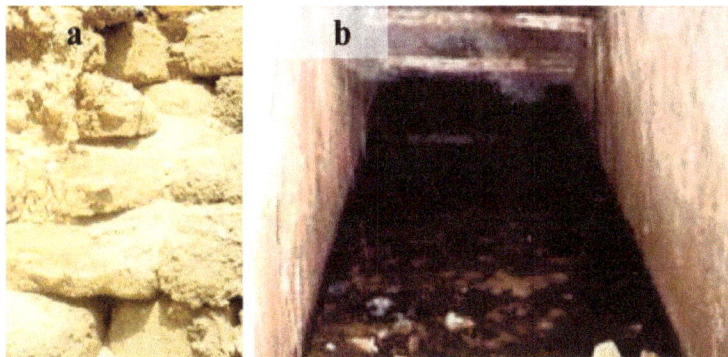

Figure 11. Examples of destruction types in Hawarah and Al-Lahun pyramids: (a) Fall of clay plaster and cement; (b) Salt crystallization in the entrance.

water of the irrigation canal and the sub-soil composed of coarse to fine sand intercalated with sandy clay and clayey sand. Regarding major cations and anions, sodium is predominant cation followed by calcium or magnesium and magnesium or calcium ($Na^+ > Ca^{2+} > Mg^{2+}$ or $Na^+ > Mg^{2+} > Ca^{2+}$), (**Table 2**). High sodium concentration is possibly due to leaching processes of clay and shale. Concerning major anions, sulfide is the mostly predominant anions followed by chloride or bicarbonates and bicarbonates or chloride ($SO_4^{2-} > Cl^- > HCO_3^-$ or $SO_4^{2-} > HCO_3^- > Cl^-$). High sulfide concentration is possibly due to leaching processes of clay and shale present, the effect of sulfide fertilizer and the infiltration of high saline excess irrigation water.

Table 2. Hydrogeochemical data of the observation wells and irrigation water in the study area.

Well No.	E.C µM hos/cm	TDS (mg/l)	Units	Ca^{2+}	Mg^{2+}	Na^+	K^+	Sum. Cat.	CO_3^-	HCO_3^-	SO_4^{2-}	Cl^-	Sum. An.
			mg/l	300	97.28	16,000	150		0	100.8	23,000	8875	
P1	70,600	48472.7	meq/l	14.97	8.00	695.68	3.84	722.49	0.00	1.65	478.86	250.28	730.79
			e%	2.07	1.11	96.29	0.53		0.00	0.23	65.53	34.25	
			mg/l	220	188.5	15,000	150		0	378	22,000	7543.8	
P2	70,100	45291.2	meq/l	10.98	15.50	652.20	3.84	682.52	0.00	6.20	458.04	212.73	676.97
			e%	1.61	2.27	95.56	0.56		0.00	0.92	67.66	31.42	
			mg/l	320	486.4	30,000	300		0	226.8	18,500	34,613	
P3	138,600	84332.3	meq/l	15.97	40.00	1304.40	7.68	1368.05	0.00	3.72	385.17	976.07	1364.96
			e%	1.17	2.92	95.35	0.56		0.00	0.27	28.22	71.51	
			mg/l	750	361.7	39,500	300		0	126	35,500	37,275	
P4	179,100	113,750	meq/l	37.43	29.75	1717.46	7.68	1792.31	0.00	2.07	739.11	1051.16	1792.33
			e%	2.09	1.66	95.82	0.43		0.00	0.12	41.24	58.65	
			mg/l	440	273.6	20,500	150		0	176.4	27,000	13,313	
P5	94,700	61764.3	meq/l	21.96	22.50	891.34	3.84	939.63	0.00	2.89	562.14	375.41	940.44
			e%	2.34	2.39	94.86	0.41		0.00	0.31	59.77	39.92	
			mg/l	280	152	6320	77		6.0001	189	10,000	3328.1	
P6	30,000	20257.6	meq/l	13.97	12.50	274.79	1.97	303.24	0.20	3.10	208.20	93.85	305.35
			e%	4.61	4.12	90.62	0.65		0.07	1.01	68.18	30.74	
			mg/l	220	103.4	2200	74		0	138.6	4580	652.31	
P7	11,300	7898.97	meq/l	10.98	8.50	95.66	1.89	117.03	0.00	2.27	95.36	18.40	116.02
			e%	9.38	7.26	81.74	1.62		0.00	1.96	82.19	15.85	
			mg/l	550	136.8	10,500	150		0	113.4	14,500	6922.5	
P8	50,100	32,816	meq/l	27.45	11.25	456.54	3.84	499.07	0.00	1.86	301.89	195.21	498.96
			e%	5.50	2.25	91.48	0.77		0.00	0.37	60.50	39.12	
			mg/l	320	115.5	4000	63		0	113.4	6550	2218.8	
P9	20,400	13,324	meq/l	15.97	9.50	173.92	1.61	201.00	0.00	1.86	136.37	62.57	200.80
			e%	7.94	4.73	86.53	0.80		0.00	0.93	67.91	31.16	
			mg/l	100	206.7	2500	18		0	970.2	4800	532.5	
P10	13,100	8642.31	meq/l	4.99	17.00	108.70	0.46	131.15	0.00	15.90	99.94	15.02	130.85
			e%	3.80	12.96	82.88	0.35		0.00	12.15	76.37	11.48	
			mg/l	100	115.5	1150	44.5		0	630	2100	399.38	
P11	6370	4224.39	meq/l	4.99	9.50	50.00	1.14	65.63	0.00	10.33	43.72	11.26	65.31
			e%	7.60	14.47	76.19	1.74		0.00	15.81	66.95	17.24	
			mg/l	60	6.08	157.5	7		0	277.2	220.5	44.375	
Irrig. Water	1080	634.055	meq/l	2.99	0.50	6.85	0.18	10.52	0.00	4.54	4.59	1.25	10.39
			e%	28.46	4.75	65.09	1.70		0.00	43.75	44.20	12.05	

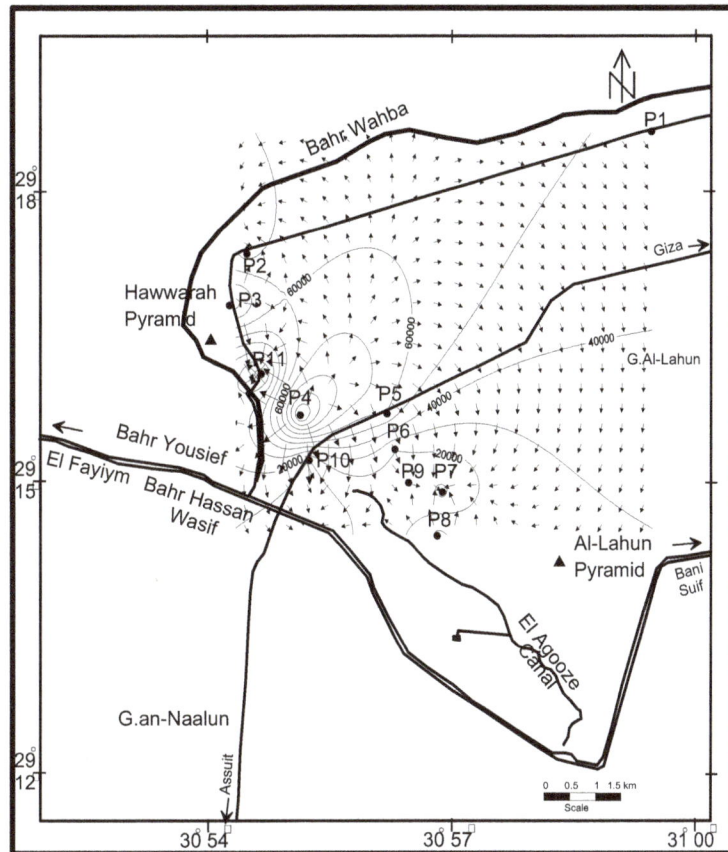

Figure 12. Salinity distribution map of sub-soil water in the study area (2008).

5. Reasons Resulted in the Waterlogging Problem

The major causes of waterlogging in the study area include excessive irrigation water supplies, seepage losses from canals, impeded sub-surface drainage and lack of proper land development. The movement of seepage water downstream has resulted in the accumulation of surface ponds, particularly in the cultivated low areas [23]. Further, litho-units of the shallow aquifer which cover major portion of the northeast El Fayoum command area are also responsible to waterlogging. These units are very deep, imperfectly to poorly drained and underlined by impervious clay or limestone which prevent downward of irrigation water. Overall, waterlogging problems in most of irrigation command areas can also be attributed to the absence drainage canal networks in the study area.

6. Solution of the Waterlogging Problem

In order to protect the antiquities from further accelerated deterioration, lowering of groundwater to safe levels was needed as follows:
- Main drainage canal should be constructed in the periphery of the old cultivated land and around the pyramids to dewatering the excess irrigation water (**Figure 13**).
- Secondary small and deep cut drainage canals network may be constructed in the all the study area and connected to the main drainage canal to dewater the excess irrigation water (**Figure 13**).
- Convert irrigation system in the new reclaimed lands from traditional (flood irrigation) to other advanced irrigation methods as shower and drip.
- Continuous monitoring of water levels to show the early detection of decline in subsoil water after constructing the drainage canals network, and hence taking the necessary corrective actions to prevent any negative impacts on groundwater level.

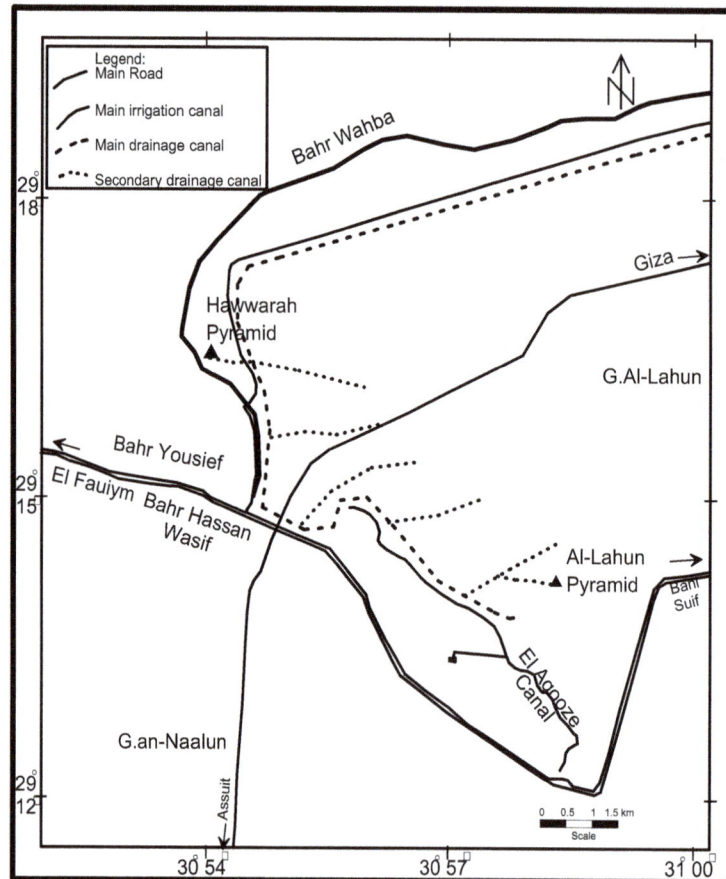

Figure 13. The proposed drainage canals network in the study area.

7. Conclusion

Reliable assessment of waterlogged areas forms a critical element in the irrigation command area development program as well as in the famous Egypt's Pharaonic monument pyramids. In the present study, eleven observation wells were constructed to measure periodic water levels in order to evaluate the impact of canal irrigation. The subsoil water moves from the high level land located in the middle of the concerned area towards the low land (old cultivated land) located in the periphery of the new reclaimed land and the two ancient pyramids (Hawarah and Al-Lahun) causing serious waterlogging and damage of the old cultivated land. The major causes of waterlogging include excessive irrigation water supplies, seepage losses from canals, impeded sub-surface drainage and lack of proper land development. Further, litho-units of the shallow aquifer which cover major portion of the northeast El Fayoum command area are also responsible to waterlogging. These units are very deep, imperfectly to poorly drained and underlined by impervious clay or limestone which prevent downward of irrigation water as well as the absence drainage canal networks in the study area. In order to protect the antiquities from further accelerated deterioration, lowering of groundwater to safe levels was needed by suggesting the construction of drainage canals network in the periphery of the old cultivated land and around the pyramids to dewatering the excess irrigation water in the study area.

References

[1]   El-Sheikh, A.E. (2004) Water Budget Analysis of the Quaternary Deposits for the Assessment of the Water Logging Problem in El Fayoum Depression. Ph.D. Thesis, Faculty of Science, Al-Azhar University, Egypt, 356 p.

[2]   Hoffman, G.J. and Durnford, D.S. (2000) Drainage Design for Salinity Control. In: Skaggs, R.W. and van Schilfgaarde, J., Eds., *Agricultural Drainage. Agronomy No.*: 38. American Society of Agronomy, Madison, 579-614.
        Khouri, N. (1998) Potential of Drydrainage for Controlling Soil Salinity. *Canadian Journal of Civil Engineering*, **25,**

195-205.

[3]   Barrett-Lennard, E. (2002) Restoration of Saline Land through Revegetation. Agric. *Water Manage*, **53**, 13-26. http://dx.doi.org/10.1016/S0378-3774(01)00166-4

[4]   Gowing, J.W. and Wyseure, G.C.L. (1992) Dry-Drainage a Sustainable and Cost-Effective Solution to Waterlogging and Salinization. *Proceedings of 5th International Drainage Workshop*, Vol. 3, ICID-CIID, Lahore Pakistan, 6.26-6.34.

[5]   Moore, G.A. and McFarlane, D.J. (1998) Waterlogging. In: Moore, G., Ed., *Soil Guide—A Handbook for Understanding and Managing Agricultural Soils*, Agriculture Western Australia, South Perth,

[6]   Brahmabhatt, V.S., Dalwadi, G.B., Chhabra, S.B., Ray, S.S. and Dadhwal, V.K. (2000) Land Use/Land Cover Change Mapping in Mahi Canal Command Area, Gujarat, Using Multitemporal Satellite Data. *Journal of the Indian Society of Remote Sensing*, **28**, 221-232. http://dx.doi.org/10.1007/BF02990813

[7]   Dutta, D., Sharma, J.R., Bothale, R.V. and Bothale, V. (2004) Assessment of Waterlogging and Salt Affected Soils in the Command Areas of All Major and Medium Irrigation Project in India. Technical Document, Regional Remote Sensing Service Centre, Jodhpur, 1-74.

[8]   Choubey, V.K. (1997) Detection and Delineation of Waterlogging by Remote Sensing Techniques. *Journal of the Indian Society of Remote Sensing*, **25**, 123-135. http://dx.doi.org/10.1007/BF03025910

[9]   Wildman, W.E. (1982) Detection and Management of Soil Irrigation Management in Mendoza. In: Menenti, M. (Ed.), Argentina. Remote Sensing in Evaluation and Management of irrigation. Institute Nacional de Cienciay Tenicas Hidricas Mendoza, Argentina, 37-58.

[10] Beadnell, H.J.L. (1905) The Topography and Geology of the Fayum Province of Egypt. Survey Department of Egypt, Cairo, 101.

[11] Said, R. (1962) The Geology of Egypt. Elsevier, Amsterdam.

[12] Said, R. (1981) The Geological Evolution of the River Nile. Springer, Berlin.

[13] Tamer, A. M. (1968) Subsurface Geology of the Fayoum Region. M.Sc. Thesis, Alexandria University, Alexandria.

[14] Vondra, C.F. (1974) Upper Eocene Transitional and Near-Shore Marine Qasr El Sagha Formation, Fayum Depression, Egypt. *Annals of the Geological Survey of Egypt*, **4**, 74-94.

[15] Abdel Hafez, N.A. (1991) Geological and Geochemical Studies on Pleistocene Old Lacustrine in El Fayoum Depressions, Western Desert Egypt. *Minia Sc. Bull.*, **4**, 145-162.

[16] Swedan, A.H. (1992) Stratigraphy of the Eocene Sediments in El Fayoum Area. *Annals of the Geological Survey of Egypt*, **18**, 157-166.

[17] Kusky, T.M., Yahia, M.A. and Ramadan, T. (2001) Notes on the Structural and Neotectonic Evolution of El-Faiyum Depression, Egypt: Relationships to Earthquake Hazards. *Egyptian Journal of Remote Sensing and Space Sciences*, **2**, 1-12.

[18] Timothy, M.K., Talaat, M.R., Mahmoud, M.H. and Safwat G. (2011) Structural and Tectonic Evolution of El-Faiyum Depression, North Western Desert, Egypt Based on Analysis of Landsat ETM+, and SRTM Data. *Journal of Earth Science*, **22**, 75-100. http://dx.doi.org/10.1007/s12583-011-0159-8

[19] Kusky, T.M. and Ramadan, T. (2002) Structural Controls on Neo-Proterozoic Mineralization in the SE Desert, Egypt: An Integrated Field, Landsat TM, and SIR-C/X SAR Approach. *Journal of African Earth Sciences*, **35**, 107-121. http://dx.doi.org/10.1016/S0899-5362(02)00029-5

[20] Pinińska, J. and Hemdan, E. (2008) Geomechanical Study of Building Materials of the Hawara Pyramid (Fayoum, Egypt). *Geologija*, **50**, 126-130.

[21] Hermina, M., Klitzsch, E. and List, F. (1987) Stratigraphic Lexicon and Explanatory Notes to the Geologic Map of Egypt 1:500,000. Continental Oil Company (CONOCO), Cairo.

[22] Chebotarev I. (1955) Metamorphism of Natural Waters in the Crust of Weathering. *Geochimica et Cosmochimica Acta*, **8**, 137-170.

[23] El Bastawesy, M., Ali, R., Faid, A. and El Osta, M.M. (2013) Assessment of Waterlogging in Agricultural Megaprojects in the Closed Drainage Basins of the Western Desert of Egypt. *Hydrology and Earth System Sciences*, **17**, 1493-1501.

7

Water Cycle and Irrigation Expansion: An Application of Multi-Criteria Evaluation in the Limestone Coast (Australia)

Zahra Paydar, Yun Chen

CSIRO Land and Water, Canberra, Australia
Email: mahbooeman@yahoo.com

Abstract

A summary of the current understanding of the hydrological system in the Limestone Coast in Australia is presented. The regional water balance analysis indicated about 90% of the water is lost through evapotranspiration. Irrigation consumes a substantial part of the groundwater and returns up to half of the water inflow to the aquifer. A multi-criteria analysis using fuzzy quantifiers and analytical hierarchy process was applied for future irrigation expansion. The results showed a total of 94,632 ha of land, considered suitable for irrigation where groundwater is not over-allocated or over used. This model showed some advantages over the conventional multi-criteria evaluation methods as it avoids arbitrary decisions on criteria weightings.

Keywords

Land Suitability, Regional Water Balance, Irrigation Expansion, Multi-Criteria Evaluation

1. Introduction

The Limestone Coast has some of the most productive land in South Australia (**Figure 1**) and supports a diverse industry base. The economy, environment and community are linked to its water resources with groundwater being the main source. The irrigation industry is the most significant user of groundwater with about 80,000 ha under irrigation; though forestry is considered as a large water user with more than 140,000 ha of forest and substantial recharge reduction. The regional water balance has been altered over the years due to a variety of factors such as climate, clearing of natural vegetation, planting of low water use crops, irrigation, drainage construction and expansion of plantation forestry.

One important aspect of the water management in the region is the unsustainable groundwater use for irriga-

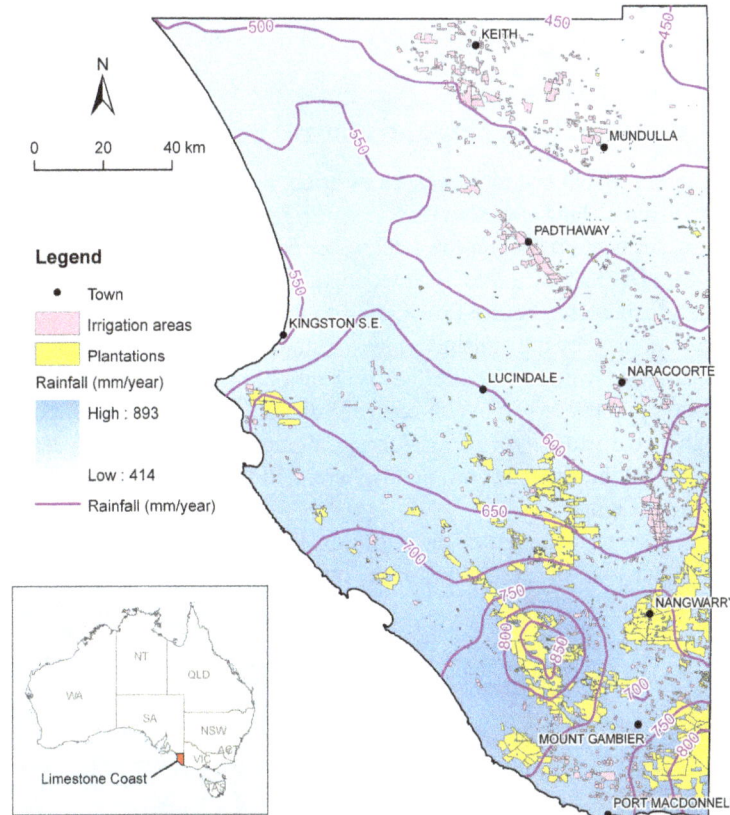

Figure 1. Limestone Coast location map and long term average (1961-1990) rainfall contours with current location of irrigation and plantation areas.

tion which will also affect the total availability of groundwater. The watertable over some parts of the region has declined over the last 30 - 40 years because of drier climate and groundwater extraction for irrigation. Uncontrolled developments of land use systems that reduce recharge and affect the availability of groundwater in the region is another issue in water management (e.g. expansion of plantation forestry). The challenge for the region is to develop an integrated water management regime at the system level that allows a balanced use of water resources taking into account the needs of all groundwater users within sustainable limits [1]. While there have been numerous studies on individual aspects of water balance; e.g. recharge [2]-[6], plantation water use [7]-[9], irrigation water use [10]-[14], there is a lack of an integrated approach to water balance to compare the relative impact of water use components in the region.

Future intensification of agriculture is likely to rely on the availability of water of suitable quality, in localities of suitable climate and soils. A spatial analysis considering bio-physical factors for irrigation expansion (land suitability and water availability) in the region would help with the future planning of such expansions. Spatial multi-criteria decision making (MCDM) is a useful approach to serving this purpose. In general, a GIS-based MCDM involves a set of geographically defined basic units and a set of evaluation criteria represented as map layers. The problem is to combine the criterion maps according to the attribute values and decision maker's preferences using a set of decision rules so as to rank each unit with an overall score. A number of spatial-based multi-criteria evaluation methods have been applied to land suitability assessment [15]-[24]. The application of advanced MCDM has shown to improve the credibility, transparency and analytic rigour of landuse and water management decisions [19]. Previous studies have assessed limitations of land for primary production in the study area [25] [26]). However, the results are only useful to identify the factor which is the primary limitation in each mapping unit because the overall classification for a land was derived essentially based on the most limiting attribute and not considering the weight of individual attributes or evaluating the resultant uncertainties.

In this paper, we present current understanding of the water cycle in a regional water balance analysis as well as a multi-criteria analysis for future expansion of irrigation in the region as steps towards the development of an integrated water management regime.

2. Materials and Methods

2.1. A Conceptual Model of Hydrology and Control Volume

Many processes at different spatial and temporal scales affect the overall water balance of a hydrological system (**Figure 2**) including different land use activities (e.g. irrigation, forestry, dryland agriculture, water bodies, towns) with different water needs affecting the total water use as well as groundwater recharge occurring from the rainfall and irrigation. Drains play a role in collecting all the surface water as well as being fed by the deeper drains tapping into the groundwater, before flowing out to the sea. At the same time, exchanges between different aquifers take place, but they are not considered here for a regional water balance analysis.

In a water balance analysis, we must define the control volume and a time frame for which the calculations are carried out and assumptions are valid. The control volume, for this analysis, consists of the area from the surface to the watertable (the vadose zone) considering all flows occurring between this zone and outside the zone (*i.e.* fluxes to and from the groundwater, atmosphere, and the sea). The boundaries used for the water balance analysis are shown in **Figure 1**.

A simplified conceptual representation of the hydrological components of the water supply for the region is shown in **Figure 3**, following our understanding of the hydrological system in the region.

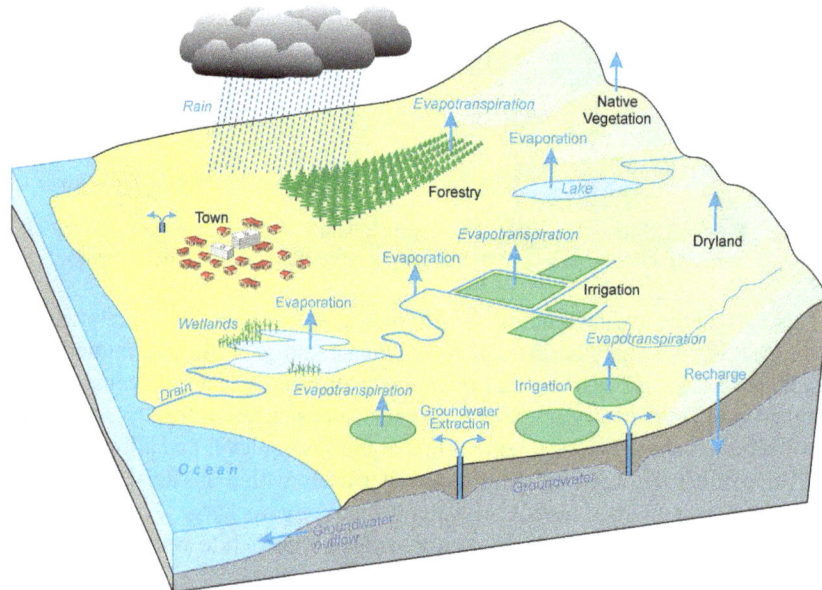

Figure 2. Schematic representation of the water cycle processes in the Limestone Coast.

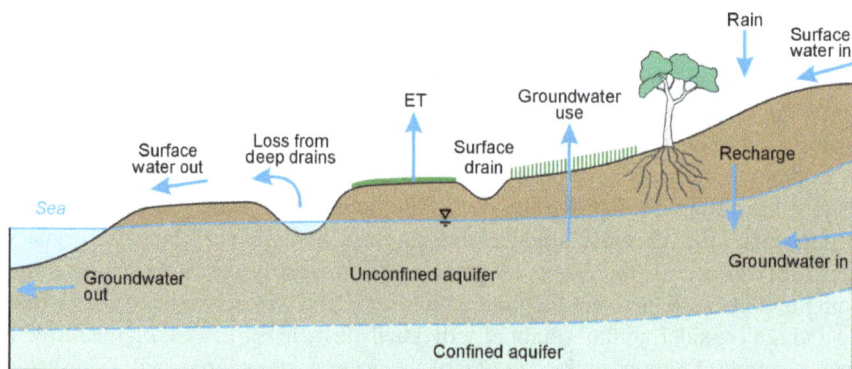

Figure 3. A simplified conceptual model of the hydrology Limestone Coast.

2.1.1. Regional Water Balance Analysis

A regional water balance can be written as:

$$\text{Inflows} - \text{Outflows} = \Delta S \qquad (1)$$

where Inflows refer to all incoming water to the region and Outflows refer to all the outgoing water flowing out of the system. ΔS refers to the change in the region water storage. When averaged over a long period, ΔS can be neglected (assumed zero).

At the regional scale, the main inflow is rain (P) which recharges (Re) the underground aquifer(s) as a proportion not being used for evapotranspiration (ET) of crops and natural vegetation or intercepted by trees and plants. Inflows also include surface inflows to the region (Q_{in}) and groundwater pumped to the surface for use (GW_{use}). The Outflows consist of mainly the ET from different land uses as well as the flow of drains consisting of surface drains (Q_{o-s}) and deeper drains collecting groundwater or fed by springs before flowing out to the sea (Q_{o-gw}).

Equation (1) thus can be written as:

$$P + Q_{in} + GW_{use} - Re - ET - Q_{o-s} - Q_{o-gw} = 0 \qquad (2)$$

The intention was to estimate an average annual water balance for the region within a period covering the years 1995 to 2004. Where data did not exist for some components, estimates based on assumptions about these average conditions were made (e.g. average land use). The steps followed for estimation of these components are described in the following sections.

2.1.2. Climate Data

Mean annual climate surfaces for 1961-1990, the reference period used by the Bureau of Meteorology, from QDNR (Queensland Department of Natural Resources) SILO daily data [27] were used for the analysis. The dataset provides interpolated data for a 5 km by 5 km grid across Australia. Areal potential evapotranspiration (ET_0) data calculated from the QDNR gridded data set (daily temperatures and solar radiation) were used in the analysis. Using these climate surfaces, the long-term average rainfall and potential evapotranspiration for the region were calculated and scaled to the water accounting period using climate data for a few stations in the region [28]. For example, average annual rainfall and ET_0 after 1995 were 95% and 98% of the long term averages. These figures were used to scale the long term averages to the period of our analysis.

2.1.3. Surface Water and Drains

The catchment contains a small number of well-defined flow paths that exist primarily as ephemeral creeks. All the data were obtained from the Department of Water Land and Biodiversity Conservation (DWLBC) surface flow archive (http://e-nrims.dwlbc.sa.gov.au/swa/map.aspx).

There are two types of drains flowing out to the sea: surface drains and ground water-fed system of drains which are deeper drains collecting groundwater or are fed by springs before they flow out of the region and to the sea.

2.1.4. Recharge to Groundwater (Re)

Here we refer to recharge as the water moving vertically through the soil, and adding to the groundwater storage. We used the mean annual values of recharge calculated by [6] and updated by [29] mostly with the watertable fluctuation method using the existing water level observation network. These values, ranging from 15 to 200 mm/year, are shown in **Figure 4** for all groundwater management zones in the region.

2.1.5. Estimating Evapotranspiration (ET)

For estimating average annual ET in the catchment, land use data for the study area were categorised into 5 classes: irrigated areas, dryland agriculture, plantation forestry, natural vegetation and wetlands/water bodies. In the absence of actual ET measurements different methods of ET estimation were used for each class.

2.1.6. Irrigation Areas

For areas under irrigation, ET estimation requires specific crop water use (ET_c) calculations. These were calculated mainly following FAO 56 methodology [30] which is based on the relationship between potential reference crop evapotranspiration (ET_0) and a crop factor (Kc) as in Equation (3) (See [28]).

Recharge (mm/year)

● Town

☐ < 80
☐ 80 - 100
☐ 100 - 110
☐ 110 - 120
☐ 120 - 130
☐ 130 - 140
☐ 140 - 165
■ >165

KEITH

MUNDULLA

PADTHAWAY

KINGSTON S.E.

LUCINDALE   NARACOORTE

NANGWARRY

N

0   20   40 km

MOUNT GAMBIER

PORT MACDONNELL

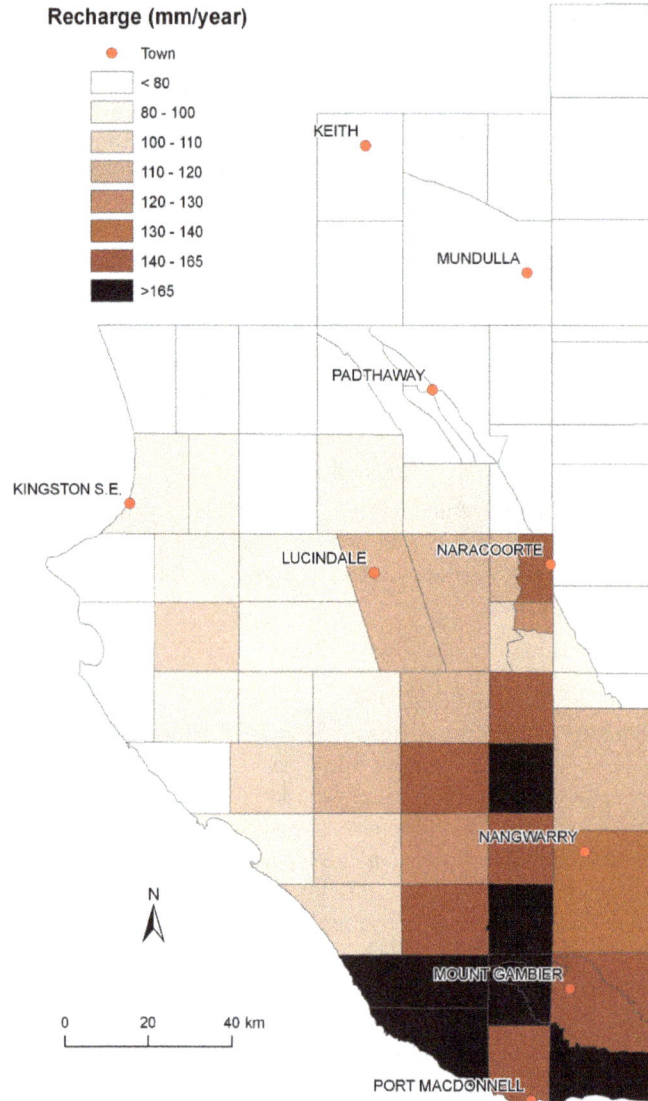

Figure 4. Recharge estimates of the region (Based on [6] and [29]).

$$ET_c = ET_0 * Kc \tag{3}$$

Monthly ET_0 values in mm were obtained from the climate data together with the information reported on land use and crop mixes under irrigation [10]-[13] to calculate ET_c according to Equation (3). Monthly Kc values for different crops (mainly pasture, vines, lucerne, cereals, fruit, vegetables and potatoes) were taken from published values and local studies [14] [25] [31]. ET_c values were converted to volumes (ML) of water considering the areas of crops in each year and averaged over the years of analysis.

2.1.7. Dryland Agriculture

For these areas, an estimate of ET was made based on average empirical values suggested by [32] based on the work of Fu [33].

Actual evapotranspiration is calculated in the model using the following equation from [33]:

$$\frac{ET}{P} = 1 + \frac{ET_0}{P} - \left(1 + \left(\frac{ET_0}{P}\right)^w\right)^{1/w} \tag{4}$$

where:

ET = actual evapotranspiration

ET_0 = potential evapotranspiration

P = rainfall

w = a model parameter

A single-parameter hyperbolic function (Equation (4)) interpolates between dry (rainfall limited) and wet (energy limited) total evaporation rates. The value of this parameter (w) describes the influence of catchment land characteristics and vegetation on actual evapotranspiration. In a water balance study of over 270 Australian catchments [32], it was found an average value of 2.55 for w parameter to give good ET prediction (by Fu's model) against the observed data. We used this average value in Equation (4) for ET estimation in dryland areas.

2.1.8. Plantation Forestry

An average area for forestry during the water accounting period was estimated as 110,000 ha. Water use by tree plantations in the region has been studied mostly on shallow groundwater areas and in closed canopy [7] [9] with a reported average water use of 944 mm/year. To consider an overall average water use (ET) by forestry, we need to take into account the relative areas on shallow water table (<6 m depth), the mix of different trees (pine plantation vs. blue gum) and the whole crop cycle water use (*i.e.* there are periods before canopy closure and after harvest). Based on these factors and previous experience, an average annual ET for the plantation forestry was estimated to be around 740 mm/year [R. Benyon, Pres. Comm.].

2.1.9. Natural Vegetation

There are large areas of land under nature conservation and natural vegetation. They are highly variable in vegetation composition, structure and spatial arrangement which are likely to affect their water use. For this land use we considered expert opinion on long term average water use of this natural vegetation to be almost the same as the average rainfall for the region [28].

2.1.10. Wetlands and Water Bodies

For water bodies including lakes and reservoirs, Kc was assumed 0.7 and for the wetlands a value of 1.1 was considered for large wetlands [34].

2.2. Spatial Multi-Criteria Analysis

A spatial multi-criteria assessment used here implements fuzzy quantifiers and analytical hierarchy process (AHP) in ArcGIS environment [35]. Topography, groundwater and landscape attributes and derivatives which affect irrigated landuse were selected as criteria and presented as spatial layers. The AHP was then employed to weight each criterion. The weighted averaging operator for selected values of fuzzy quantifiers was applied to derive several evaluation scenarios showing alternative potential suitability maps.

2.2.1. Suitability Classification

The suitability classes were adapted from the Food and Agricultural Organization system [36]. They consist of four levels describing land with high potential (S1) to unsuitable (N) as in **Table 1**.

2.2.2. Evaluation Criteria

Groundwater, topography, and soil landscape attributes, considered important in irrigation, were selected mostly based on the available spatial data which were collected and compiled in previous studies [37] [38]. The Fifteen attributes including groundwater salinity, depth to water table, salinity, susceptibility to waterlogging, susceptibility to acidity, inherent fertility, boron toxicity, sodicity, root zone depth, water holding capacity, surface condition, steepness, subsoil structure, surface rockiness and surface texture were considered [24]. They are further grouped into five categories/objectives: 1) waterlogging and salinity; 2) soil fertility; 3) chemical barriers to root growth: 4) soil depth and water storage; and 5) topography and soil physical conditions (**Figure 5**). The threshold values for each of the four suitability classes of most criteria were determined based on the synthesis of main irrigated crops in the region [38]. Criteria indicating the impact of groundwater salinity were based on the general distribution of groundwater salinity for the unconfined aquifer through the area and taking into account the soil salinity associated with groundwater [25].

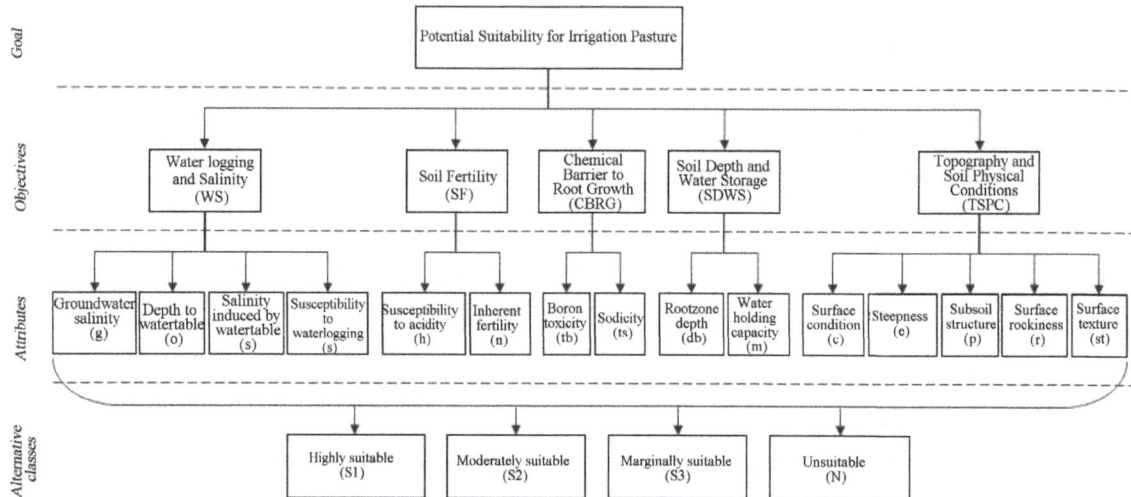

Figure 5. The hierarchical structure used in this study.

Table 1. Land suitability classification.

Class	Definition
S1	Highly suitable: land having no significant limitations for sustained applications to irrigated cropping, or only minor limitations that will not significantly reduce the productivity.
S2	Moderately suitable: land having limitations that are moderately severe for sustained application to irrigated cropping, and may reduce the productivity marginally.
S3	Marginally suitable: land with limitations that are severe for sustained application to irrigation cropping, and as such reduce productivity significantly but is still marginally economical.
N	Unsuitable: land with extreme limitations which appear to preclude sustained application to irrigation cropping.

2.2.3. Criterion Maps

This procedure links the classified evaluation criteria to the spatial data. The original Soil Landscapes dataset is in vector format (polygon shapefile) [37]-[39]. Each polygon of the dataset is classified into four suitable classes. Where there are several land uses in a polygon, each with a different suitability classification, a single overall suitable class was identified based on the following two rules:

1) Assign a class level which represents the largest proportion of areas in the polygon;

2) Take the lower or lowest class level where multiple classes are contributing equally in a polygon (covering same proportions of the area).

All layers have four classes representing different suitability levels (**Table 1**) based on the threshold values assigned to them in **Table 2**.

In the study area, a key issue in the management of landuse and irrigation practices is to reduce accessions to the water table thus minimising the risk of waterlogging and salinisation due to rising water table levels. Therefore, two critical criteria from the Waterlogging and Salinity objective, groundwater salinity and soil salinity, were identified as key factors which constrain the expansion of irrigated pasture. These two spatial data layers were used as masks to exclude areas where groundwater or soil is classified as unsuitable before determination of criteria weights.

2.2.4. Criterion Weights

Criterion weights at both objective and attribute levels were derived using the pair-wise comparison which is the basic measurement mode of the AHP procedure. The procedure greatly reduces the conceptual complexity of a problem since only two components are considered at a time. This approach required the expert's judgment for relative importance of one evaluation factor (objective and criterion) against another. The method employs an underlying semantic scale with values from 1 to 9 to rate the relative importance for two elements. The available values for the comparison are the member of the set: {9, 8, 7, 6, 5, 4, 3, 2, 1, 1/2, 1/3, 1/4, 1/5, 1/6, 1/7, 1/8, 1/9},

Table 2. Pair-wise comparison matrix of objectives and calculated weights.

Objective	Waterlogging and Salinity	Soil Fertility	Chemical Barriers to Growth	Soil Depth & Water Storage	Topography & Soil Conditions	Weight
Waterlogging and Salinity	1	2	3	1/2	2	0.238
Soil Fertility	1/2	1	2	1/2	1/2	0.140
Chemical Barriers to Root Growth	1/3	1/2	1	1/4	1/3	0.077
Soil Depth & Water Storage	2	2	4	1	2	0.365
Topography & Soil Physical Conditions	1/2	2	3	1/2	1	0.180

with 9 representing absolute importance and 1/9 the absolute triviality [40] [41]. The comparison matrix at objective level is presented in **Table 2** which assigns a numerical value showing relative importance of each objective. In this study, "Soil Fertility" objective has been regarded slightly more important than "Chemical Barriers to Growth"; hence a value of 2 has been assigned to the corresponding matrix position (**Table 2**). The transpose position automatically gets the reciprocal value, in this case 1/2. At the attribute level, a comparison matrix taking the same format of **Table 2** was constructed for each of the objectives by comparing associated attributes. Based on these matrices, the relative weights for objectives and criteria were derived by a GIS-based model called FLOWA [20] [21] [42].

2.2.5. Order Weights

After the weights related with criterion maps were input into the FLOWA, the module specified a linguistic quantifier to generate a set of weights, and therefore, computing the overall evaluation by means of the weighted averaging (OWA) combination function. Several resultant maps representing different scenarios were obtained. The suitability map, which was derived from the moderately optimistic scenario, was then overlaid by the current land use map. It was further compared with water availability data by excluding areas identified as having a negative water balance (*i.e.* groundwater over use or allocation) [29] to show the extent of potential areas for irrigation expansion.

3. Results and Discussion

3.1. Regional Water Balance

Calculations of recharge resulted in a mean annual volume of 1110 GL, while average rainfall was calculated as 11,942 GL. For the surface inflows, a total of 17.52 GL was calculated while the total outflow from the region consists of surface drains with total flow of 106,400 ML and groundwater-fed drains with total of 97,315 ML.

Annual *ET* estimates are summarised in **Table 3**. Dryland is by far the largest consumer of water (*ET*) of all land use classes (7687 GL). Total *ET* from irrigated land is estimated at 404 GL which compares reasonably well with the average reported irrigation water use in the region [43] of 448 GL which includes irrigation losses as well as *ET*. Considering all components of annual regional water balance comprising all water entering the surface (inflows) and leaving the region (outflows), **Table 4** gives a summary of the water balance. The difference (balance) between inflows and outflows is relatively small (42 GL). This is equivalent to 0.33% water balance error. This, without considering uncertainties and some possible compensating errors, is an indication that the system overall is in balance and our estimates roughly describe the physical characteristics of the hydrology of the region.

The results show that a large proportion (90%) of all the water entering the region leaves the surface as evapotranspiration. Of the remaining water, 8.8% goes to the groundwater as recharge, while less than 2% leaves the region as surface water in drains or creeks.

Of the total annual evapotranspiration of 11,193 GL from the region, irrigation accounts for only 3.6% and plantation forestry for 7% total *ET*. Not all of this water use comes from the groundwater and in our analysis we have not separated the sources of water use (*i.e.* rainfall and groundwater).

Table 3. Average annual *ET* estimates from each land use type.

Land use	Area (ha)	*ET* (GL)
Irrigated land	65,877	404
Plantation forestry	110,000	814
Natural vegetation	268,438	1532
Wetland/marshes	36,306	513
Dryland agriculture	1,536,058	7687
Water bodies	29,720	243
Other areas-residential	45,857	
Total-study area	2,092,257	11,193

Table 4. Annual regional water balance components.

Inflows (GL/year)				Outflows (GL/year)				
Rain	Surface	GW extraction Unconfined confined		*ET*	Surface	GW-fed drains	Recharge	Other uses
11,942	23	573	30	11,193	106	97	1110	20
		Total inflows = 12,568				Total outflows = 12,526		

The average annual volume of water applied for irrigation is 400 GL/yr or 605 mm/yr which, when added to the average rainfall, yields a total of 1175 mm/yr. Based on the first order approximations, if only 630 mm/yr is evapotranspired from these lands, then the rest (545 mm/yr) is recharge. Thus, on average (across all soil, crop and irrigation system types) about half (46%) of the total water applied is returned back to the aquifer. This simplified analysis does however neglect the timing of rainfall, irrigation application, crop water use and fallow periods. In reality, a higher proportion of irrigation water returning to the aquifer will be expected under flood systems and lighter soils and much lower rates under more-efficient systems.

It should be noted that this regional (vadose zone) water balance gives an overall picture of the hydrology of the system without giving a balance of the groundwater. It also does not consider the spatial distribution of these components as they are averaged in space and time with rough estimates and assumptions. For a comprehensive and integrated approach for studying impact of different scenarios (e.g. climate, land use change) on water resource management, a fully dynamic model of the surface and groundwater of the region is needed which is beyond the scope of this study. Nevertheless, the water balance results when considered with the results of spatial analysis could help in better understanding of the system and making decisions on expansion of irrigation in the region.

3.2. Spatial Analysis

The resultant suitability map (**Figure 6**) shows the extent and distribution of the land suitability classes. There is about 13% (217,542 ha) of total area being classified as highly suitable (S1), while moderately and marginally suitable classes represent 42% (728,452 ha) and 5% (87,863 ha) of land respectively. The suitability map when overlaid with present landuse map (**Figure 6**) revealed that, in the study area, 25% (16,325 ha) of present irrigated land falls in the highly suitable class, while approximately 46% (30,265 ha) occurs in moderately suitable areas. There is only over 4% (2932 ha) under marginally suitable areas and more than 25% (16,755 ha) exists under unsuitable regions which consists of irrigated modified pasture, irrigated cropping (e.g. hay and silage) and irrigated perennial horticulture (e.g. vine fruits). They are mainly located in the north of the region where groundwater salinity is classified as greater than 2000 mg/L. Highly suitable land (S1) should be retained for irrigated agriculture since limitations to irrigated cropping in S1 land can be overcome by standard management practices. Policies should be considered which protect this land from unnecessary subdivision for urban or rural residential use or other uses such as roads. Limitations to irrigated pasture on moderately suitable land (S2) need

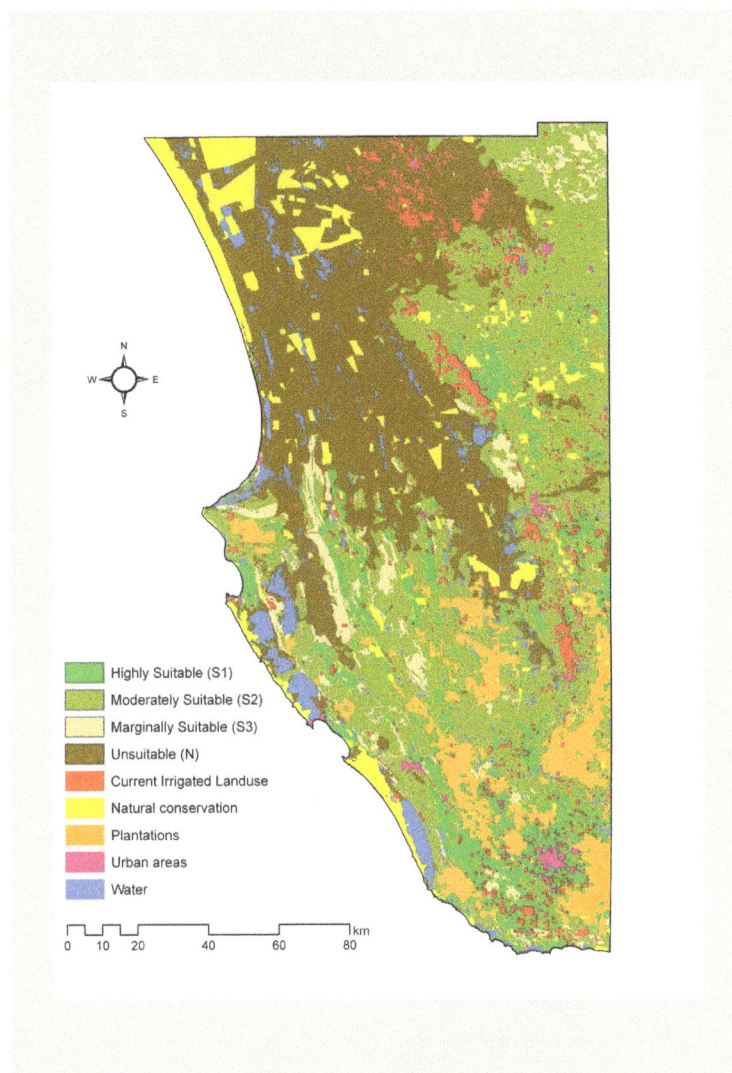

Figure 6. An evaluation map of irrigation suitability derived from WLC approach and overlaid with present landuse map.

to be recognised because a decline in productivity may occur and a range of landuse problems may develop if this land is used and managed inappropriately.

Currently, less than 8% of the total highly suitable lands have been used for irrigation practice. A large proportion of highly suitable land, located in the south part of the region in particular, has a great potential in cultivating more irrigated agriculture. However, not all suitable areas can be considered for irrigation as groundwater resource is limited and its management and allocation planning has been the subject of many reports (e.g. [6] and [29]). When water availability was considered as a limiting factor (**Figure 7**) where the groundwater allocation or use was larger than the recharge, the total highly suitable areas for irrigation expansion is estimated around 94,632 ha (S1 class) mostly around centre, south and western side of the region. In planning any expansion of irrigation areas, other important factors, such as economics and environmental aspects of such plans must also be considered. Future studies should also link a water balance analysis of land use change (*i.e.* crop-soil-water balance model) to a spatial multi-criteria analysis to enable estimation of future water use and recharge under proposed scenarios of irrigation expansion. A fully dynamic groundwater model could better describe the groundwater balance of each management zone for future scenarios. Such an analysis, when coupled with economic analysis, could give a more comprehensive view of the water cycle and its spatial distribution in the region for current and future water management planning.

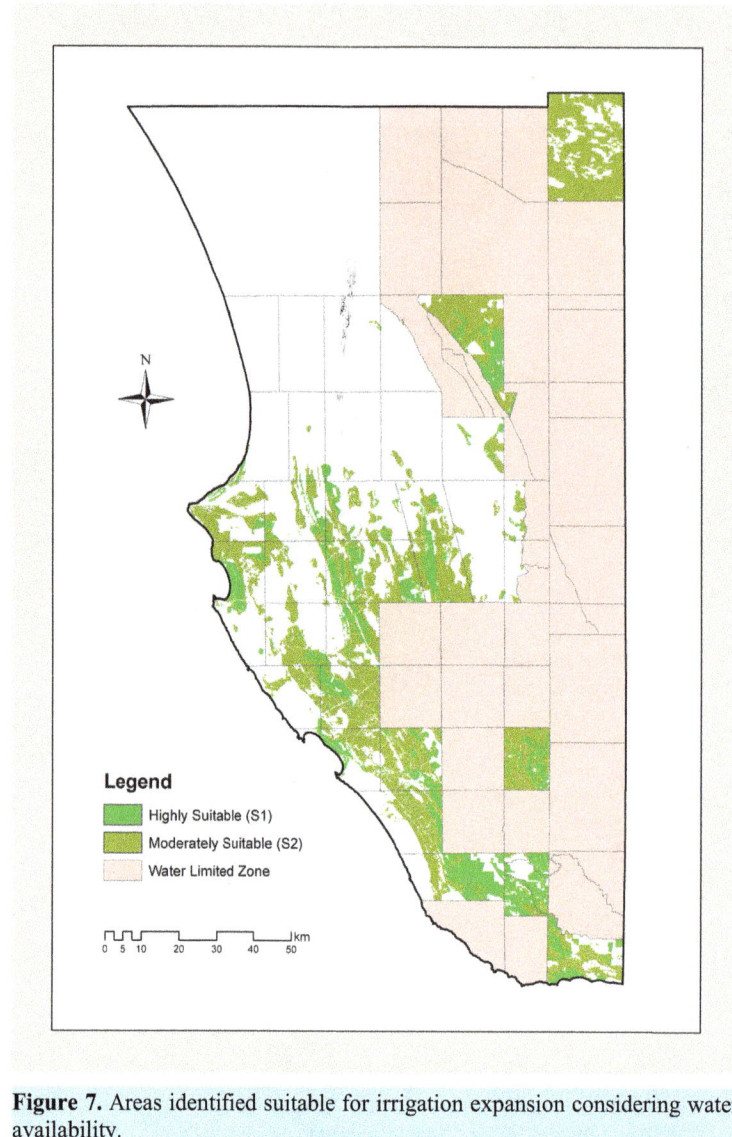

Figure 7. Areas identified suitable for irrigation expansion considering water availability.

4. Conclusions

At the regional level, our analysis showed that a substantial (~90%) part of the water input to the region is consumed through either plant transpiration or evaporation from water bodies, wetlands and soils. Irrigation only contributes to about 4% of this total *ET*, though it consumes a substantial part of the groundwater extraction in the region. Plantation forestry is considered as a "water use" activity and its share of the total regional *ET* is around 7%. Around 9% of the total rainfall is recharged to the aquifer and is the major contribution to this huge groundwater resource while less than 2% leaves the region as surface water in drains or creeks.

The regional water balance gives an indication of the relative contribution of different components of the water cycle at the system level and as such it does not show the spatial and temporal variability of these terms.

A spatial fuzzy multi-criteria evaluation approach has been applied for spatial suitability assessment for irrigation. The resultant evaluation has revealed that about 70% of existing irrigated cropping is located in highly suitable and moderately suitable lands in the eastern half of the region. But vast areas of highly suitable land exist where irrigation can be expanded. Expansion in irrigated agriculture is likely to present new demand for water resources and as such, the availability of groundwater in potential suitable areas should be considered when planning for expansion. When areas identified as water limited were excluded from the potential suitable areas, the total highly suitable areas for irrigation showed a maximum of 94,632 ha distributed mainly in the center, south and western part of the region. The resultant potential irrigation maps are intended to provide a regional view of areas potentially suitable for irrigation. Future planning should consider water balance changes resulting from land use change (through modeling) as well as economic and environmental assessment of future land use scenarios.

The use of a multi-criteria evaluation model in this study showed some advantages over the conventional methods as it avoids arbitrary decisions on assigning related weights to the criteria. It allows exploring and visualizing a wide range of different multi-criteria decision strategies. When (in future) linked to water balance models, it would facilitate a better understanding of the patterns of alternative land use and could give a comprehensive picture of irrigation land use impact on future water resources management and development planning.

Acknowledgements

This work has been carried out as part of a suite of activities funded by the Cooperative Research Centre for Irrigation Futures (CRC IF) in the Systems Harmonisation Program. The funding of the CRC IF is gratefully acknowledged. We also wish to acknowledge those who have provided inputs to this study. In particular we acknowledge: DWLBC staff at Mt. Gambier for providing spatial data, information and reports.

Drs. Richard Benyon and Judy Eastham (former CSIRO) for providing insight to the estimation of water use by plantation forestry and natural vegetation and Mr Peter Briggs (CSIRO) for providing long-term climate surfaces (rainfall and potential ET) and Mr Heinz Buettikofer (CSIRO) for the graphics and GIS support.

References

[1] South East Natural Resource Consultative Committee (SENRCC) (2003) South East Natural Resource Management Plan. A Plan for the Integrated Management of Natural Resources in the South East, Consultation Report.

[2] Hopton, H., Schmidt, L., Stadter, F. and Dunkley, C. (2001) Forestry, Land Use Change and Water Management: A Green Triangle Perspective. In: Nambiar, E.S.K. and Brown, A.G., Eds., *Plantations, Farm Forestry and Water*. Water and Salinity Issues in Agroforestry No. 7. Agriculture, Fisheries and Forestry—Australia; Joint Venture Agroforestry Program; CSIRO Forestry and Forest Products. Rural Industries Research and Development Corporation, Canberra.

[3] Colville, J.S. and Holmes, J.W. (1972) Watertable Fluctuations under Forest and Pasture in a Karstic Region of Southern Australia. *Journal of Hydrology*, **17**, 61-80. http://dx.doi.org/10.1016/0022-1694(72)90066-2

[4] Allison, G.B. and Hughes, M.W. (1978) The Use of Environmental Chloride and Tritium to Estimate Total Recharge to an Unconfined Aquifer. *Australian Journal of Soil Research*, **16**, 181-195. http://dx.doi.org/10.1071/SR9780181

[5] Herczeg, A.L. and Leaney, F.W.J. (1993) Estimates of Regional Recharge to a Karst Aquifer: Naracoorte Ranges, SA. Centre for Groundwater Studies Report No. 53, Adelaide.

[6] Brown, K.G., Harrington, G.A. and Lawson, J. (2006) Review of Groundwater Resource Condition and Management Principles for the Tertiary Limestone Aquifer in the South East of South Australia. Department of Water, Land and Biodiversity Conservation, Report DWLBC 2006/02.

[7] Benyon, R.G. and Doody, T.M. (2004) Water Use by Tree Plantations in South East South Australia. CSIRO Forestry and Forest Products, Mount Gambier SA, Technical Report No. 148.

[8] Smith, M.K. (1974) Throughfall, Stemflow and Interception in Pine and Eucalypt Forests. *Australian Forestry*, **36**, 190-197. http://dx.doi.org/10.1080/00049158.1972.10675584

[9] Benyon, R.G. (2002) Water Use by Tree Plantations in the Green Triangle: A Review of Current Knowledge. Glenelg Hopkins Catchment Management Authority, Hamilton and CSIRO Forestry and Forest Products, Mt Gambier.

[10] Binks, B. (2000) Profile of the South East Irrigation Industry. South East Catchment Water Management Board, South Australia.

[11] Kelly, R. and McIntyre, N. (2005) Water Allocation and Use in the South East 2003-2004. Annual Water Use Report, Department of Water, Land and Biodiversity Conservation, South Australia.

[12] Kelly, R. and Laslett, D. (2002) Water Allocation and Use in the South East 2001-2002. Report to the South East Catchment Water Management Board, South Australia.

[13] Kelly, R. and Laslett, D. (2003) Water Allocation and Use in the South East 2002-2003. Report to the South East Catchment Water Management Board, South Australia.

[14] Skewes, M. (2006) Definition of Net Irrigation Requirements in the South East of South Australia. Report to the South Australian Department of Water Land and Biodiversity Conservation. Irrigated Crop Management Services, Rural So-

lutions SA.

[15] Malczewski, J. (2006) Ordered Weighted Averaging with Fuzzy Quantifiers: GIS-Based Multicriteria Evaluation for Land-Use Suitability Analysis. *International Journal of Applied Earth Observation and Geoinformation*, **8**, 270-277. http://dx.doi.org/10.1016/j.jag.2006.01.003

[16] Pereira, J.M.C. and Duckstein, L. (1993) A Multiple Criteria Decision-Making Approach to GIS-Based Land Suitability Evaluation. *International Journal of Geographical Information Systems*, **7**, 407-424. http://dx.doi.org/10.1080/02693799308901971

[17] Bojorquez-Tapia, L.A., Diaz-Mondragon, S. and Ezcurra, E. (2001) GIS-Based Approach for Participatory Decision-Making and Land Suitability Assessment. *International Journal of Geographical Information Science*, **15**, 129-151. http://dx.doi.org/10.1080/13658810010005534

[18] Joerin, F., Theriault, M. and Musy, A. (2001) Using GIS and Outranking Multicriteria Analysis for Land-Use Suitability Assessment. *International Journal of Geographical Information Science*, **15**, 153-174. http://dx.doi.org/10.1080/13658810051030487

[19] Malczewski, J. (2004) GIS-Based Land-Use Suitability Analysis: A Critical Overview. *Progress in Planning*, **62**, 3-65. http://dx.doi.org/10.1016/j.progress.2003.09.002

[20] Chen, Y., Khan, S. and Paydar, Z. (2009) To Retire or Expand? A Fuzzy GIS-Based Spatial Multi-Criteria Evaluation Framework for Irrigated Agriculture. *Irrigation and Drainage*, **59**, 174-188.

[21] Chen, Y., Yu, J. and Khan, S. (2010) Spatial Sensitivity Analysis of Multi-Criteria Weights in GIS-Based Land Suitability Evaluation. *Environmental Modelling & Software*, **25**, 1582-1591. http://dx.doi.org/10.1016/j.envsoft.2010.06.001

[22] Chen, Y., Yu, J. and Khan, S. (2013) The Spatial Framework for Weight Sensitivity Analysis in AHP-Based Multi-Criteria Decision Making. *Environmental Modelling & Software*, **48**, 129-140. http://dx.doi.org/10.1016/j.envsoft.2013.06.010

[23] Yu, J., Chen, Y., Wu, J. and Khan, S. (2011) Cellular Automata Based Spatial Multi-Criteria Land Suitability Simulation for Irrigated Agriculture. *International Journal of Geographical Information Science*, **25**, 131-148. http://dx.doi.org/10.1080/13658811003785571

[24] Chen, Y. and Paydar, Z. (2012) Evaluation of Potential Irrigation Expansion Using a Spatial Fuzzy Multi-Criteria Decision Framework. *Environmental Modelling & Software*, **38**, 147-157. http://dx.doi.org/10.1016/j.envsoft.2012.05.010

[25] PIRSA Rural Solutions (2000) Annual Irrigation Requirement for Horticultural Crops. Technical Report 263, Irrigated Crop Management Services.

[26] SA Government (2000) Sustainable Development in South East South Australia. Project Report and Prototype Information System. Collaborative Project of South Australian State and Local Government.

[27] Jeffrey, S.J., Carter, J.O., Moodie, K.B. and Beswick, A.R. (2001) Using Spatial Interpolation to Construct a Comprehensive Archive of Australian Climate Data. *Environmental Modelling & Software*, **16**, 309-330. http://dx.doi.org/10.1016/S1364-8152(01)00008-1

[28] Paydar, Z., Chen, Y., Xevi, E. and Buettikofer, H. (2009) Current Understanding of the Water Cycle in the Limestone Coast Region. Water for a Healthy Country National Research Flagship, CSIRO. http://www.irrigationfutures.org.au/newsDownload.asp?ID=1027&doc=CRCIF-TR0309-web.pdf

[29] Latcham, B., Carruthers, R., Harrington, G.A. and Harvey, D. (2007) A New Understanding on the Level of Development of the Unconfined Tertiary Limestone Aquifer in the South East of South Australia. Department of Water, Land and Biodiversity Conservation, Report DWLBC 2007/11.

[30] Allen, R.G., Pereira, L.S., Raes, D. and Smith, M. (1998) Crop Evapotranspiration; Guidelines for Computing Crop Water Requirements. FAO Irrigation and Drainage Paper 56, Food and Agriculture Organization of the United Nations, Rome.

[31] Christiansen, J.E. and Hargreaves, G.H. (1969) Irrigation Requirements from Evaporation. *Transaction of the International Commission on Irrigation and Drainage*, **23**, 569-596.

[32] Zhang, L., Hickel, K., Dawes, W.R., Chiew, F.H.S., Western, A.W. and Briggs, P.R. (2004) A Rational Function Approach for Estimating Mean Annual Evaporation. *Water Resources Research*, **40**, Article ID: W02502. http://dx.doi.org/10.1029/2003WR002710

[33] Fu, B.P. (1981) On the Calculation of the Evaporation from Land Surface. *Scientia Atmospherica Sinica*, **5**, 23-31.

[34] Allen, R.G. (1998) Predicting Evapotranspiration Demands for Wetlands. *ASCE Wetlands Engineering and River Restoration Conference*, Denver, 20-29 March 1998. http://www.kimberly.uidaho.edu/water/papers/wetlands/Allen_Predicting_ET_demands_for_wetlands_ASCE_1998.pdf

[35] Chen, Y., Khan, S. and Paydar, Z. (2007) Irrigation Intensification or Extensification Assessment Using Spatial Modelling in GIS. In: Oxley, L. and Kulasiri, D., Eds., *MODSIM* 2007 *International Congress on Modelling and Simulation*, Modelling and Simulation Society of Australia and New Zealand, Christchurch, December 2007, 1321-1327.

[36] FAO (Food and Agriculture Organisation of the United Nations) (1976) A Framework for Land Evaluation. FAO Soils Bulletin 32, FAO, Rome.

[37] DWLBC—Department of Water, Land and Biodiversity Conservation, South Australia (2002) Soil and Land Information—Soils of South Australian's Agricultural Lands. [CD ROM]

[38] DWLBC—Department of Water, Land and Biodiversity Conservation, South Australia (2002) Soil and Land Information—Assessing Agricultureal Land. [CD ROM]

[39] NRC (Natural Resources Council) (2000) Sustainable Development in South East South Australia. Project Report, Collaborative Project of South Australian State and Local Government.

[40] Saaty, T.L. (1980) The Analytical Hierarchy Process. McGraw Hill, New York.

[41] Saaty, T.L. and Vargas, L.G. (1991) Prediction, Projection and Forecasting. Kluwer Academic Publisher, Dordrecht, 251 p. http://dx.doi.org/10.1007/978-94-015-7952-0

[42] Chen, Y., Yu, J. and Xevi, E. (2009) A GIS-Based Sensitivity Analysis of Multi-Criteria Weights. In: Anderssen, R.S., Braddock, R.D. and Newham, L.T.H., Eds., 18*th World IMACS Congress and MODSIM09 International Congress on Modelling and Simulation*, Modelling and Simulation Society of Australia and New Zealand and International Association for Mathematics and Computers in Simulation, Cairns, 13-17 July 2009, 3137-3143.

[43] EconSearch (2006) Profile of Irrigated Activity in the South East catchment Region, 2003/04. Report Prepared for South East Catchment Water Management Board. EconSearch, Unley.

8

Frequent Breaches in Irrigation Canals in Sindh Pakistan

Riaz Bhanbhro[1,2], Nadhir Al-Ansari[1*], Sven Knutsson[1]

[1]Department of Civil, Environmental and Natural Resources Engineering, Luleå University of Technology, Luleå, Sweden
[2]Quaid-e-Awam University of Engineering, Science & Technology, Nawabshah, Pakistan
Email: *nadhir.alansari@ltu.se

Abstract

Every year, a number of breach failures occur in Irrigation Canals in Sindh. Those failures cause displacement of thousands of people, destruction of properties, land, and damage to costly crops that worth millions Rupees. In addition to that, breach failures also can cause water shortages when the failure occurs during the peak demand period. There are various causes of embankment failures which include overtopping, internal erosion, structural defects and piping. State of art for breach failures is presented in this paper and suggestions for prediction of breaches in context to canals in Sindh are discussed. Seepage and slope stability analyses are recommended for a long-term breach predictions.

Keywords

Canal Breaches; Embankment Failures; Long Term Breach Prediction; Irrigation System Sindh

1. Introduction

Sindh is the province of Pakistan and part of Indus Basin system. There are three barrages and 14 main canals which irrigate about 5 million hectares of area in Sindh [1]. Each year, breach failures in canals in Sindh are reported resulting loss of people, property, land and crops from flooding. Irrigation network map is shown in **Figure 1**. Embankments are constructed to withstand against water for several purposes including irrigation, supply and flood defense to protect people, land, crops, and property [2] [3]. These embankments can survive up to some limited safety levels and are subject to decay with time and might fail due to several triggering mechanisms [4] [5]. Most frequent embankment failures are caused by overtopping and internal erosion [6] structural de-

---

*Corresponding author.

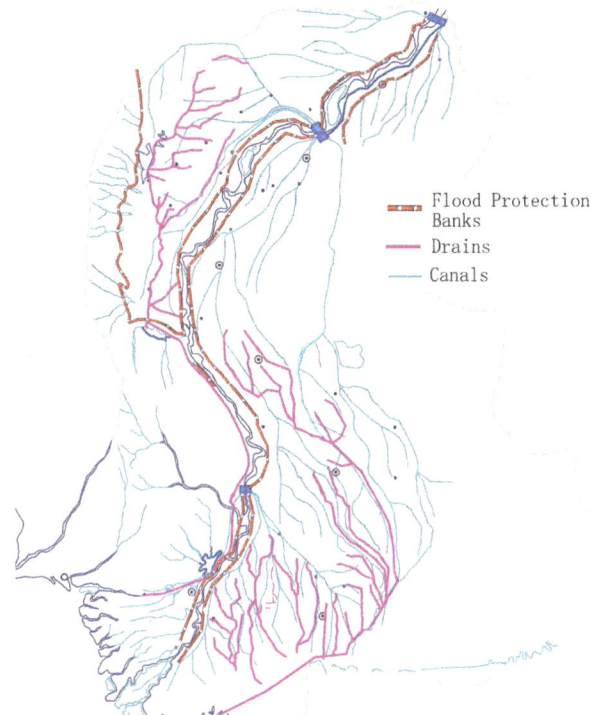

Figure 1. Irrigation network in Sindh, Pakistan [13].

fects and piping [2]. Breach process and breach rate can depend upon various factors including hydraulic conditions, materials, geometry, compaction, water content, density strength etc. [7] [8].

There is a need to improve understanding of breach failure warning particularly for canals in Sindh. This can be achieved by considering factors (and causes), causing the breach failures. Many researches carried out previously on this subject and had developed models for prediction of erosion rates, time to failure and modes of failure. Other studies on this problem were conducted using laboratory experiments and developing field models, which are summarized by [2] [9] [10].

However, there is need to conduct such type of studies for canals in Sindh towards safety and prediction of breaches.

This article presents the state of art study for breach failures in context to canals in Sindh. Attempt have been made to develop some guidelines for long-term breach predictions prior to failure. Various factors are discussed that can predict upcoming failure and some suggestions are given that can be included in Bund Manual. Bund Manual is official document for construction and regulation of the irrigation systems in Sindh [12]. Some preliminary basic laboratory tests are recommended to find out the strength of embankment materials in order to conduct seepage analysis and perform slope stability analysis

2. Breaches in Canals in Sindh

Every year, a number of breaches in canals of Sindh are reported, causing floods and displacement of thousands of people, causalities of people and animals, destruction of crops of worth of millions of rupees and property losses. A view of breach in Mirwah canal in Sindh is shown in **Figure 2** and some of recent breach failures are summarized in **Table 1**.

3. Common Causes of Breach Failures

There are several reasons involved for canal breaches in irrigation canals of Sindh. Breach data presented in **Table 1** show that the majority of failures occurred during flood seasons or when water is being released at full supply levels in canals or even in excessive quantity. Generally, embankments of canals in Sindh are built as homogenous usually constructed with earth (clay, silt and sand). Most frequent embankment failures are caused

Figure 2. View of breach along Mir Wah Canal, Sindh [11].

Table 1. Some of recent breach failures in Canals in Sindh.

Canal and its location	Description and Damages	Year	Source
Rohri Canal At RD 723-724	Widened up to 200 ft with flow around 5000 cusec from breach point. Caused displacement of 40,000 people, more than 25 villages were completely flooded and thousands of acres of cotton crops were destroyed. Flow in canal was around 13000 cusec before breach.	May 2012	The Express Tribune (2012) [14]
Rohri Canal Near Kumb Daron	People from 40 villages were displaced, 400 houses destroyed. Thousands of acres of crops were damaged.	September 2007	Rajput (2012)[15]
Rohri Canal RD 752	50 villages were submerged under water and crops were destroyed.	July 2008	Rajput (2012) [15]
Rohri Canal Near Kesana mori	Failure occurred during peak demand season and there was already water shortage. Huge quantity of water wasted followed by damages to many villages and crops.	2007	Rajput (2012) [15]
Rohri Canal RD 839	Caused 2 months of emergency in the vicinity and Tando Adam city was submerged in water.	1990	Rajput (2012) [15]
Mir Wah Canal	A 100 ft breach in Mirwah canal. It destroyed 20 houses along with damages to date palm orchards and flooding 500 acres of agricultural land. **Figure 2**	June 2011	Dawn (2011) [11]
Mir Wah Canal RD240	100 ft breach was reported. Thousands of acres of land were inundated with water causing sever damages to crops of sugarcane, jute and cotton. 400 houses nearby were flooded and 100 mud houses were damaged.	July 2010	The Express Tribune (2010) [16]
Jamrao Canal RD 164-165	Crops of rice and cotton standing on 400 acres of land were washed away. More than 1000 families of 10 villages were affected by the breach	July 2012	The News (2012) [17]
Jamrao Canal	The gushing water from 300 ft wide breach in Jamrao canal not only inundated agricultural land but also caused damage to wildlife sanctuary washing away jackals, rabbits and perished the eggs of crocodiles.	May 2011	Business Recorder (2011) [18]
Nusrat Canal	Similarly a breach in Nusrat canal was recorded by Sama TV on 22 September 2010 which caused around 200 acres of flood damaging crops.	September 2010	Saama TV (2010) [19]
Twin jamrao Canal Mile 33		July 2012	Rajput (2012) [15]

by overtopping, internal erosion [6], structural defects and piping [2].

3.1. Internal Erosion and Piping

Internal erosion in embankments can be originated by several factors such as backward erosion, leakage because of hydraulic fracturing, slope instability, high permeability region due to improper compaction, internal instability, hydraulic gradient, soil type and degree of saturation [6] [20]. Piping can be originated when hydraulic gradient of seepage is greater than hydraulic gradient of that soil [21].

Apart from earlier mentioned failures for internal erosion and piping, uncontrolled animal activity has always been a core issue of the canals in Sindh, where burrowing animals [2] [20] [22] like rats, earth worms burrow and dug the tunnels and holes [22] [23]. These holes cause weakening of embankments (**Figure 3**). This ultimately

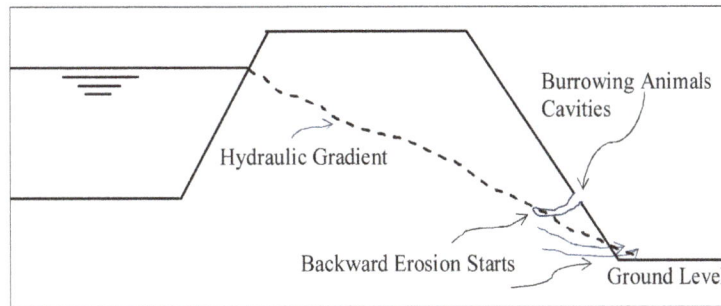

Figure 3. Some observed scenarios in canal systems in Sindh.

originates the flow when such cavities and holes come in contact with hydraulic gradient leading to breach failure by developing internal erosion and piping.

Most of the embankments for main canals in Sindh had deteriorated due to wear and natural ground levels had been lowered due to several reasons including agricultural activities. Due to lowering of ground levels near toe of embankments, hydraulic gradients are exposed to space. Since, the process that flow of water though embankment is natural as water stays against embankment almost every time, it start to seep and with time it develops small pathways for water to flow out.

Seepage energy can cause the finer particles to flow from upstream to downstream of embankment, removal of particles cause increase in diameter of pipe resulting collapse of crest roof [2]. Embankments fail due to wash out of its materials by flow of water [2] followed by development of internal erosion. The procedure of internal erosion and piping involves four stages; initialization of erosion, continuation, progression to enlarge leakage and failure/breach formation [6].

3.2. Overtopping

Water can overtop the embankments due to several reasons. Those reasons can be the settlement of embankments, erosion of top surface of embankment and high flood levels in the canal. High monsoon rainfalls can also take active part to erode top most layer of the embankments [24]. Deposition of sediments in the bed of a canal may cause water to overtop even canal runs at designed capacity. Non cohesive embankments usually have surface erosion when they are subject to overtopping whereas cohesive embankments have head cut erosion, highly compacted non-cohesive soils might also erode in head cut [2]. Several tests were conducted by [25] to describe breach process during overtopping. Initially erosion starts from downstream side making some micro rill erosion on downstream slopes and as a result, series of rills which gradually becomes master rill, eventually initiating main headcut resulting widening of canal. Head cut gradually migrate to upstream crest of embankment ultimately causing breach.

3.3. Frequent Drying and Wetting

Sindh is situated in subtropical region where summers are hot and winters are cold.

It is expected to observe climate changes with rise in temperatures and drier weather. Drier summers and wet winters possibly result in larger changes of pore water pressure and moisture content in clay soils that causes more common slope serviceability problems and damages [26].

One reason for drying and wetting can be because of climate changes i.e. summers and winters. Another reason of frequent drying and wetting within the embankment is due to the fluctuation of the water flow in canals causing change in moisture content (**Figure 4**). Drying and wetting within embankment can reduce or increase water content and degree of saturation which are also factors that attribute to initialize flow path enlargement that causes erodibility [6]. Higher cycles of moisture in old clays can cause problems like shrink and swell [26]. Tension cracks are likely to develop during long dry periods for high plasticity clay soils [24]. Series of tests were conducted by [27] and it indicated that drying and wetting can reduce long-term shear strength for high plastic clays.

3.4. Trees

Cutting down of trees can attribute towards breach failure. Tree roots are subject to decomposition upon cut-

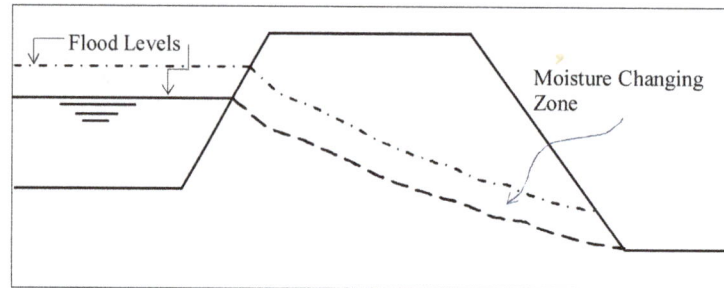

Figure 4. Moisture changing zone due to change in flood level in canal.

ting that can leave root holes [22]. These root holes can act as seepage flow path, that might lead to the progress-sion of erosion ultimately that leads to failure. On the other hand, there are also positive effects of tree planting on the embankments on the stability and maintenance. It can control erosion and also provide protection against burrowing animals [28]. However, in windy areas plantation of trees can be harmful due to chances of breaking of trees.

3.5. Structural Defects

Small structures (modules/conduits etc.) constructed within embankment of the main canals for off taking water courses can cause the leakage. This may be due to weak bond between the hydraulic structure and the embank-ment. Chances are high for initialization of leakage in cases when canal runs at designed full supply level or even in excess. Structural defects might include differential settlements followed by cracks and weaken layers that originate internal erosion and piping which causes failure [2].

3.6. Sand boils/Bubbles

According to McCook [21] sand boils are phenomenon related to piping and that may develop at downstream side of the embankments, which might lead to failures due to sudden unexpected water levels. However, sand boils do not cause breach failures always; they can be immediately treated by providing sand bags. Water leaks from foundation of embankment through a sand stratum towards downstream side and forms bubbling spring. In such underground leaks, water streams flows like fountain and carries sand along its flow. These water leaks look like small volcanoes emitting water and sediments [2] [12]. Saturation of foundation materials may cause slope instability and sliding [2]. Such type of leak that led to breach was observed at a location mile 13/4 of the left bank bund near Moro in 1942 [12]. Chances of breach are higher when underground flow of water is muddy; similarly less if the water flow is clear.

3.7. Exposed Wave Action in Coastal Areas

Islam [22] stated that in coastal areas tidal waves might damage the embankments. Hydraulic loads are gradually applied on toes and slopes due to action of tidal waves. On the other hand, cyclonic storms in coastal areas if occurs recurrently causes enormous hydraulic loads by acting on water surface. These ultimately hit the em-bankments, and if there is overtopping, the embankment gets destroyed. High velocity flow in coastal areas of-ten causes erosion of banks by undermining.

3.8. Other Causes of Failures

High energy flow through canals can cause significant erosion near banks of channel that may lead to sliding in-side the canal followed by bank failure. Regular monsoon rains create surface runoff that causes sheet erosion and formation of rills on the poor protected embankments that can weaken the banks leading to failures [22]. Earth quakes, maintenance and control against local weaknesses, vegetation can also sometimes be a cause of breaching [2]. Islam [22] summarizes various other factors that lead to failures. They are:
- Human interference factor: This is most common where the people living near by banks of canals use embank-ments as their main travel path and crest serving as rural communication road between villages.
- Heavy movement during dryer seasons: This can cause decaying of embankment.

- Uncontrolled cattle movements/grazing slowly affect the shoulders and toes of banks.
- Public cuts: These cuts are frequently observed, and they weaken the embankments which lead to slow and continued erosion [22] [23]. These cuts can lead to a breach during heavy floods or storm. The cuts are usually made for personal benefits of yielding by individuals owning crops nearby.
- Other factors: Apart from those factors mentioned above, poor planning and maintenance, faulty design and construction, proper supervision, inappropriate soil materials, proper compaction [2] [22] [23] are also factors that contribute to a breach or failure.

4. Prediction of Breach Failures in Embankments of Canals

It is very difficult to predict upcoming breach for any embankment/levee, although it is not a new technique. There can be several factors involved that indicate the upcoming failures [10]. Extensive studies have been made on modeling of breach failures; almost all of them represented the breach formation methods, models and to evaluate risks related to breach failures. These studies included the laboratory tests and modeling for overtopping breach, internal erosion, piping failure and breach propagation which are summarized by [2] [9] [10]. However, very fewer studies had been conducted to predict a breach failure long time before its failure. Short term breach prediction might include the surveillance and monitoring for any seepage, leaks, and erosion from the embank- ments at regular basis.

Long-Term Breach Predicts

This method might involve developing an equation of physical characteristics or data of earlier failures [10]. That data may include a complete previous flood record, earlier failure histories of the location [10], discharge during previous failures, the height of embankment etc. Previous records of failures might provide some hints towards expected upcoming breach. For example, it had been observed that many breach failure had been reported between locations of RD 600 to RD 900 on both sides along Rohri Canal in the past few years. All these failures were reported during monsoon/high flood seasons. Depending on any mode of the failure, it still can be linked to expect the upcoming failure.

There can be several reasons for failure, improper maintenance/surveillance, poor design, high permeable material used for the construction of embankments in this area, and creep settlements cannot be neglected. Although these failure reasons are still unanswered at the moment for the said canal, and a comprehensive study of area is required. A broader idea, which might predict breach long time before it takes place might include a comprehensive preliminary study of the site. The study should include, checking the free board according to high flood level design, detailed soil investigation/classification, hydraulic conductivity within embankment, compaction ratio, seepage paths to verify if the seepage gradients are exposed to outer slope, monitoring the pore water pressure within embankment, and to analyzing slope stability after performing laboratory tests. Upon receipt of these results, an expected failure can be predicted and avoided accordingly.

5. Remedial Measures/Suggestions

5.1. Heavy Surveillance

Continuous surveillance should be performed at regular basis, especially during monsoon/flood seasons. Surveillance may include, monitoring of free board, water levels, and downstream seepage/boils. Visual surveillance and measurement is a common method to detect internal erosion and piping in embankments. However, in most cases it is recognized at the stage when erosion has already been initiated and pipe has progressed, this can also depend on several factors [6]. Immediate measures should be taken to stop the piping once it is detected as usually pipes are the last visible stage followed by failure if not treated. Special attention should be paid to surveillance during high flooding in canals as most of piping cases through embankments are due to high levels [6].

5.2. Operation and Maintenance

A canal should be continuously monitored throughout its lengths. Weak locations should be determined and filled accordingly, the free boards should be maintained throughout according to its designed capacity and high flood levels. If any canal is subject to sedimentation, it should be dredged in order to maintain desired discharges. Any cuts and rills on the embankment due to rain should properly be treated.

Decayed trees along embankments should be properly removed. Scouring of berms is usually observed at canal curves. Measures should be taken to minimize the erosion at such places; this can be done by providing rip raps on upstream side. Canals should be properly operated from head regulators i.e. sudden draw down and sudden over flow can cause severe damages to canal embankments. If there is any breach in any off taking canal from the main canal, it should not be closed immediately as it may increase the flow in the main canal that ultimately can cause breaches in main canal as well [29].

5.3. Proper Compaction

Compaction can significantly improve the engineering properties of earthen materials [30]. Degree of compaction has significant effects on the erodibility of material used for the construction of embankment. Soil erodibility is the main material property in defining the piping and internal erosion of earthen embankments [31]. All soils are susceptible to erosion [20], measures should be taken to reduce the erodibility of soils. This can be achieved by increasing the degree of compaction. Presently compaction dry density ratio of 85% is used for construction and rehabilitation works along canals in Sindh. However, any compaction dry density ratio less than 90% is very poorly compacted that appears to be fast erodible irrespective of any soil type [6]. Poorly compacted soils (less than 95%) are more likely influence to erosion; 95% - 98% of standard compaction density ratio is neutral whereas 98% or greater than 98% are less likely to pipe or erosion [6] [32] [33]. Series of laboratory tests were conducted by [30] [34] [35], concluded that soils which are compacted on wet side of optimum moisture content are less erodible as compared to dry side of optimum water content. Also the rate of change of erodibili- ty versus dry density is quicker on dry side as compared to wet side. **Figure 5** shows and illustration of same dry density at points a and b; however, the only difference is water content. Many upcoming failures can be avoided by selecting proper compaction ratios in construction or rehabilitation works; it is recommended that at least 95% degree of compaction on wet side should be used. Apart from strength, improved compaction can also offer more resistance to burrowing animals to create tunnels and holes [36].

5.4. Soil Texture and Gradation

Soil gradation is one of key factors for determination erodibility [30]. Texture and gradation of soils used in construction of canals should be analyzed. As seepage through embankments can be high if poorly graded or gap graded soils are used. Study should be conducted to find soil properties related to gradation curves, remedial measures should be taken accordingly. Materials which are used in construction or rehabilitation works along canals should be used as high plastic of liquid limit 50% or more in the downstream zone of embankments [6], as it can increase time for expected development of breach.

5.5. Bio-Tech Reinforcements and Crest Sloping

Islam [22] defines that embankment erosions can be reduced by vegetation and plantation on embankments. Severe erosions of monsoon rainfall can be significantly reduced by surface vegetation; this surface vegetation can be usually a mat of well-maintained grasses. This continuous mat of grass works as bio tech reinforcements for embankments against runoff erosion due to overtopping or heavy rain falls. Providing crest sloping (**Figure 6**) can assist drainage to flow towards the upstream side, it helps in reducing the velocity of overtopping waters; usually it should be provided with slope of 1 in 20 [37].

5.6. Detailed Laboratory Investigations

To find stability of slopes of embankments and calculate the factors of safety, some advanced laboratory investigations are recommended. Once the laboratory data are available, then, weak zones and potential embankment failures can be detected. There is strong need to perform laboratory investigations as up to date there is no available data for the laboratory investigations for the canals in Sindh in literature. These laboratory investigations may include, 1) liquid limits and plastic limits; 2) gradation curves; 3) degree of compaction at present level; 4) strain rates upon application of loads; 5) hydraulic conductivity; 6) particle shape and mineralogy; and 7) shear strength parameters of soils. For determination of strength parameters, both direct shear and triaxial tests should be performed.

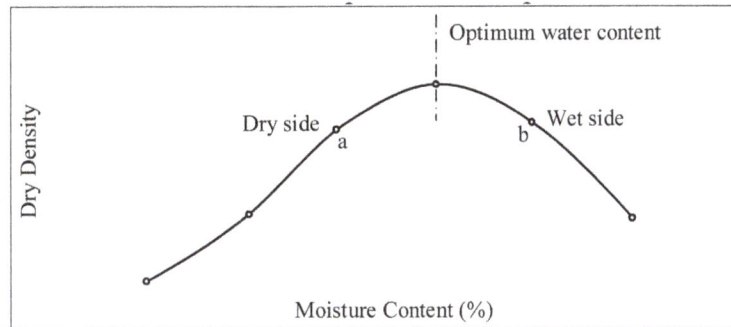

Figure 5. Dry density vs. water content.

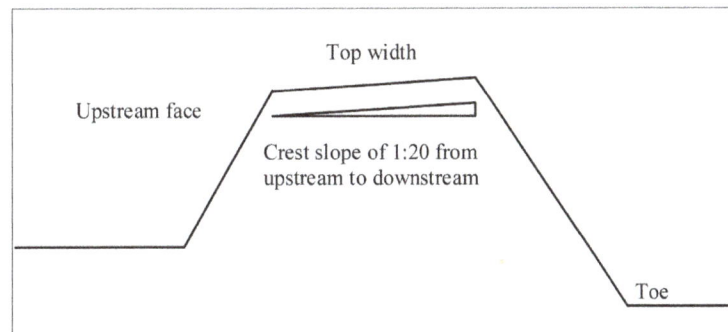

Figure 6. Top width section, crest slope, after [37].

5.7. Seepage Analysis

Complete study of seepage paths should be carried out. The saturation gradient in present construction or rehabilitation works of canals is assumed theoretically as suggested in Bund Manual [12], which is the official document for irrigation and drainage department, Government of Sindh. However, seepage gradient can vary tremendously depending on the compaction ratio, material gradation, and soil types.

So it is very necessary to conduct the study for seepage analysis in order to find relatively more realistic seepage gradient for the future constructions and for the existing canals. Embankment materials and its engineering properties can vary along its lengths. In order to perform the analysis for seepage gradient; material properties should be determined by taking several samples along its length to determine material gradation and hydraulic conductivity. Seepage analysis can be performed by using finite element method, and there are various computer programs available that can perform the seepage analysis. Computer programs can reduce enormous time and money [38].

5.8. Slope Stability Analysis

Slopes have tendency to slide due to the effect of gravitational and/or other forces. Failure mode of breach at location RD 723 of Rohri canal was due to sliding [15]. There could have been several failures which are not reported as most of failures modes of breaches are not reported. The engineering properties of materials used in construction are controlled by compaction and methods of construction. However, these properties can change with time due to several factors (seepage pressures, deterioration, draw down effects and creep etc.). It is thus important to analyze the slopes for stability of the present conditions. The analyses can be performed for these critical conditions such as 1) long term conditions; 2) effective normal and principal stresses within embankments; 3) effects of sudden draw down or vice versa; 4) seismic disturbances if any 5) creep; and 6) for safety factors. Safety measures can be taken accordingly after receipt of the results from stability analysis.

6. Conclusions/Discussions

Many beach failures occur every year in Canals in Sindh. There are various reasons for failures including overtopping, internal erosion and piping, structural defects and high energy flows in canals. Prediction of breaches is not a new technique; it requires several factors and reliable data. It can be further divided as short term prediction and long term prediction. Short term breach prediction may include the monitoring and surveillance for any seepage or erosion along the embankments. Long term breach prediction may include a detailed study about the location, materials, construction methods, rate of erosions, compaction, hydraulic conductivity and strength parameters, then to develop an equation to correlate the upcoming failure. Some initial laboratory studies are recommended for canals in Sindh in order to conduct detailed seepage studies and slope stability analysis, which might predict weak zones in embankments that can reduce potential breach failures.

Acknowledgements

The research presented was carried out as a part of "Swedish Hydropower Centre—SVC". SVC has been established by the Swedish Energy Agency, Elforsk and Svenska Kraftnät together with Luleå University of Technology, The Royal Institute of Technology, Chalmers University of Technology and Uppsala University.

References

[1]   Steenbergen, F. (2014) Water Charging in Sindh, Pakistan—Financing Large Canal Systems.
       http://www.metameta.nl/wp-content/uploads/2012/02/MetaMeta_Watercharging_Sindh.pdf

[2]   Wu, W., Altinakar, M., Al-Riffai, M. and Bergman, N. (2011) Earthen Embankment Breaching. *Journal of Hydraulic Engineering*, **137**, 1549-1564. http://dx.doi.org/10.1061/(ASCE)HY.1943-7900.0000498

[3]   Mohamed, M., Samuels, P., Morris, M. and Ghataora, G. (2002) Improving the Accuracy of Prediction of Breach Formation through Embankment Dams and Flood Embankments. *Proceedings International Conference, Fluvial Hydraulics*, Louvain-la-Neuve, Belgium, 663-673.

[4]   Foster, M., Fell, R. and Spannagle, M. (2000) The Statistics of Embankment Dam Failures and Accidents. *Canadian Geotechnical Journal*, **37**, 1000-1024. http://dx.doi.org/10.1139/t00-030

[5]   Allsop, W., Kortenhaus, A., Morris, M., *et al.* (2007) Failure Mechanisms for Flood Defense Structures. FLOODsite Report.

[6]   Fell, R., Wan, C.F., Cyganiewicz, J. and Foster, M. (2003) Time for Development of Internal Erosion and Piping in Embankment Dams. *Journal of Geotechnical and Geoenvironmental Engineering*, **129**, 307-314.
       http://dx.doi.org/10.1061/(ASCE)1090-0241(2003)129:4(307)

[7]   Hunt, S., Hanson, G. and Temple, D. (2006) Breach Widening Observations Related to Clay Core Earthen Embankment Tests. FLOODsite project report T04_06_01, April. http://hikm.ihe.nl/floodsite/data/Task4/pdf/failmechs.pdf

[8]   Hanson, G., Cook, K. and Hunt, S. (2005) Physical Modeling of Overtopping Erosion and Breach Formation of Cohesive Embankments. *Transactions of the American Society of Agricultural Engineers*, **48**, 1783-1794.
       http://dx.doi.org/10.13031/2013.20012

[9]   Wahl, T.L. (2007) Laboratory Investigations of Embankment Dam Erosion and Breach Processes. CEA Technologies inc., dam safety interest group, CEATI report no. T032700-0207A, 60 p.

[10]   Morris, M. and Hassan, A. (2002) Breach Formation through Embankment Dams & Flood Defence Embankments: A State of the Art Review. HR Wallingford, IMPACT Project Workshop, Wallingford.

[11]   Dawn (2013) Canal Breach Inundates 500 Acres.
       http://dawn.com/2011/06/12/canal-breach-inundates-500-acres

[12]   Bund Manual (2008) 4th Edition, Irrigation and Power Department, Government of Sindh, Pakistan.

[13]   Sindh Irrigation and Drainage Authority (SIDA) (2013)
       http://www.sida.org.pk/download/maps/IrrigationNetworkPlanMap.jpg

[14]   The Express Tribune (2013) More than 30000 People Have Been Displaced by Breach in Rohri Canal.
       http://tribune.com.pk/story/378786/more-than-30000-people-have-been-displaced-by-breach-in-rohri-canal

[15]   Rajput, M.I. (2012) Translated from Sindhi Book (Saal 2011 wari barsatibod ae sindh ja soor). Sindh National Academy (Trust), Hyderabad.

[16]   The Express Tribune (2013) 400 Houses Damaged by Canal Breach in Khairpur.
       http://tribune.com.pk/story/26853/400-houses-damaged-by-canal-breach-in-khairpur

[17]   The News (2013) Breach in Jamrao Canal Inundates Several Villages.
       http://www.thenews.com.pk/Todays-News-2-121534-Breach-in-Jamrao-canal-inundates-several-villages

[18]   Business Recorder (2013) More than 300 ft Jamrao Canal Breach Damages Wildlife Sanctuary.

http://www.brecorder.com/pakistan/general-news/14397-more-than-300ft-jamrao-canal-breach-damages-wildlife-sanct uary.html

[19] Saama, T.V. (2013) Naushero Feroze's DKM Submerged as Nusrat Canal Breaches. http://www.samaa.tv/nadeemmaliklive/Naushero-Feroze%E2%80%99s-DKM-submerged-as-Nusrat-Canal-breaches-2 3769-1.html

[20] Hanson, G., Tejral, R., Hunt, S. and Temple, D. (2010) Internal Erosion and Impact of Erosion Resistance. 30*th Annual USSD Conference*, Sacramento, 12-16 April 2010, 773-784.

[21] McCook, D. (2004) A Comprehensive Discussion of Piping and Internal Erosion Failure Mechanisms. *Proceedings of the* 2004 *Annual Association of State Dam Safety Officials*, 1-6.

[22] Islam, M.N. (2000) Embankment Erosion Control: Towards Cheap and Simple Practical Solutions for Bangladesh. 2*nd International Conference on Vetiver* (*ICV*2), 2000, 307-321.

[23] Hoque, M.M. and Siddique, M.A.B. (1995) Flood Control Projects in Bangladesh: Reasons for Failure and Recommendations for Improvement. *Disasters*, **19**, 260-263.

[24] Ismail, F., Mohamed, Z. and Mukri, M. (2008) A Study on the Mechanism of Internal Erosion Resistance to Soil Slope Instability. *Electronic Journal of Geotechnical Engineering*, 13.

[25] Hahn, W., Hanson, G.J. and Cook, K.R. (2004) Breach Morphology Observations of Embankment Overtopping Tests. *Joint Conference on Water Resource Engineering and Water Resources Planning and Management*, Minneapolis, 30 July-2 August 2000, 1-10.

[26] Clarke, D. and Smethurst, J. (2010) Effects of Climate Change on Cycles of Wetting and Drying in Engineered Clay Slopes in England. *Quarterly Journal of Engineering Geology and Hydrogeology*, **43**, 473-486. http://dx.doi.org/10.1144/1470-9236/08-106

[27] Rogers, L.E. and Wright, S.G. (1986) The Effects of Wetting and Drying on the Long-Term Shear Strength Parameters for Compacted Beaumont Clay. Center for Transportation Research, Bureau of Engineering Research, University of Texas at Austin, Texas.

[28] Lammeranner, W. and Meixner, H. (2008) Woody Plants on Floods Protection Levees: A Contradiction? The Use of Vegetation to Improve Slope Stability. 2*nd International Conference on Ground Bio-Engineering*, Beijing, 14-18July 2008.

[29] Rajput, M.I. (2010) Translated from Sindhi Book (Sindhu jee Saar). Sindh National Academy (Trust), Hyderabad.

[30] Hanson, G. and Hunt, S. (2006) Determining the Erodibility of Compacted Soils for Embankment Dams. *Proceedings of the 26th Annual USSD Conference*, San Antonio, 1-6 May 2006, 311-320.

[31] Wan, C.F. and Fell, R. (2004) Investigation of Rate of Erosion of Soils in Embankment Dams. *Journal of Geotechnical and Geoenvironmental Engineering*, **130**, 373-380. http://dx.doi.org/10.1061/(ASCE)1090-0241(2004)130:4(373)

[32] Foster, M. and Fell, R. (1999) A Framework for Estimating the Probability of Failure of Embankment Dams by Internal Erosion and Piping Using Event Tree Methods. University of New South Wales, Kensington.

[33] Foster, M. and Fell, R. (2000) Use of Event Trees to Estimate the Probability of Failure of Embankment Dams by Internal Erosion and Piping. *Proceedings of the* 20*th International Congress on Large Dams*, Beijing, International Commission on Large Dams (ICOLD), Paris, Question 76, **1**, 237-260.

[34] Hanson, G. and Hunt, S. (2007) Lessons Learned Using Laboratory JET Method to Measure Soil Erodibility of Compacted Soils. *Applied Engineering in Agriculture*, **23**, 305. http://dx.doi.org/10.13031/2013.22686

[35] Wahl, T.L., Lentz, D.J. and Feinberg, B.D. (2011) Physical Hydraulic Modeling of Canal Breaches. Hydraulic Laboratory Report HL-2011-09, Bureau of Reclamation, US Department of the Interior, Denver.

[36] Sjoerd, W.D. (2004) Effects of Soil Compaction. http://pubs.cas.psu.edu/freepubs/pdfs/uc188.pdf

[37] Hughes, A.K. (1981) The Erosion Resistance of Compacted Clay Fills in Relation to Embankment Overtopping. Ph.D. Thesis, Newcastle University, Newcastle.

[38] Giglou, A.N. and Zeraatparvar, A. (2012) Seepage Estimation through Earth Dams. *Journal of Basic and Applied Scientific Research*, **2**, 7861-7865.

9

Assessment of Groundwater Quality in Central India

Shabya Choudhary[1], Shobhana Ramteke[1], Keshaw Prakash Rajhans[1], Pravin Kumar Sahu[1], Suryakant Chakradhari[1], Khageshwar Singh Patel[1*], Laurent Matini[2]

[1]School of Studies in Chemistry/Environmental Science, Pt. Ravishankar Shukla University, Raipur, India
[2]Department of Exact Sciences, E.N.S., Marien Ngouabi University, Brazzaville, Congo
Email: *patelks_55@hotmail.com

Abstract

The groundwater is widely used for irrigation of rice crops. The overuse of groundwater causes depletion of the water quality (*i.e.* enormous increase in conductivity, hardness and ion and metal contents, etc.) in several regions of the country and world. In this work, the quality of the groundwater in the densestrice cropping area, Saraipali, Chhattisgarh, Central India is discussed. The water is sodic in nature with extremely high electrical conductivity. The mean concentration (n = 30) of F^-, Cl^-, NO_3^-, SO_4^{2-}, NH_4^+, Na^+, K^+, Mg^{2+}, Ca^{2+} and Fe in the water was 1.2 ± 0.2, 98 ± 31, 46 ± 15, 56 ± 9, 19 ± 4, 206 ± 25, 9.2 ± 2.3, 39 ± 6, 114 ± 19 and 1.7 ± 0.6 mg/L, respectively. The sources of the contaminants are apportioned by using the factor analysis model. The suitability of the groundwater for the drinking and irrigation purposes is assessed.

Keywords

Groundwater, Indices, Sources

1. Introduction

The urban groundwater has emerged as one of the world's most challenging issues due to large users and contamination with chemicals of geogenic and anthropogenic origins [1]. The quality of available groundwater was degraded enormously by enhancing conductivity, alkalinity, hardness and contaminant levels [2]-[15]. Hence, in this work, the groundwater quality of the rice growing area, Saraipali block, Mahasamund, Chhattisgarh, India was selected for the assessment and rating.

---

*Corresponding author.

2. Materials and Methods

2.1. Study Area

Saraipali (21.33°N 83.0°E) is a block in Mahasamund district, Chhattisgarh state, India, including 299 town and villages inclusive of Saraipali town with population of ≈0.3 million. The rice is a main crop of the area with use surplus amount of groundwater to take the multiple crops in a year. The water is hard and become turbid on the storage due to precipitation of the metals *i.e.* Mg, Ca and Fe into oxides and hydroxides. The health problems (*i.e.* tiredness, diarrhea, stone formation in kidney and spleen, etc.) in the residence of the studied area due to intake of the groundwater were marked. Therefore, in the present work, the water quality assessment of Saraipali area was chosen.

2.2. Sample Collection

The groundwater samples were collected from 30 locations of the town and nearby villages, **Figure 1**. The water was collected in the post monsoon period, January, 2014 in a 1-L cleaned polyethylene bottle by using established methodology [16]. The bottle was ringed thrice with the sampling water prior to collection and filled up to the mouth with the water. The physical parameters *i.e.* pH, temperature (T), electrical conductivity (EC), reduction potential (RP) and dissolved oxygen (DO) were measured at the spot.

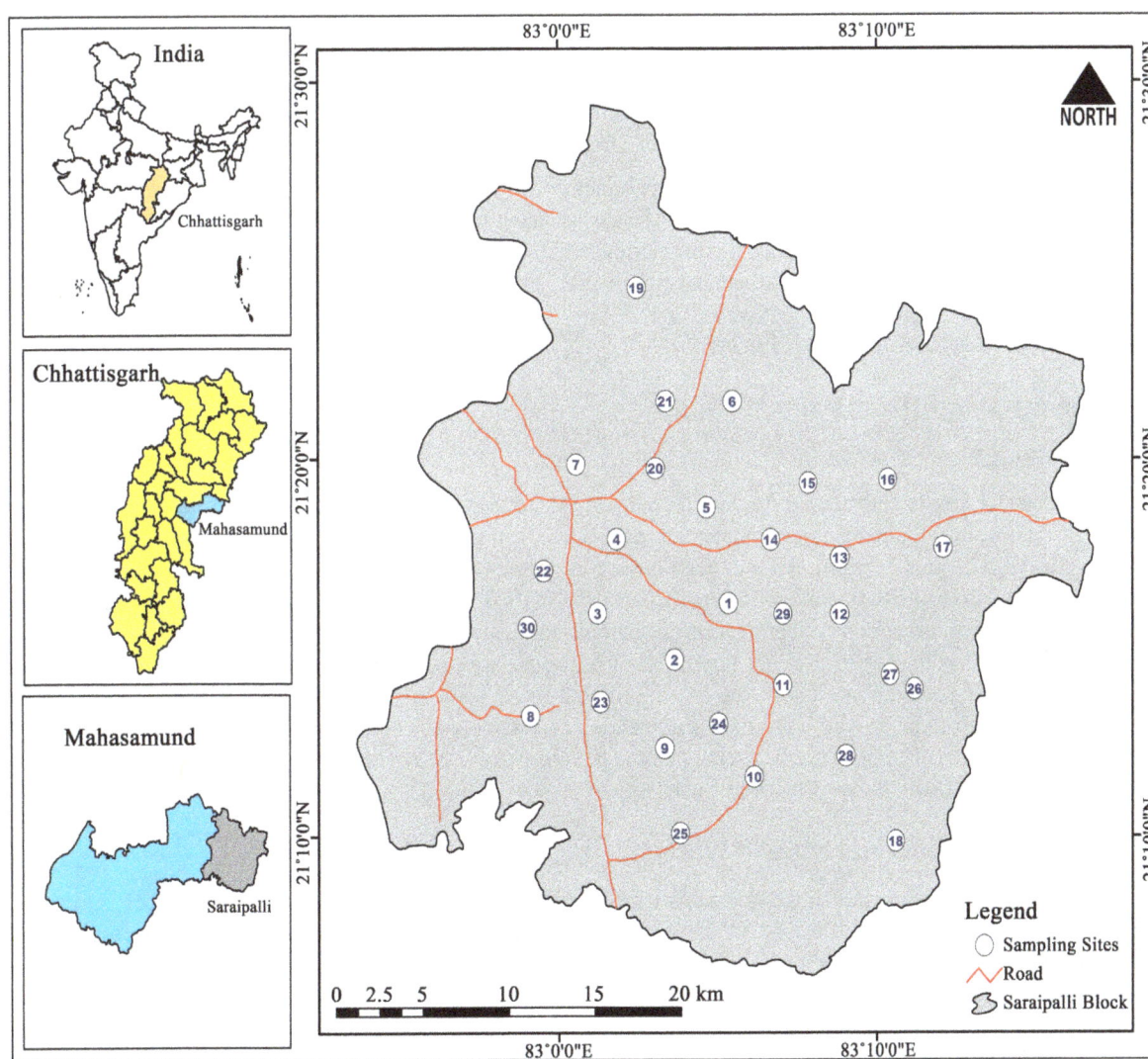

Figure 1. Representation of sampling locations in Chhattisgarh, India.

2.3. Analysis

The Hanna water analyzer kits was used for the measurement of the physical parameters. The total dissolved solid (TDS) value was determined by evaporation method by prior filtering the water through glass fiber with subsequent drying at the constant weight [16]. The total hardness (TH) and total alkalinity (TA) values were analyzed by titration methods [17]. The Metrohm ion meter-781 was used for monitoring of F^- by using the buffer in a 1:1 volume ratio. The Dionex ion chromatography-1100 was used for the quantification of the ions. Multivariate statistical model *i.e.* factor analysis (FA) was used for the source apportionment of ions and metals [18]. The statistical software STATISTICA 7.1 was employed for the multivariate statistical calculations.

The various water quality indices *i.e.* sodium adsorption ratio (SAR), sodium hazard (SH) and water quality index (WQI) were used for rating of the water quality. The weighed arithmetic method was employed for computation of the WQI of the groundwater by using four parameters *i.e.* pH, DO, EC and TDS [19] [20]. The following equations were used for calculation of the indices.

$$SAR = \left[Na^+\right] \Big/ \sqrt{\left\{\left(\left[Ca^{2+}\right]+\left[Mg^{2+}\right]\right)\Big/2\right\}}$$

$$SH = \left(\left\{[Na]+[K]\right\}\Big/\left\{[Na]+[K]+[Mg]+[Ca]\right\}\right) \times 100$$

The equivalent concentrations of cations were used.

$$WQI = \sum q_n W_n \Big/ \sum W_n$$

$$q_n = 100\left(V_n - V_{io}\right)\Big/\left(S_n - V_{io}\right)$$

q_n = Quality rating of the *n*th water quality parameter.
V_n = Estimated value of the *n*th parameter of a given water.
S_n = Standard permissible value of the *n*th parameter.
V_{io} = Ideal value of the *n*th parameter of pure water (*i.e.* 0 for all other parameters except pH and dissolved oxygen (7.0 and 14.6 mg/L, respectively).
W_n = Unit weight for the *n*th parameter.

3. Results and Discussion

3.1. Geology

Chhattisgarh basin is characterized by rocks belonging to Proterozoic aged sandstone, limestone, and dolomite, conglomerate, etc. Siliciclastic-carbonates are deposited in muddy shelf and platformer environment, indicative of more stable tectonic condition. Its deposition is controlled by several cycles of transgressions and regressions. The Proterozoic grouprocks are found to spread over the studied area. The gypsum minerals are found to be more intense than calcareous minerals, containing both toxic and precious elements at traces.

The physical characteristics of 30 tube well of Saraipali area is summarized in **Table 1**. The depth of tube well (n = 30) is moderate, ranging from 24 - 63 m with mean value of 32 ± 2 m. The ionic contamination of the water was found to be related with the depth profile of the tube wells and increased as the depth profile was increased (r = 0.59). The age of tube wells was ranged from 7 - 25 Yr with mean value of 17 ± 2 Yr. The water quality was also found to be influenced by the age of tube wells.

3.2. Physical Characteristics of Water

The chemical characteristics of the groundwater are presented in **Table 2**. The T, DO, RP and pH value of water (n = 30) was ranged from 19°C - 22°C, 4.8 - 5.4 mg/L, 117 - 238 mV and 6.2 - 8.3 with mean value of 20.9°C ± 0.3°C, 5.1 ± 0.1 mg/L, 187 ± 9 mV and 6.88 ± 0.13, respectively. In some locations, the water was found to be slightly acidic due to higher Cl^- and NO_3^- contents. The EC, TDS, TA and TH value of water was ranged from 785 - 4589 µS/cm, 651 - 2836 mg/L, 159 - 610 mg/L and 186 - 864 mg/L with mean value of 1946 ± 363 µS/cm, 1411 ± 221 mg/L, 352 ± 45 mg/L and 355 ± 58 mg/L, respectively. The EC value was mainly contributed by the ions *i.e.* Na^+, K^+, Cl^-, NO_3^- and SO_4^{2-} (r = 0.93).

Table 1. Geophysical characteristics of tube well and groundwater during January, 2014.

S. No.	Location	Age, Yr	Depth, m	T, °C	pH	EC, μS/cm	RP, mV	DO, mg/L
1	Joganipalidipa	22	30	22	7.1	1169	200	5.2
2	Joganipali	10	30	22	6.2	1776	187	5.3
3	Kejuan	18	33	21	6.9	966	170	5.2
4	Harratar	13	27	21	7.2	1433	212	5.0
5	Kutela	15	24	21	7.1	1099	139	5.4
6	Bastisaraipali	19	27	22	7.0	2097	165	5.0
7	Madhopali	17	27	22	7.0	1190	238	5.1
8	Parsada	16	24	22	7.2	1127	186	5.3
9	Telidipa	12	27	21	6.8	888	180	5.0
10	Lukapara	7	63	21	6.8	3770	187	4.8
11	Lakhanpali	21	33	20	6.8	1209	218	5.3
12	Barihapali	10	48	20	6.8	2545	191	4.9
13	Mokhaputka	25	33	21	6.6	2467	181	4.9
14	Kumhardipa	17	36	22	6.5	1375	214	5.1
15	Saraipali	20	30	22	6.7	1100	183	5.0
16	Paterapali	15	33	22	6.9	4589	161	5.3
17	Balsi	25	33	22	6.5	1928	172	5.1
18	Kendudhar	24	30	21	7.2	1910	219	5.1
19	Bichhiyan	22	33	21	7.0	4082	205	5.2
20	Sagarpali	18	39	20	6.8	3666	194	5.1
21	Amarkot	22	24	21	6.6	1080	188	5.0
22	Mohda	20	27	20	6.3	1888	163	5.4
23	Navrangpur	18	33	20	7.1	1251	172	5.3
24	Patsendri	16	36	20	7.1	2730	194	5.2
25	Bonda	17	36	20	7.0	3094	201	5.1
26	Girsa	15	33	19	8.3	2045	117	5.1
27	Jambahlin	20	27	20	6.9	1806	213	5.1
28	Baitari	15	30	21	6.8	1340	226	5.4
29	Chattigirhola	16	33	21	7.0	785	157	5.2
30	Echchhapur	18	30	20	7.1	1968	170	5.2

Table 2. Chemical characteristics of groundwater during January, 2014, mg/L.

S. No.	TDS	TA	TH	F$^-$	Cl$^-$	NO$_3^-$	SO$_4^{2-}$	NH$_4^+$	Na$^+$	K$^+$	Ca^{2+}	Mg^{2+}	Fe
1	748	353	210	0.8	27	22	27	13	156	9.5	57	34	2.4
2	1183	298	318	0.9	92	29	44	15	246	5.5	99	39	3.8
3	857	286	243	0.6	18	21	69	12	118	6.0	75	30	2.4
4	896	420	306	1.2	36	28	31	14	163	6.5	101	31	0.5
5	651	286	207	0.8	18	18	38	11	125	4.0	68	22	1.1
6	1310	311	330	1.3	129	18	53	17	218	17.0	101	42	0.7
7	1028	335	246	0.9	27	14	42	31	146	5.5	75	31	1.1
8	1071	237	246	1.0	42	104	34	7	118	6.5	75	31	0.4
9	978	347	408	1.6	23	29	39	9	102	8.5	130	47	2.1
10	2588	585	693	1.8	190	120	40	31	311	4.0	226	74	0.4

Continued

11	906	280	258	0.8	36	32	69	7	163	4.5	86	26	0.8
12	1731	384	501	1.9	125	23	57	26	254	9.5	164	53	0.9
13	1868	317	471	1.8	134	67	79	19	233	3.0	153	51	1.5
14	1554	170	186	0.8	65	163	38	23	175	7.0	62	18	1.2
15	805	213	222	0.7	51	15	36	11	156	14.0	73	23	2.7
16	2836	464	864	2.2	374	42	47	12	351	5.5	286	88	0.6
17	1646	183	471	1.9	166	34	46	13	251	11.0	156	48	1.4
18	1106	573	276	0.8	42	22	79	15	260	36.0	83	36	0.6
19	2626	543	513	1.7	254	120	68	29	311	5.0	151	72	1.9
20	2207	610	438	1.6	231	68	42	33	317	13.5	138	52	0.4
21	1212	244	348	1.2	36	22	88	17	155	8.5	117	34	7
22	1960	159	327	1.1	120	153	100	13	248	5.5	112	30	6.9
23	963	268	222	1.0	47	25	85	12	131	8.0	70	26	2.1
24	1854	360	543	1.8	161	32	39	28	282	13.0	179	56	1.1
25	1948	329	552	1.9	231	28	35	19	257	20.0	182	57	1.2
26	2022	372	231	1.1	116	21	140	57	226	6.0	75	25	0.3
27	1097	433	354	1.2	47	21	43	18	179	5.0	117	36	1.1
28	945	402	219	0.8	34	26	61	18	152	12.0	70	25	3.1
29	792	244	216	0.9	35	31	46	12	122	8.0	73	21	1.5
30	956	549	228	0.7	47	31	60	16	260	7.5	73	26	0.9

3.3. Chemical Characteristics of Water

The concentration of F^-, Cl^-, NO_3^-, SO_4^{2-}, NH_4^+, Na^+, K^+, Mg^{2+}, Ca^{2+} and Fe was ranged from 0.6 - 2.2, 18 - 374, 14 - 163, 27 - 140, 7.0 - 57, 102 - 351, 3.0 - 36, 18 - 88, 57 - 286 and 0.3 - 7.0 mg/L with mean value of 1.2 $\pm$ 0.2, 98 $\pm$ 31, 46 $\pm$ 15, 56 $\pm$ 9, 19 $\pm$ 4, 206 $\pm$ 25, 9.2 $\pm$ 2.3, 39 $\pm$ 6, 114 $\pm$ 19 and 1.7 $\pm$ 0.6 mg/L, respectively. Among them, Na^+ showed the highest content followed by Ca^{2+} and Cl^-. The highest ionic content was marked at locations lying close to at the highway junctions and water reservoirs due to their increased mineralization in the groundwater, **Figure 2**.

3.4. Source

The correlation coefficient matrix of the water variables are shown in **Table 3**. Among them, ions *i.e.* F^-, Cl^-, Na^+, Mg^{2+} and Ca^{2+} were found to be well correlated, showing origin from the common sources. The molar ratio of [Na^+]/[Cl^-] was ranged from 1.5 - 11 with mean value of 5 $\pm$ 1, indicating both geogenic and anthropogenic origins of Na in the water.

The FA model showed the extraction of six factors with account for 84.04% of total variance, **Table 4**. Factor-1 accounts for 39.27% of the total variance with strong positive loadings of TH, Ca^{2+}, Mg^{2+}, F^-, Cl^-, EC and TDS; related to hardness depending on the weathering of fluoride bearing materials such as CaF_2. Factor-2 explains 14.79% of the total variance with high positive loading of SO_4^{2-}, correlated to evaporation of the water. Factor-3 explains 9.06% of the total variance with high positive loading of alkalinity in opposition to Fe. Factor-4 accounts for 8.32% of the total variance with a negative loading of DO. Factor 5 explains 6.87% of the total variance with a negative loading of the variable Age of the tube wells. Factor-6 accounts for 5.74% of the total variance with a high positive loading of NO_3^-, indicating agricultural impacts in the water.

3.5. Water Quality

The value of TA, TH, Mg, Ca and Fe content was found to be higher than recommended value of 120, 200, 30, 75 and 0.30 mg/L, respectively [19] [20]. The value of SAR, SH and WQI was ranged from 1.8% - 28%, 19% - 84% and 86% - 713% with mean value of 6.6% $\pm$ 1.7%, 50% $\pm$ 5% and 275% $\pm$ 60%, respectively. The

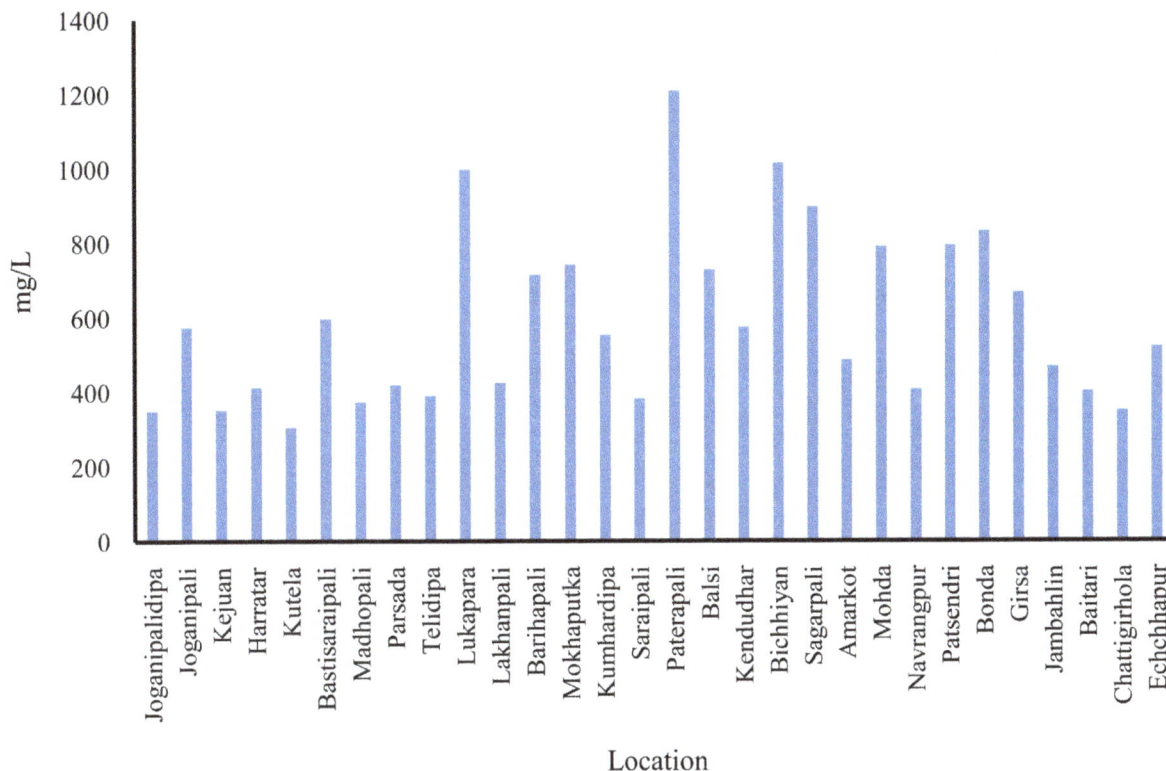

Figure 2. Spatial variations in sum of total concentration of the ions.

Table 3. Correlation coefficient matrix of elements in water.

	F^-	Cl^-	NO_3^-	SO_4^{2-}	NH_4^+	Na^+	K^+	Ca^{2+}	Mg^{2+}	Fe
F^-	1									
Cl^-	0.81	1								
NO_3^-	0.15	0.29	1							
SO_4^{2-}	−0.11	−0.01	0.03	1						
NH_4^+	0.24	0.32	0.11	0.40	1					
Na^+	0.65	0.86	0.29	0.09	0.42	1				
K^+	−0.02	0.02	−0.26	−0.02	−0.05	0.18	1			
Ca^{2+}	0.91	0.85	0.17	−0.14	0.15	0.73	−0.05	1		
Mg^{2+}	0.88	0.86	0.18	−0.17	0.19	0.75	0.01	0.93	1	
Fe	−0.20	−0.20	0.13	0.31	−0.25	−0.18	−0.12	−0.15	−0.21	1

Table 4. Eigenvalues and factor loadings of groundwater.

Variable	Factor-1	Factor-2	Factor-3	Factor-4	Factor-5	Factor-6
Age	−0.10	0.02	−0.18	−0.20	**−0.83**	0.04
Depth	0.41	0.06	0.19	0.57	0.32	0.32
T	−0.06	−0.67	−0.20	−0.01	−0.11	0.09
pH	−0.09	0.55	0.64	−0.12	0.14	−0.30
EC	**0.88**	0.14	0.31	0.02	−0.03	0.25
RP	−0.15	−0.59	0.33	0.22	−0.35	0.30
DO	−0.08	0.03	0.09	**−0.90**	0.07	0.10
TDS	**0.85**	0.30	0.10	0.07	0.02	0.40

Continued

TA	0.43	0.10	**0.76**	0.04	−0.05	0.05
TH	**0.97**	−0.02	0.09	0.07	0.14	0.00
F⁻	**0.91**	0.08	0.04	0.15	0.11	−0.07
Cl⁻	**0.91**	0.10	0.07	−0.02	−0.05	0.19
NO_3^-	0.21	−0.02	−0.17	−0.08	0.10	**0.85**
SO_4^{2-}	−0.03	**0.87**	−0.15	−0.12	−0.21	0.08
NH_4^+	0.18	0.67	0.44	0.32	0.06	0.33
Na^+	0.29	0.66	0.40	0.31	0.00	0.36
K^+	−0.08	−0.10	−0.04	0.40	−0.69	−0.28
Ca^{2+}	**0.96**	−0.01	0.06	0.07	0.16	−0.01
Mg^{2+}	**0.95**	−0.05	0.19	0.06	0.06	0.04
Fe	−0.24	−0.01	**−0.76**	0.06	−0.33	0.15
Eigenvalue	7.85	2.96	1.81	1.66	1.37	1.15
% Total variance	39.27	14.79	9.06	8.32	6.87	5.74
Cumulative %	39.27	54.05	63.11	71.43	78.30	84.04

classification of groundwater was grouped on the basis of SH values, excellent (<20%), good (20% - 40%), permissible (40% - 60%), doubtful (60% - 80%) and unsuitable (>80%). It means the water of the studied area was found to be sodic and hard in nature, being unsuitable for the drinking purposes. They could be used for the irrigation purposes but prolonged excessive extraction of the water may cause adverse impacts in rice yields in near future.

4. Conclusion

The groundwater of Saraipali area is deteriorated rapidly due to its excessive extraction for the irrigation purposes. The water is sodic and hard in nature. The values of EC, TH, TA, Na, Mg, Ca and Fe were observed to be above reported permissible limits. The water is seemed to be unsuitable for the drinking purposes due to high mineralization of the bed-rock elements in the aquifer. The water could be used for the irrigation of the new varieties rice crops required less water with lower ripping life.

Acknowledgements

We are thankful to the Pt. Ravishankar Shukla University, Raipur for awarding scholarship to one of the author *i.e.* SC.

References

[1] Kulkarni, H., Shah, M. and Vijay Shankar, P.S. (2015) Shaping the Contours of Groundwater Governance in India. *Journal of Hydrology: Regional Studies*, **4**, 172-192. http://dx.doi.org/10.1016/j.ejrh.2014.11.004

[2] Machiwal, D. and Jha, M.K. (2015) Identifying Sources of Groundwater Contamination in a Hard-Rock Aquifer System Using Multivariate Statistical Analyses and GIS-Based Geostatistical Modeling Techniques. *Journal of Hydrology: Regional Studies*, **4**, 80-110. http://dx.doi.org/10.1016/j.ejrh.2014.11.005

[3] Basavarajappa, H.T. and Manjunatha, M.C. (2015) Groundwater Quality Analysis in Precambrian Rocks of Chitradurga District, Karnataka, India Using Geo-Informatics Technique. *Aquatic Procedia*, **4**, 1354-1365. http://dx.doi.org/10.1016/j.aqpro.2015.02.176

[4] Verma, S., Mukherjee, A., Choudhury, R. and Mahanta, C. (2015) Brahmaputra River Basin Groundwater: Solute Distribution, Chemical Evolution and Arsenic Occurrences in Different Geomorphic Settings. *Journal of Hydrology: Regional Studies*, **4**, 131-153. http://dx.doi.org/10.1016/j.ejrh.2015.03.001

[5] Hallett, B.M., Dharmagunawardhane, H.A., Atal, S., Valsami-Jones, E., Ahmed, S. and Burgess, W.G. (2015) Mineralogical Sources of Groundwater Fluoride in Archaen Bedrock/Regolith Aquifers: Mass Balances from Southern India and North-Central Sri Lanka. *Journal of Hydrology: Regional Studies*, **4**, 111-130.

[6]   Banerjee, A. (2015) Groundwater Fluoride Contamination: A Reappraisal. *Geoscience Frontiers*, **6**, 277-284. http://dx.doi.org/10.1016/j.gsf.2014.03.003

[7]   Rosin, K.G., Kaur, R., Singh, S.D., Singh, P. and Dubey, D.S. (2013) Groundwater Vulnerability to Contaminated Irrigation Waters—A Case of Peri-Urban Agricultural Lands around an Industrial District of Haryana, India. *Procedia Environmental Sciences*, **18**, 200-210. http://dx.doi.org/10.1016/j.proenv.2013.04.026

[8]   Venkateswaran, S. and Deepa, S. (2015) Assessment of Groundwater Quality Using GIS Techniques in Vaniyar Watershed, Ponnaiyar River, Tamil Nadu. *Aquatic Procedia*, **4**, 1283-1290. http://dx.doi.org/10.1016/j.aqpro.2015.02.167

[9]   Adnan, S. and Iqbal, J. (2014) Spatial Analysis of the Groundwater Quality in the Peshawar District, Pakistan. *Procedia Engineering*, **70**, 14-22. http://dx.doi.org/10.1016/j.proeng.2014.02.003

[10]   Banerjee, S., Das, B., Umlong, I.M., Devi, R.R., Kalita, H., Saikia, L.B., Borah, K., Raul, P.K. and Singh, L. (2011) Heavy Metal Contaminants of Undergroundwater in Indo Bangla Border Districts of Tripura, India. *International Journal of ChemTech Research*, **3**, 516-522.
http://sphinxsai.com/Vol.3No.1/chem_jan-mar11/pdf/CT=80(516-522)%20JM11.pdf

[11]   Kumar, A., Narang, S., Mehra, R. and Singh, S. (2015) Assessment of Radon Concentration and Heavy Metal Contamination in Groundwater Samples from Some Areas of Fazilka District, Punjab, India. Indoor and Built Environment.

[12]   Borah, J. (2011) Monitoring Fluoride Concentration and Some Other Physico-Chemical Properties of Groundwater of Tinsukia District, Assam, India. *International Journal of ChemTech Research*, **3**, 1339-1342.

[13]   Singaraja, C., Chidambaram, S., Anandhan, P., Prasanna, M.V., Thivya, C., Thilagavathi, R. and Sarathidasan, J. (2014) Geochemical Evaluation of Fluoride Contamination of Groundwater in the Thoothukudi District of Tamilnadu, India. *Applied Water Science*, **4**, 241-250. http://dx.doi.org/10.1007/s13201-014-0157-y

[14]   Singaraja, C., Chidambaram, S., Anandhan, P., Prasanna, M.V., Thivya, C. and Thilagavathi, R. (2013) A Study on the Status of Fluoride Ion in Groundwater of Coastal Hard Rock Aquifers of South India. *Arabian Journal of Geosciences*, **6**, 4167-4177. http://dx.doi.org/10.1007/s12517-012-0675-6

[15]   Ghosh, S., Chakraborty, S., Roy, B., Banerjee, P. and Bagchi, A. (2010) Assessment of Health Risks Associated with Fluoride-Contaminated Groundwater in Birbhum District of West Bengal, India. *Journal of Environmental Protection Science*, **4**, 13-21. http://aes.asia.edu.tw/Issues/JEPS2010/GhoshS2010.pdf

[16]   APHA (2005) Standard Methods for the Examination of Water and Wastewater. 21st Edition, AWWA, WEF and APHA, Washington DC.

[17]   Nollet, L.M.L. and De Gelder, L.S.P. (2007) Handbook of Water Analysis, 2nd Edition, CRC Press, Boca Raton. https://www.crcpress.com/Handbook-of-Water-Analysis-Second-Edition/Nollet-De-Gelder/9780849370335

[18]   Shrestha, S. and Kazama, F. (2007) Assessment of Surface Water Quality Using Multivariate Statistical Techniques: A Case Study of the Fuji River Basin, Japan. *Environmental Modelling and Software*, **22**, 464-475. http://dx.doi.org/10.1016/j.envsoft.2006.02.001

[19]   BIS (2003) Indian Standard Drinking Water Specifications (IS 10500:1991), Ed. 2.2 (2003-2009), Bureau of Indian Standard, New Delhi.
http://www.indiawaterportal.org/sites/indiawaterportal.org/files/drinking_water_standards_is_10500_1991_bis.pdf

[20]   WHO (2011) Guidelines for Drinking Water Quality. 4th Edition, World Health Organization, Geneva. http://apps.who.int/iris/bitstream/10665/44584/1/9789241548151_eng.pdf

10

The Improvement of the Quality of Irrigation Water Contaminated with Heavy Metals in the Borg El Arab, Egypt

Mohamed Kamel Fattah[1], S. A. E. Abdelrazek[2]

[1]Environmental Studies and Researches Institute, Sadat City University, Sadat, Egypt
[2]Soil, Water and Environment Research Institute, Agriculture Research Center, Cairo, Egypt
Email: kamelhydro90@yahoo.com

Abstract

The aim of this study is to estimate the effects of the accumulation of harmful heavy metals in the irrigation water resulting from the proximity of the various activities of various water: sewage, artesian wells and industrial activities, particularly Borg El Arab Industrial area and take Nile water l Control (Bahig canal). The concentrations of these heavy metals in samples drawn from different distances from the source of irrigation in summer 2011 that contain the highest concentrations, followed by the winter of 2012. These results have shown that the amount of heavy metals in the samples at a distance of 50 meters is more than quantity of 5000 m by a 30 to 35 twice in water for irrigation, also the amount of heavy metals in the samples at a distance of 50 meters has values less than the permissible limits and disappeares completely at a distance of 5000 meters and that for different sources.

Keywords

Artificial Water, Sludge Water, Pollution, Heavy Metals, Water Quality Improving

1. Introduction

Survey study was carried out an area located at the satellite of Burg El-Arab. The studied area constitutes a part of the eastern section of the north-western coastal of A.R.E 48 km, and west of Alexandria-Marsa Matroh road it lies approximately between latitudes 30°45'N and 30°55'N and longitudes 29°30'E and 29°50'E, The study area covers the total area 5000 fed dens map. The area shows many differences in elevation and relief the elevation varies between zero and more than 40 M above sea level. The hydrochemical study reveals the quality of water that is suitable for drinking, agriculture and industrial purposes and helps in understanding the change in quality due to rock-water interaction or any type of anthropogenic influence. The rapid extension of industrial developed countries was accompanied by many problems of pollution especially due to wastewater discharged from factories. Heavy metals produced by different industrial processes are the most harmful pollutes in these wastes [1].

In general, wastewater could be divided into water as the bulk volume and three other categories included: suspended solids, colloidal materials, and dissolved materials. [2] showed that the potential problems due to the use of wastewater on cropland include ground and surface water contamination, pathogens, odors, and heavy metals entry into the food chain that industrial liquid wastes are more various and more concentrated and contain certain various acids, alkalis chemical contaminants oil, coarse, solids, and other constituents. The dissolved nutrients (phosphate, ammonium, nitrate, potassium, sodium, calcium, sulfates, etc). toxic wastes heavy metals are mostly from industry; Cu, Zn, Hg, Pb, Cd, Cr, Co, As etc.) and non biodegradable organic chemicals variation of groundwater quality in an area is a function of physical and chemical parameters that are greatly influenced by geological formations and anthropogenic activities [3].

In Egypt, many investigators warned of the residual effects of factories pollution of the biosphere [4].

2. Material and Methods

In winter 2012, the **Investors Union in Burg El-arab area** has been treated the Artificial water before mixing with Nile water and the same with Sludge water in mary mina station with **E-JUST Japans Tokyo Technology**, (E-JUST, Tokyo Institute of Technology, Report in heavy metals and salts treatment in Burg-El-Arab area 2012) The objective of this work was to investigate the contamination of water in the vicinity of factories region of Burg El-Arab by heavy metals and improving the quality of this irrigation water. Cetro-Cal compound from AGRI-CO reduced salinity and heavy metals in artesian water. Agricultural area around Burg-El Arab Artificial water, sludge water, artesian water and Nile water were selected for the current study (**Figure 1**).

In July 2011 and January 2012, samples of irrigation water samples were carried according to standard methods [5]. Total soluble salts pH by pH meter and EC according to [6] **Table 1**. The analysis chemical properties of the non polluted water (Bahig Canals) at (50 - 5000 m) and Polluted water (Artificial water, sludge water, artesian water) at (50 - 5000 m). The concentration of Fe, Cu, Zn, Mn, Pb, Cd, Co and Ni were determined by atomic absorption spectrophotometer Model 238 Perkin-Elme [7], BOD (biochemical oxygen demand) [8]-[11].

3. Results and Discussion

3.1. Major Ions Distribution

The results of chemical analyses of the collected water samples (**Table 1**) showed that, (Bahig (fresh water) calcium is the most dominant cation followed by sodium and magnesium ($Ca^{2+} > Na^+ > Mg^{2+}$). These elements show wide range of chemical concentration, calcium concentration ranges from 68 to 36 mg/l, sodium concentration ranges from 27 to 23 mg/l, and magnesium concentration ranges from 18 to 14 mg/l. Low concentrations represent low water salinity dominated at the Bahig area.

El-Rawaisate (artesian water), sodium is the most dominant cation followed by calcium and magnesium and ($Na^+ > Ca^{2+} > Mg^{2+}$). These elements show wide range of chemical concentration, sodium concentration ranges from 136 to 134 mg/l, calcium concentration ranges from 102 to 98 mg/l, and magnesium concentration ranges from 22 to 27 mg/l. Low concentrations represent low water salinity dominated at the El-Rawaisate area.

Mary Mina (sewage water), sodium is the most dominant cation followed by magnesium and calcium ($Na^+ > Mg^{2+} > Ca^{2+}$), These elements show relatively high values of chemical concentration, sodium concentration ranges from 301 to 439 mg/l, calcium concentration ranges from 156 to 148 mg/l, and magnesium concentration ranges from 88 to 99 mg/l. high concentrations represent high water salinity dominated at Mary Mina area. While some groundwater samples show increasing of calcium over magnesium ($Na^+ > Ca^{2+} > Mg^{2+}$) as a result of cation exchange processes

Burg-El-Arab Industrial Zone (artificial water), sodium is the most dominant cation followed by magnesium and calcium ($Na^+ > Mg^{2+} > Ca^{2+}$), These elements show relatively high values of chemical concentration, sodium concentration ranges from 441 to 382 mg/l, calcium concentration ranges from 150 to 144 mg/l, and magnesium concentration ranges from 86 to 73 mg/l. high concentrations represent high water salinity dominated at

Table 1. Chemical properties of irrigation water of BAHIG area (Nile water) and BURG-EL ARAB area (Artesian, Sludge[*] and Artificial water).

Distance (m)	Seasons	pH	E.C $dS \cdot m^{-1}$	Cation epm				Anion epm			SAR	Quality classes
				Na	K	Ca^2	Mg^2	HCO_3	Cl	SO_4^{2-}		
							Nile water					
50	S	7.9	0.81	1.19	0.13	3.40	1.5	2.81	2.91	0.50	0.76	C2-S1
	W	7.6	0.78	1.11	0.08	2.90	1.2	2.16	2.81	0.32	0.75	C2-S1
250	S	7.8	0.66	1.28	0.12	2.25	1.2	2.65	1.72	0.48	0.98	C2-S1
	W	7.5	0.61	1.12	0.09	2.12	1.1	2.42	1.61	0.40	0.89	C2-S1
500	S	7.9	0.91	1.23	0.14	2.15	1.2	2.31	1.81	0.60	0.95	C2-S1
	W	7.4	0.89	1.05	0.11	2.11	0.9	1.95	1.72	0.50	0.86	C2-S1
1000	S	7.8	0.66	1.52	0.21	1.52	1.5	2.81	1.42	0.52	1.24	C2-S2
	W	7.3	0.61	1.31	0.13	1.31	1.3	2.25	1.32	0.48	1.15	C2-S2
1500	S	7.7	0.72	1.45	0.23	2.77	1.5	2.61	2.88	0.46	0.99	C2-S1
	W	7.4	0.62	1.21	0.11	1.99	1.4	1.64	2.70	0.37	0.93	C2-S1
5000	S	7.5	0.69	1.24	0.29	1.82	1.6	2.62	1.91	0.45	1.04	C2-S2
	W	7.2	0.63	1.01	0.21	1.09	1.2	1.53	1.67	0.31	0.94	C2-S1
							Artesian water					
50	S	8.8	2.53	7.1	0.15	5.12	1.9	5.94	5.91	2.42	3.80	C3-S3
	W	8.2	1.96	6.35	0.06	4.99	1.4	4.47	5.42	1.91	3.55	C3-S4
250	S	8.5	2.92	9.5	0.29	7.84	3.8	2.50	14.9	4.03	3.86	C3-S3
	W	7.9	1.82	7.24	0.09	7.15	2.9	1.57	12.81	3.00	3.19	C3-S4
500	S	8.4	2.71	9.66	0.19	9.89	2.6	5.30	12.12	4.92	3.86	C3-S3
	W	7.9	1.61	8.15	0.11	8.96	2.0	4.98	10.33	3.91	3.48	C3-S4
1000	S	8.3	2.34	10.5	0.21	7.91	3.9	5.10	13.21	4.21	4.32	C3-S4
	W	7.8	1.72	8.41	0.09	7.82	2.8	3.93	11.51	3.68	3.68	C3-S4
1500	S	8.3	2.42	7.67	0.12	4.92	1.8	4.90	5.10	4.51	8.02	C3-S4
	W	8.0	1.36	9.92	0.08	4.81	1.1	4.65	9 .35	1.91	5.77	C3-S4
5000	S	8.2	2.31	7.10	0.13	4.91	2.3	5.80	6.20	2.44	3.76	C3-S3
	W	8.0	1.14	6.09	0.07	3.96	1.9	3.69	6.12	2.21	3.56	C2-S4
							Sludge water[*]					
50	S	7.3	3.80	13.10	0.31	7.81	7.3	4.32	14.9	9.30	4.76	C4-S4
	W	7.1	3.72	12.81	0.21	6.71	6.9	3.91	13.82	8.90	4.91	C4-S4
250	S	7.3	3.70	17.00	0.25	6.29	7.1	2.12	27.49	1.10	6.56	C4-S4
	W	7.0	3.69	14.12	0.12	6.11	6.8	1.94	24.30	1.00	5.59	C4-S4
500	S	6.8	3.70	18.17	0.29	7.08	7.1	4.47	24.67	3.50	6.83	C3-S4
	W	6.2	3.61	16.98	0.15	6.92	6.9	3.62	23.91	3.42	6.49	C4-S4
1000	S	6.5	3.5	19.00	0.25	7.36	6.3	4.21	28.00	0.70	7.28	C3-S4
	W	6.2	3.31	18.28	0.12	6.31	6.0	3.91	26.29	0.51	7.37	C3-S4
1500	S	7.2	3.5	19.0	0.28	7.40	6.9	4.79	28.60	2.19	7.12	C4-S4
	W	6.9	3.36	18.91	0.19	6.31	6.2	3.82	26.58	1.21	7.56	C3-S4
5000	S	6.9	3.60	19.0	0.25	7.41	8.3	3.21	27.80	3.95	6.79	C4-S4
	W	6.5	3.27	18.82	0.18	6.36	6.9	3.38	25.90	2.98	7.29	C3-S4

Continued

						Artificial water						
50	S	6.2	3.6	19.00	0.29	7.55	7.2	4.24	24.20	5.60	6.99	C4-S4
	W	6.0	3.41	18.69	0.25	6.99	6.8	4.18	23.50	5.05	7.11	C3-S4
250	S	6.1	3.6	19.00	0.25	8.33	9.4	3.98	23.50	9.50	6.38	C4-S4
	W	6.0	3.21	18.72	0.23	7.13	8.2	3.18	22.40	8.70	6.76	C3-S4
500	S	6.7	4.0	19.20	0.34	9.29	9.9	3.23	25.80	9.70	6.21	C4-S4
	W	6.4	3.92	18.89	0.28	7.11	8.4	2.19	23.70	8.79	6.79	C4-S4
1000	S	7.3	3.9	18.00	0.35	10.31	8.9	4.48	24.80	8.28	5.83	C4-S4
	W	7.1	3.81	17.01	0.29	9.21	8.5	3.51	23.71	7.79	5.71	C4-S4
1500	S	6.4	3.8	20.43	0.36	7.25	7.5	4.78	23.36	7.40	7.51	C4-S4
	W	6.2	3.79	19.61	0.21	6.11	6.4	3.81	22.21	6.31	7.84	C4-S4
5000	S	6.8	3.3	16.65	0.36	7.20	6.1	4.21	23.50	2.60	6.45	C3-S4
	W	6.5	3.49	15.42	0.29	6.36	5.9	3.41	22.53	2.30	6.22	C3-S4

[*]Refers to output of human uses; W: Winter; S: Summer; C1: Low salinity; C2: Medium salinity; C3: High salinity; C4: Very high salinity; S1: Low alkalinity; S2: Medium alkalinity; S3: High alkalinity; S4: Very high alkalinity; Epm: equivalents per million.

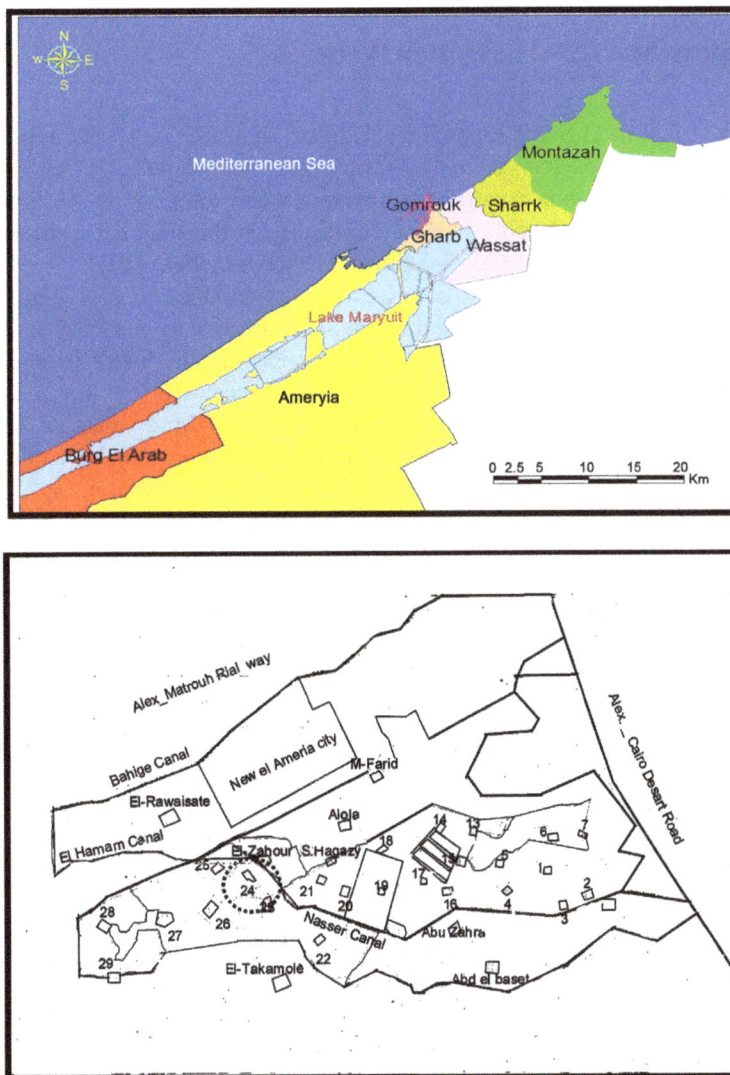

Figure 1. Location map of the study area.

Mary Mina area. While some groundwater samples show increasing of calcium over magnesium (Na$^+$ > Ca^{2+} > Mg^{2+}) as a result of cation exchange processes

Chemical distribution of anions shows that bicarbonate is the most dominant anion followed by chloride and sulphate anions of the irrigated water sample of low salinity contents (HCO$_3^-$ > Cl$^-$ > SO$_4^{2-}$), bicarbonate concentration ranges from 171 mg/L to 177 ppm, chloride concentration ranges from 103 mg/l to 67 mg/l, sulphate concentration ranges from 24 to 22 mg/l at Bahig (fresh water) area.

El-Rawaisate (artesian water), shows that chloride is the most dominant anion followed by bicarbonate and sulphate anions of the irrigated water sample of low salinity contents (Cl$^-$ > HCO$_3^-$ > SO$_4^{2-}$), chloride concentration ranges from 209 mg/l to 220 mg/l, bicarbonate concentration ranges from 170 mg/L to 158 ppm and sulphate concentration ranges from 116 to 117 mg/l.

Mary Mina (sewage water), (artesian water), shows that bicarbonate is the most dominant anion followed by chloride and sulphate anions of the irrigated water sample of low salinity contents (Cl$^-$ > SO$_4^{2-}$ > HCO$_3^-$), chloride concentration ranges from 859 mg/l to 834 mg/l, bicarbonate concentration ranges from 268 to 124 mg/L and sulphate concentration ranges from 116 to 115 mg/l

Burg-El-Arab Industrial Zone (artificial water), shows that bicarbonate is the most dominant anion followed by chloride and sulphate anions of the irrigated water sample of low salinity contents (HCO$_3^-$ > Cl$^-$ > SO$_4^{2-}$), bicarbonate concentration ranges from 360 mg/L to 378 mg/L, chloride concentration ranges from 209 mg/l to 220 mg/l, sulphate concentration ranges from 116 to 115 mg/l.

3.2. Geochemical Classification of Irrigation Water

Trilinear Diagram

The main purpose of the trilinear diagram proposed by [12] is to show different water types in a particular area. The diamond-shaped field of this diagram consists of two equal triangular fields. Generally, water which appears in **the upper triangle has secondary salinity properties**, where (SO$_4^{2-}$ + Cl$^-$) exceed (Na$^+$ + K$^+$) and the characteristic water types are Ca and Mg chlorides and sulphates. On the other hand, those which appear in **the lower triangle are considered to have primary alkalinity properties where** (CO$_3^{2-}$ + HCO$_3^-$) exceed (Ca^{2+} + Mg^{2+}), and the characteristic water types are Na and K carbonates and bicarbonates. **Figure 2** shows two trilinear plots of major ions in water samples collected from the study area in summer 2011 and winter 2012.

-Bahig area (fresh water) and El-Rawaisate area (artesian) appear in **the lower triangle are considered to have primary alkalinity properties where** (CO$_3^{2-}$ + HCO$_3^-$) exceed (Ca^{2+} + Mg^{2+}), and the characteristic water types are Na and K carbonates and bicarbonates. While, Burg-El-Arab Industrial Zone (artificial water), Mary Mina (**Sludge water**) samples appear in the upper triangle of the diamond-shaped field, the dominant water types are **Ca and Mg chlorides and sulphates**. Water in these areas enriched in Mg^{2+} than water in the Bahig area (fresh water) and El-Rawaisate area (artesian)) indicating its precipitation from water as it moves downgradient. This process is responsible for the dominance of Mg-rich carbonate deposits and cement within the Quaternary clastic sediments of the study area. These deposits affect ground-water quality and may create perched conditions in some areas.

3.3. Heavy Metals in Irrigation Water

Generally,in Burg–El-Arab enriched with Fe, followed by Mn, Pb, Zn, Cu, Co, Ni, and Cd. [13] fond that the soluble heavy metals (ppm) in wastewater were as follows: Fe (34.0 and 56.0) > Mn (9.60 and 17.90) > Zn (5.70 and 6.80) > Pb (5.70 and 4.30) > Cu (4.60 and 3.90) > Co (2.50 and 3.70) > Ni (2.50 and 1.90).Concentration of Fe, Mn, Zn, Cu, Pb, Cd, Co and Ni in water samples from Burg –El-Arab canals (**Table 2**), revealed that the factories liquid wastes, sludge , artesian discharged in irrigation water.

The concentration of metals decreased by getting far away from the sources of pollution and could be arranged, descending between distance at 50m (in summer 201) Nile water (Bahig canal) (control), as following: Fe (0.3 [1] 1.23) > Mn (0.06 - 0.65) > Zn (0.08 - 0.52) > Cu (0.0 [1] 0.16) > Pb, Cd, Co, Ni. Generally the liquid wastes were enriched with these metals and were much higher than the maximum concentration recommended by [14] for waste water for irrigation. Also, levels of the heavy metals in water samples were generally higher than which are recorded for fresh water in several studies, since the reported values (ppm) were in the range of 0.11 - 0.77 for Fe, 0.01 - 0.11 for Mn, 0.01 -0.09 for Zn, 0.0 [1] 0.12 for Cu, 0.01 - 0.06 for Pb, 0.02 - 0.04 for Cd, 0.0 [1] 0.06 for Coand 0.01 - 0.02 for Ni [15]-[18]. Stated that the concentrations of heavy metals in unpol-

Table 2. Heavy metals in irrigation water of samples collected at different distance from Bahig area (Nile water) and BURG-EL ARAB area (Artesian, Sludge* and Artificial water).

Distance (Meter)	Seasons	Fe	Mn	Zn	Cu	Pb	Cd	Co	Ni
					ppm (mg/L)				
Nile water									
50	S	1.23	0.65	0.52	0.16	Nd	Nd	Nd	Nd
	W	1.23	0.71	0.52	0.16	Nd	Nd	Nd	Nd
250	S	0.71	0.43	0.31	0.07	Nd	Nd	Nd	Nd
	W	0.93	0.61	0.31	0.07	Nd	Nd	Nd	Nd
500	S	0.62	0.25	0.21	0.09	Nd	Nd	Nd	Nd
	W	0.73	0.26	0.21	0.09	Nd	Nd	Nd	Nd
1000	S	0.58	0.07	0.06	0.07	Nd	Nd	Nd	Nd
	W	0.71	0.02	0.06	0.07	Nd	Nd	Nd	Nd
1500	S	0.44	0.09	0.05	0.02	Nd	Nd	Nd	Nd
	W	0.61	0.01	0.05	0.02	Nd	Nd	Nd	Nd
5000	S	0.31	0.06	0.08	0.01	Nd	Nd	Nd	Nd
	W	0.21	0.01	0.08	0.01	Nd	Nd	Nd	Nd
Artesia									
50	S	2.81	0.95	0.90	0.31	Nd	Nd	Nd	Nd
	W	2.82	0.91	1.0	0.35	Nd	Nd	Nd	Nd
250	S	0.91	0.58	0.20	0.21	Nd	Nd	Nd	Nd
	W	1.66	0.81	0.26	0.24	Nd	Nd	Nd	Nd
500	S	0.82	0.35	0.16	0.11	Nd	Nd	Nd	Nd
	W	0.93	0.36	0.17	0.16	Nd	Nd	Nd	Nd
1000	S	0.78	0.17	0.90	Nd	Nd	Nd	Nd	Nd
	W	0.81	0.12	Nd	Nd	Nd	Nd	Nd	Nd
1500	S	0.74	0.09	0.7	Nd	Nd	Nd	Nd	Nd
	W	0.91	0.8	Nd	Nd	Nd	Nd	Nd	Nd
5000	S	0.17	0.06	Nd	Nd	Nd	Nd	Nd	Nd
	W	0.21	0.06	Nd	Nd	Nd	Nd	Nd	Nd
Sludge water*									
50	S	3.69	1.32	1.15	0.64	0.62	0.46	0.3	0.82
	W	3.77	1.36	1.19	0.68	0.66	0.61	1.20	0.92
250	S	1.81	0.68	0.49	0.38	0.34	0.30	0.22	0.47
	W	1.86	0.71	0.49	0.41	0.39	0.34	0.32	0.28
500	S	1.31	0.45	0.33	0.27	0.26	0.24	0.15	0.47
	W	1.36	0.46	0.34	0.32	0.27	0.26	0.27	0.50
1000	S	1.17	0.27	0.23	0.07	0.17	0.14	0.10	0.47
	W	1.13	0.32	0.23	0.12	0.21	0.16	0.17	0.32
1500	S	0.83	0.17	0.16	0.08	0.16	0.08	0.04	0.32
	W	0.91	0.18	0.16	0.09	0.23	0.09	0.19	0.10
5000	S	0.17	0.09	0.08	0.04	0.04	0.3	.0.01	0.09
	W	0.21	0.09	0.10	0.05	0.04	0.06	0.06	0.06
Artificial water									
50	S	7.38	2.63	2.29	1.28	1.24	0.91	0.54	1.56
	W	7.53	2.72	2.38	1.36	1.32	1.21	0.95	1.68

Continued

250	S	3.61	1.35	0.97	0.75	0.68	0.59	0.44	1.09
	W	3.72	1.42	0.98	0.81	0.78	0.68	0.61	0.95
500	S	2.61	0.90	0.66	0.53	0.51	0.48	0.29	0.90
	W	2.72	0.92	0.67	0.64	0.54	0.52	0.49	0.95
1000	S	2.15	0.53	0.45	0.31	0.33	0.27	0.18	0.59
	W	2.26	0.64	0.46	0.23	0.42	0.31	0.31	0.61
1500	S	1.60	0.33	0.31	0.16	0.23	0.15	0.07	0.17
	W	1.89	0.35	0.32	0.18	0.45	0.18	0.19	0.19
5000	S	0.35	0.17	0.15	0.08	0.07	0.06	0.06	0.09
	W	0.37	0.18	0.19	0.09	0.08	0.12	0.09	0.12
R. g for water irrigation agriculture (FOW,1985)		5.0	0.30	2.0	0.2	-	5.0	0.05	0.20
L. S. D		0.146	0.102	0.056	0.41	0.07	0.04	0.02	0.01

S: summer; W: winter; ppm: part per million.

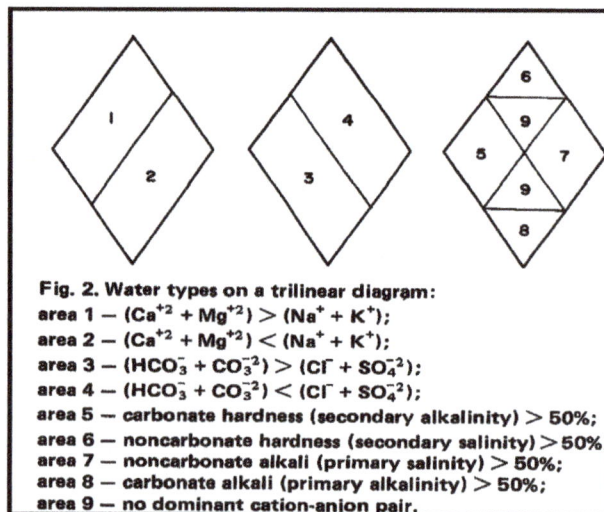

Fig. 2. Water types on a trilinear diagram:
area 1 — $(Ca^{+2} + Mg^{+2}) > (Na^+ + K^+)$;
area 2 — $(Ca^{+2} + Mg^{+2}) < (Na^+ + K^+)$;
area 3 — $(HCO_3^- + CO_3^{-2}) > (Cl^- + SO_4^{-2})$;
area 4 — $(HCO_3^- + CO_3^{-2}) < (Cl^- + SO_4^{-2})$;
area 5 — carbonate hardness (secondary alkalinity) > 50%;
area 6 — noncarbonate hardness (secondary salinity) > 50%;
area 7 — noncarbonate alkali (primary salinity) > 50%;
area 8 — carbonate alkali (primary alkalinity) > 50%;
area 9 — no dominant cation-anion pair.

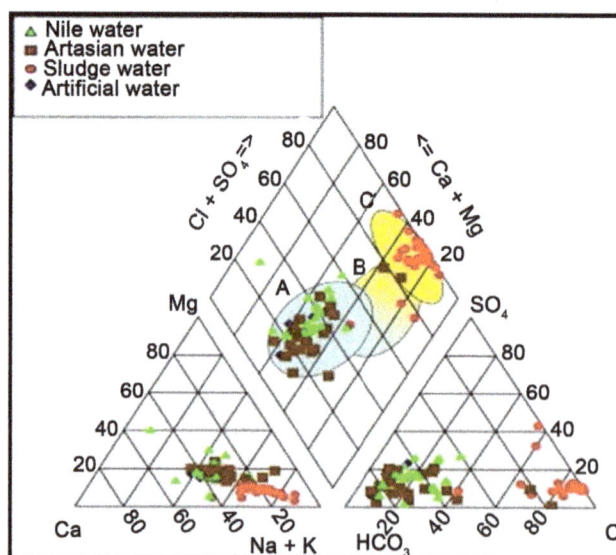

Figure 2. Piper's trilinear diagram.

luted Nile water were as follow in ppm :Fe 0.77 > Mn 0.11 > Zn 0.09 > Cu0.01 and Pb 0.01. Also, concentrations of heavy metals (ppm) in industrial wastewater were as follow: Fe 34.40 > Pb 3.80 > Mn 2.50 > Co 1.50 > Zn 0.69 > Cu 0.47 **Table 2**.

[19] found that the superphosphate fertilizer contained varying amounts of heavy metals particularly Cd, Pb and Ni and these concentrations (ug/g) were 23.1 Cd, 35.0 Pb and 60.0 Ni **Table 2**.

Superphosphate, pesticides, sludge, brick factories were the main sources of pollution by heavy metals such as Fe, Mn, Zn, Cu, Pb, Cd, Co and Ni [16] [20]-[23].

3.4. BOD$_5$ and Turbidity in Irrigation Water

[8] [9] found that BOD (biochemical oxygen demand) this test measures the oxygen required by microorganisms during the degradation of water sample increasing by increasing in sewage sludge In summer July 2011 (50 m – 5000 m) the Turbidity in Industrial water (1286 TU-1246 TU) and domestic wastes (1487 TU - 1446 TU) contribute with both inorganic and organic materials **Table 3**. Practically, all public water supplies are free from noticeable turbidity [10] and [11]. It is the clear from the present study the waste water irrigated decreased with treatment the artificial water before mixed with Nile water and the same with the sludge water collected in Mary mina station from **Tokyo Tech** and used **Cetro-cal** compound from **Agrico.co** reduced salinity and heavy metals content in artesian water.

[2] elevated levels of metals in water may be toxic they enter the food chain and could be harmful to animal and human being.

3.5. Assessment of the Environmental Impacts Resulting from Irrigation by Different Types of Water on Soil and Planting:

Irrigation water in the study area including Burg-El-Arab (Industrial water), Mary Mina (sewage water) , Bahig (fresh water) and El-Rawaisate (artesian water). According to U.S Salinity Laboratory [24] based on the Sodium Adsorption Ratio (SAR), and the specific conductance. Generally, irrigation water with low SAR is much desirable. The suitability of groundwater for agricultural purposes depends on the effect of mineral constituents of water on both plants and soil. Effects of salts on soils causing changes in soil structure, permeability and aeration indirectly affect plant growth. [25] and US Salinity Laboratory [24] proposed irrigational specifications for evaluating the suitability of water for irrigation use. There is a significant relationship between sodium adsorption ratio (SAR) values for irrigation water and the extent to which sodium is adsorbed by the soils. If water used for irrigation is high in sodium and low in calcium, the cation exchange complex may become saturated with sodium, which can destroy the soil structure owing to dispersion of clay particles [26]. SAR was computed using the equation given below [27]

$$SAR = \frac{Na}{\sqrt{(Ca + Mg)/2}}$$

where, the concentration of these cations is expressed in e.p.m. According to the U.S. Salinity Laboratory diagram, the water is divided into classes C1, C2, C3, and C4, which denote the conductance and S1, S2, S3, and S4, which denote SAR (table). The recommended classification of water for irrigation according to SAR ranges are shown in **Table 4**. If the SAR value is greater than 6 - 9, the irrigation water will cause permeability problems on shrinking and swelling of clayey soils types [28].

As shown from **Figure 3** as well as **Table 4**, and according to the foregoing discussion, it clears that:

-**Nile water** samples falls in the good water class (C2-S1, S2) where C2 reflects good water for soil of medium permeability for most plants, and S1, S2 reflect good for coarse grained permeability soils, unsatisfactory for highly clayey soil with low leaching.

-**Artesian water** falls in the moderate water class (C3-S3) where medium to highly saline water and highly sodium content. C3 reflects satisfactory for plants, having a moderate salts tolerance on soils of moderate permeability with leaching, S3 means suitable only with good drainage, high leaching and organic matter addition some chemical additives.

-**Sludge water and artificial water** fall in the bad water class C4-S4, where C4 reflects highly saline water, having satisfactory for salt tolerant crops on soils of good permeability with special leaching. S4 reflects very highly sodium and unsatisfactory.

Table 3. BOD and Total Suspended Mater in irrigation water of samples collected at different distance from BAHIG area (Nile water) and BURG-EL ARAB area (Artesian, Sludge[*] and Artificial water).

Distance (m)	Seasons	BOD mg/L	Total suspended mater TU (mg/L)
		Nile water	
50	S	7.5	151
	W	7.2	149
250	S	7.6	149
	W	7.3	146
500	S	7.6	152
	W	7.4	150
1000	S	7.6	152
	W	7.3	148
1500	S	7.5	150
	W	7.2	147
5000	S	7.8	153
	W	7.3	148
		Artesian water	
50	S	6.5	142
	W	6.1	140
250	S	6.5	150
	W	6.1	147
500	S	6.4	145
	W	6.2	143
1000	S	6.5	142
	W	6.2	140
1500	S	6.5	152
	W	6.3	146
5000	S	6.8	146
	W	6.5	142
		Sludge water[*]	
50	S	563.7	1487
	W	561.2	1482
250	S	546.5	1483
	W	542.2	1480

Continued

500	S	463.3	1389
	W	461.2	1386
1000	S	578.3	1487
	W	576.1	1485
1500	S	563.2	1453
	W	560.1	1450
5000	S	563.8	1446
		562.4	1442
Artificial water			
50	S	283.4	1286
	W	280.2	1284
250	S	279.8	1278
	W	276.3	1273
500	S	299.8	1280
	W	296.5	1276
1000	S	264.2	1269
	W	262.1	1263
1500	S	246.3	1285
	W	242.1	1282
5000	S	268.8	1246
	W	265.6	1243

S: summer; W: winter; mg/L: milligrams per liter; TU: turbidity.

Table 4. Classification and description of conductivity and sodium.

C	Degree	Description
C1	Low salinity water	Good
C2	Moderate to saline water	Good for soil of medium permeability for most plants.
C3	Medium to highly saline water	Satisfactory for plants, having a moderate salts tolerance on soils of moderate permeability with leaching.
C4	Highly saline water	Satisfactory for salt tolerant crops on soils of good permeability with special leaching.
S	Degree	Description
S1	Low sodium water	Good
S2	Medium sodium water	Good for coarse grained permeability soils, unsatisfactory for highly clayey soil with low leaching.
S3	Highly sodium water	Suitable only with good drainage, high leaching and organic matter addition some chemical additives.
S4	Very highly sodium	Unsatisfactory.

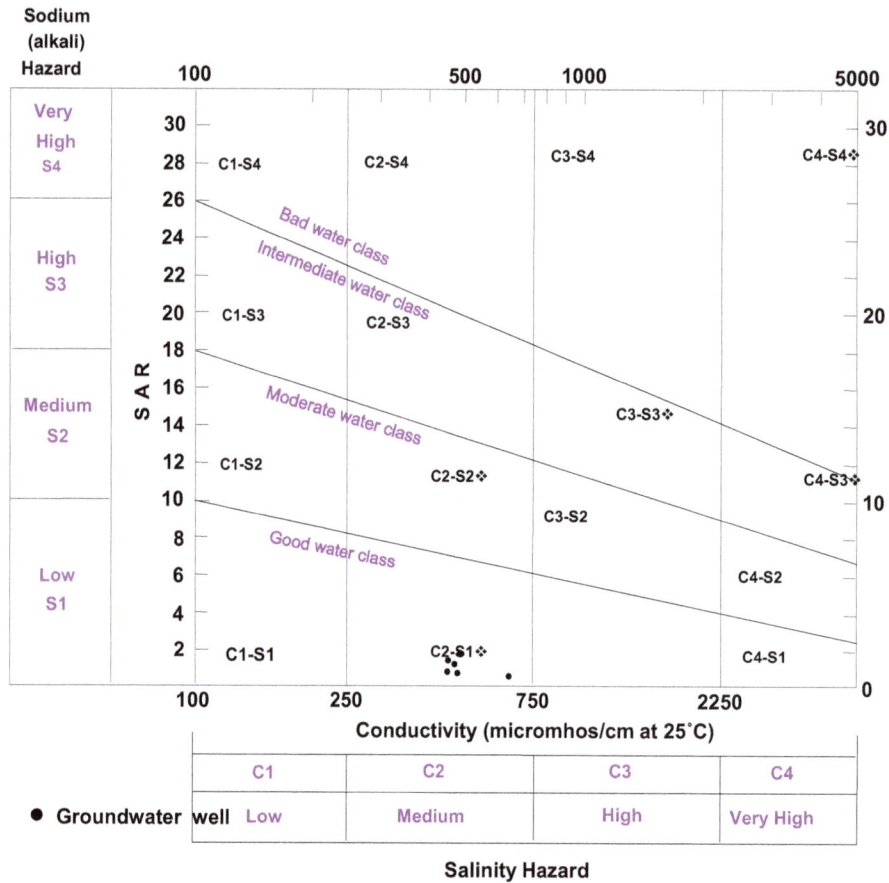

Figure 3. US salinity diagram for classification of irrigation waters summer season.

The non desirable effects resulting from irrigate by Sludge water and artificial water leading to excess concentration of sodium in irrigation water, where sodium ions tend to be absorbed by clay particles, displacing Mg and Ca ions. This exchange process of Na in water for Ca and Mg in soil reduces the permeability and eventually results in soil with poor internal drainage. Hence, air and water circulation is restricted during wet conditions and such soils are usually hard when dry. The irrigation water will cause permeability problems on shrinking and swelling of clayey soils types [28]-[31].

4. Conclusion and Recommendation

Obtained result indicated that samples of water of 2011 summer and winter of 2012 being at the least indicating that a pronounced accumulation of heavy metals were established. Those concentrations, however, had decreased by avoiding the pollution source; where samples collected at 5000 m from the source contained the lowest of heavy metals concentration. The results displayed that heavy metals contents in samples at 50 m distant from the sources of pollution exceeded that of the control 5000 m by about 30 to 35 time in irrigation water which used for grown plants in 5000 fedden. Generally, this work may express a vital point in heavy metals accumulation in the environment specially water system.

References

[1] Berrow, M.L and Reeves, G.A. (1984) Background Levels of Trace Elements in Soil and Water. International Conference, London.

[2] CAST (Council for Agricultural and Technology) (1976) Application of Sewage Sludge to Cropland; Appraisal of Potential Hazards of the Heavy Metals to Plants and Animals. Report No. 64. Coun. For Agric. Sci. and Tech., Ames, Iowa, 63 p.

[3] Belkhiri, L., Boudoukha, A., Mouni, L. and Baouz, T. (2010) Application of Multivariate Statistical Methods and Inverse Geochemical Modeling for Characterization of Groundwater—A Case Study: Ain Azel Plain (Algeria). *Geoderma*, **159**, 390-398. http://dx.doi.org/10.1016/j.geoderma.2010.08.016

[4] Abdel-Tawab, M.M. (1985) Soil Pollution as Effected by Some Industrial Waste at Helwan. El-Saff area. M.Sc. Thesis, Faculty of Agriculture, Cairo University, Giza.

[5] APHA (American Public Health Association) (1992) Stander Methods of Examination of Water and Waste Water. 18th Edition, American Public Health Association, Washington DC.

[6] Jackson, M.L. (1973) Soil Chemical Analysis. Advanced Course Ed. 2. A Manual of Methods Useful for Instruction and Research in Soil Chemistry, Physical Chemistry of Soils, Soil Fertility and Genesis Revised from Original Edition of 1965.

[7] Page, A.L., Miller, R.H. and Keeney (1982) Method of Soil Analysis. Part 2, ASA, SSSA, Madison.

[8] Mitchell, R. (1972) Water Pollution Microbiology. Wiley-Interscience, New York.

[9] Geradi, M.H. (1982) BOD Analysis: A Brief Review. Water Poll Control Association, PA Mag, 12.

[10] Sawyer, C.N. and McCarty, P.L. (1967) Chemistry for Sanitary Engineers. McGraw-Hill Book Co., New York.

[11] Holden, W.S. (1970) Water Treatment and Examination. Churchill, London.

[12] Piper, A.M. (1953) A Graphic Procedure in the Geochemical Interpretation of Water Analyses. *Transactions, American Geophysical Union*, **25**, 914-928.

[13] Shalaby, M.H., Gobran, O.A. and Raslan, M.I. (1996) Chemical Properties of Soils as Affected by Pollution of Different Wastes. *Egyptian Journal of Soil Science*, **36**, 23.

[14] FAO (1985) Water Quality for Agriculture, Irrigation Drainage. FAO, Roma.

[15] Eissa, A.M. and El-Kassas, H.I. (1999) Impact of Heavy Metals on Soil, Plant and Water at Abou-Zaable Area. *Egyptian Journal of Soil Science*, **9**, 351.

[16] Ahmed, S.M. (2001) Phytoremidiation of Some Contaminated Water with Heavy Metals in the Kafer El-Dawar Industrial Area. Ph.D. Thesis, Faculty of Agriculture, University of Alexandria, Alexandria.

[17] Abou El-Naga, S.A., El-Shinnawi, M.S., El-Swaby, M.S. and Salem, A.A. (1996) Changes in Elemental Constituents of Soil and Plants under Irrigation with Wastewater. *Menofiya Journal of Agricultural Research*, **21**, 1575.

[18] Ramadan, M.A. (1995) Studied on the Pollution of Agricultural Environment in Egypt. Ph.D. Thesis, Faculty of Agriculture, Cairo University, Cairo.

[19] Peeezzarossa, B., Lubranal, Petruzzelli, G. and Toognoni (1991) The Effect of Cd Contents on Tomato Plants. *Water, Air, & Soil Pollution*, **57**, 589.

[20] Peeezzarossa, B., Petruzzelli, G., Matorgio, F. and Tognoni (1993) Effect of Reported Phosphate Fertilization on Heavy Metals Accumulation in Soil and Plants under Protected Cultivation. *Communications in Soil Science and Plant Analysis*, **24**, 8.

[21] El-Sebaey, M.M. (1995) Studied on Chemical Pollution of Deferent Waters with Some Heavy Metals in Fayoum Governorate. Master's Thesis, Faculty of Agriculture at Moshtohor, Zagazig University, Zagazig.

[22] Omran, M.S., Fayiad, M.N. and El-Shikha, S.H. (1996) Effect of Industrial Activities on Pollution of Egyptian Soils. *Proceeding of 6th International Conference on "Environmental Protection Is a Must"*, Alexandria, 21-23 May 1996, 658-671.

[23] USDA (1997) Development of Phosphate Resource. In: *Final Environmental Impact Statement*, S.E. Ldohoi, 40 p.

[24] WHO (2006) Guidelines for Drinking-Water Quality. Recommendations, Volume 1, 3rd Edition, World Health Organization, Geneva.

[25] Wicklander, L. (1964) Chapter 4: Cation and Anion Phenomena. In: Bear, F.E., Ed., *Chemistry of the Soil*, Reinhold Publishing Co. PVT. LTD., New Delhi, 163-205.

[26] Singh, S. and Kumar, M. (2006) Heavy Metal Load of Soil, Water and Vegetables in Peri-Urban Delhi. *Environmental Monitoring and Assessment*, **120**, 79-91. http://dx.doi.org/10.1007/s10661-005-9050-3

[27] Richards, R.L., Ed. (1954) Diagnosis and Improvement of Saline and Alkali Soils. Agriculture Hand Book No. 60, US Government Printing Office, Washington DC.

[28] Saleh, A., Al-Ruwih, F. and Shehata, M. (1999) Hydrogeochemical Processes Operating within the Main Aquifers of Kuwait. *Journal of Arid Environments*, **42**, 195-209. http://dx.doi.org/10.1006/jare.1999.0511

[29] Collins, R. and Jenkins, A. (1996) The Impact of Agricultural Land Use on Stream Chemistry in the Middle Hills of the Himalayas, Nepal. *Journal of Hydrology*, **185**, 71-86. http://dx.doi.org/10.1016/0022-1694(95)03008-5

[30] Scott, C.A., Faruqui, N.I. and Raschid-Sally, L. (2004) Wastewater Use in Irrigated Agriculture: Confronting the Live-

lihood and Environmental Realities. CABI Publishing, Wallingford.
http://dx.doi.org/10.1079/9780851998237.0000

[31] Seckler, D., Amarasinghe, U., Molden, D., de Silva, R. and Barker, R. (1998) World Water Demand and Supply, 1990 to 2025: Scenarios and Issues. Research Report 19. International Water Management Institute, Colombo.

11

Irrigation Demand VS Supply-Remote Sensing and GIS Approach

Ch. Ramesh Naidu[1], M. V. S. S. Giridhar[2]

[1]GVP College of Engineering (Autonomous), Visakhapatnam, India
[2]Jawaharlal Nehru Technological University, Hyderabad, India
Email: chrnaidu@rediffmail.com, mvssgiridhar@gmail.com

Abstract

To determine the irrigation requirements of rice crop on different soils, an integrated approach is used using Remote Sensing and GIS techniques. Depending on the type of soil, climate and the crop acreage, the water requirement for paddy fields is derived. This study is focused on estimating the water demand for rice crop in Rabi season. Crop evapo-transpiration and soil percolation losses account more in rice fields especially in hot climate like Rabi season. In addition to evapo-transpiration and percolation losses, the conveyance losses are also accountable in the case of unlined canals. Satellite data is used to estimate the rice and fallow lands. In conjunction to satellite interpreted data, climate and soil data are also integrated in GIS platform. CROPWAT model is used to determine the crop evapo-transpiration (ET_c). There are 11 Water User's Associations (WUA) in the command area and under which 13 canal blocks are delineated. These blocks are again delineated in to 212 sub blocks. This study indicates that there exists a 5% to 20% of water deficiency in some WUAs and also water surplus in some WUAs varying from 15% to 40%.

Keywords

GIS, Water Users Association, Evapo-Transpiration, Remote Sensing, Irrigation

1. Introduction

The estimation of irrigation demand [1] [2] is an important component for managing the water effectively in the canal command area. To meet the present water demand, irrigation management plays an important role. One of the reasons for low efficiency is the un-accounted volume of water loss due to percolation in the rice fields and conveyance and seepage losses occurred during the water distribution through canals to fields [3] [4]. These losses are accounts in calculating total irrigation demand in addition to crop evapo-transpiration. Traditional way of calculating irrigation demand and the absence of scientific approach and adoption of new technologies is another reason for low irrigation efficiency. The selected study area lacks complete information about the command area including its water resources, distribution system details, land use, cropping patterns, soil, geology, climate and socio economic factors and also lacking regular evaluation and working knowledge regarding proper water management. This can be attributed to the dearth of spatially related information available regarding the canal network system and command area. The hard copy printed maps which are currently available are not in scale followed by lack of integration between the maps for the purpose of evaluation. The water allocation in command area is marked by blocks and sub-blocks. These blocks and sub-blocks come under the purview of Water Users' Association (WUA). Proper delineation of block boundaries is essential in order to evaluate water allocation and its usage in terms of actual or theoretical water demand. GIS technology [5] in conjunction with Remote Sensing [6] has proved to be effective for land use and water management. For effective decision making on water use, stresses the need of generation of spatial and non-spatial database in GIS platform [7]. Multispectral and multi-temporal remote sensing data are very useful as input for finding out rice crop acreage and with this, cartographic and data overlaying capability of GIS coupled with its dynamic linking ability plays a vital role in decision making process. Resulting disparities are existed in the availability of water between head-reach and tail end farms and between large and small farms that reflect some portions of the command area posing water shortage and water logging problems due to under and over utilization of canal water. This demands the calculation and analysis of irrigation water requirements of the command area in detail up to water users association, canal block level by using the latest available technologies for optimum utilization of irrigation water.

2. Materials and Methods

The selected study area is under block number five left main canal in Nagarjuna Sagar Project [8]. The area is located between 16°39'2.84" and 16°56'40.81" N latitude and 79°25'16.01" and 79°40'52.90" E longitude. The study area is bounded by Musi River in the East, Krishna River in the South, Lalbahadur canal in the North and Tungapadu vagu in the West. In the command area reserved forest occupies 18 to 20 percent. The command area extends to a total of 26,725 ha, distributed across three mandals namely Mirialguda, Dameracherla and Tripuraram. The methodology adopted for the study involves the use of satellite imagery in conjunction with soil, crop and climate data. Water losses from the field due to seepage and percolation, evapo-transpiration and conveyance losses were estimated. The crop information and statistical reports extracted from satellite imagery were integrated to crop, soil and climate data and the results were analyzed on GIS platform. The flowchart in **Figure 1** depicts overall methodology.

Since rice is dominant crop in the study area, multi-spectral and multi-temporal satellite imagery on the basis of crop calendar for Rabi season were procured from National Remote Sensing Centre (NRSC), Hyderabad, India. Imageries in the initial crop development stage and the late mid season stage were selected for the present study. This includes IRS P6 LISS III [9] [10] in the month of February and March. Survey of India (SOI) topomaps in 1:25,000 scale were collected to use as a reference. WUA maps were collected from the Walamtari, Hyderabad and integrated to canal blocks by setting the common projection and converted to GIS format. **Figure 2** and **Figure 3** shows the maps in GIS platform depicting WUAs in command area and canal distribution on different soils. Collateral data such as soil data, daily meteorological data, and climate and rainfall data were collected from different sources to study the parameters of evapo-transpiration, percolation/seepage loss and conveyance loss. Daily meteorological data was collected from the nearest and most representative meteorological station. Mandal wise daily rainfall data was collected from the District Information Center. Some essential information were collected from the field that includes crop and crop variety, first and last planting date and first and last harvesting date. The satellite imageries after being geo-referenced with respect to SOI topo maps were digitally enhanced to extract appropriate information for rice crop acreage estimates. The crop acreage reports were then generated WUA wise in GIS platform to find out the water demands for each WUA. For this study CROPWAT version 8.0 [11] [12] was used. ETc and Net Irrigation Water Requirements (NIWR) were estimated by using CropPWAT Software. The inputs used at various stages of the software are rainfall, temperature, wind speed, sunshine, sowing and harvesting dates of Rabi, crop factors, rooting depth and soil characteristics of the study area [13] [14]. The effective rainfall and the crop water requirement decide the amount of irrigation water that has to be applied. The effective rainfall is subtracted from the crop water requirement to calculate the net irrigation water requirements (NIWR). This formula is taken into consideration while running the CROP-WAT model using the daily rainfall data in the program. The net irrigation water requirements have been computed on daily and 10-day basis for each WUA and block/canal [15]. Calculation of the NIWR is carried out by pooling up sequentially the appropriate climate and rainfall data sets together with the crop files and the corresponding sowing/planting dates. In case of NIWR calculation of rice, soil data is also required. NIWR is determined by using the formula: NIWR = ETc − Re. Re is the effective rainfall. The Gross Irrigation Water Requirements (GIWR) [16] required for irrigation gift for the entire command area is equal to the sum of the net irrigation water requirements, water losses due to percolation and seepage and canal conveyance losses.

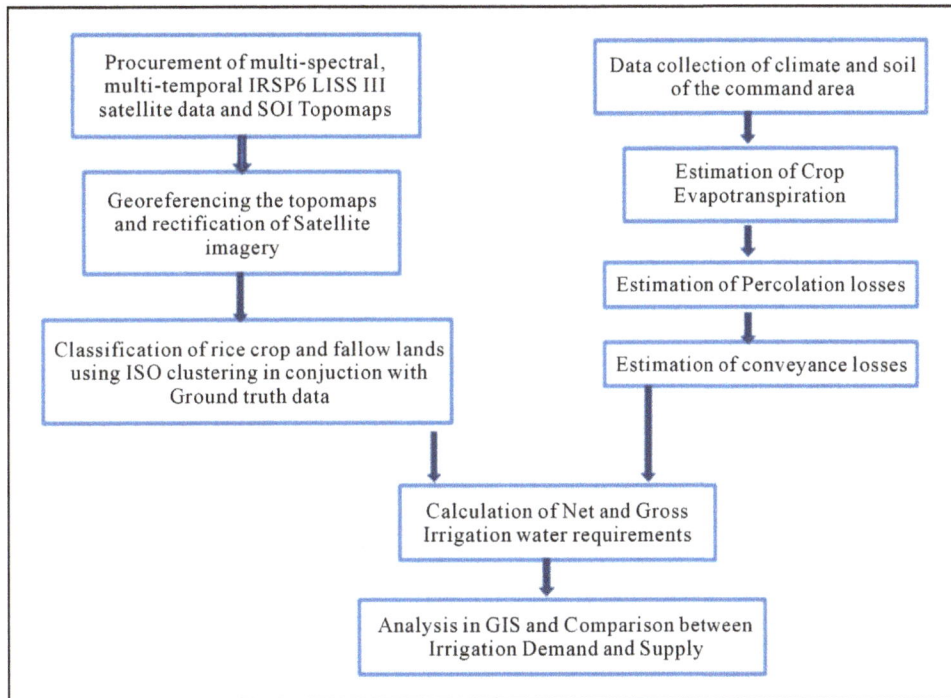

Figure 1. Flow Chart showing the study methodology.

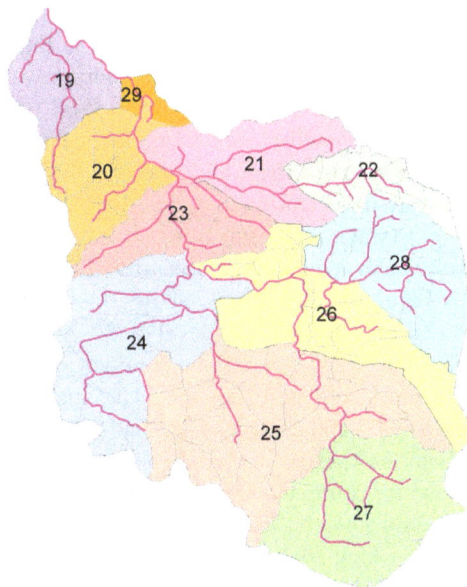

Figure 2. WUA boundaries of command area.

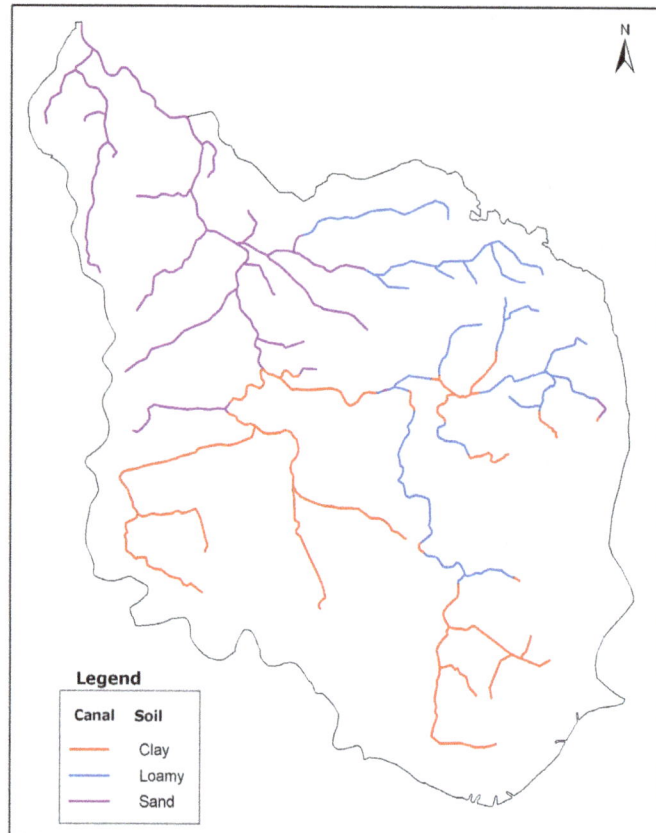

Figure 3. Canals distributed on different soils.

3. Discussion and Results

From the results, it was observed that for WUAs of Chillapuram, Venkatadripalem, Kothagudem, Mirialguda and Chinthapalle, the percolation losses accounts more than 10 million litres per ha. Where as for WUAs of Appalammagudem and Kalleyapalle the percolation losses are 5 million litres per ha. For Kondrapole and Kesavapuram WUAs, the percolation losses accounts 7 to 8 million litres per ha. The conveyance losses calculated from GIS ranging from 1 to 25 MLD (million litres per day) per km. **Figure 4** and **Figure 5** shows WUA wise irrigation demand in the year 2007 and 2009. The percolation losses were calculated using soil, crop, WUA, canal/block themes in the GIS environment using GIS software. Soil type and crop acreage under each command area unit were taken as main parameters for calculation of percolation losses.

The gross irrigation water requirements are compared with the irrigation supplies at WUA level. The study indicates that there is a deficit of water for irrigation in some canal blocks/WUAs, where in water deficiency varies from 5 to 20 percent of total crop water requirement. In some canal blocks and WUAs, water surplus varies from 15 to 40 percent. The data obviously indicates the necessity of additional canal supplies to some blocks and less canal supply to some blocks. **Figure 6** and **Figure 7** shows the WUA wise irrigation demand versus supply for the year 2007 and 2009. It is observed that at block level or WUA level, the water requirements are not only based on crop extent and also depends on the soil and meteorological conditions of the area. The net irrigation water requirements (NIWR) in the years 2007 and 2009 are 54125 and 85208 million litres. The gross water requirements calculated for the years 2007 and 2009 is 4.875 and 7.25 thousand million cubic feet (TMC). The supply to the command area in both the years is 4.35 and 5.49 thousand million cubic feet (TMC). It indicates the supply is lower than the demand for command area. The study found that there is an excess release of water in the year 2007 for Chillapuram, Kothagudem, Kondrapole, Dameracherla and Kesavapuram WUAs and deficiency of water in Venkatadripalem, Chinthapalle, Borrayapalem, Appalammagudem and Kalleyapalle WUAs. Whereas in the year 2009, the excess water is released for Dameracherla and Mirialguda WUAs and remaining WUAs have deficiency in supply of water to fields.

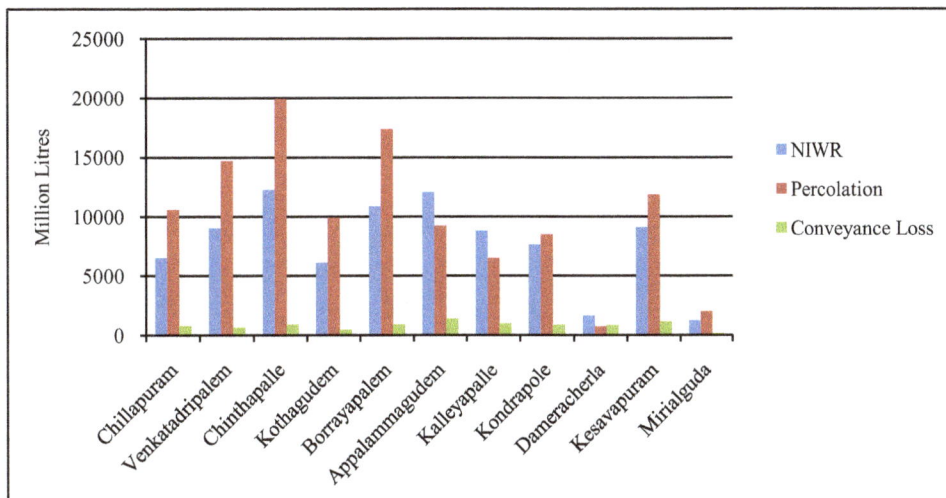

Figure 4. WUA wise irrigation demand in the year 2007.

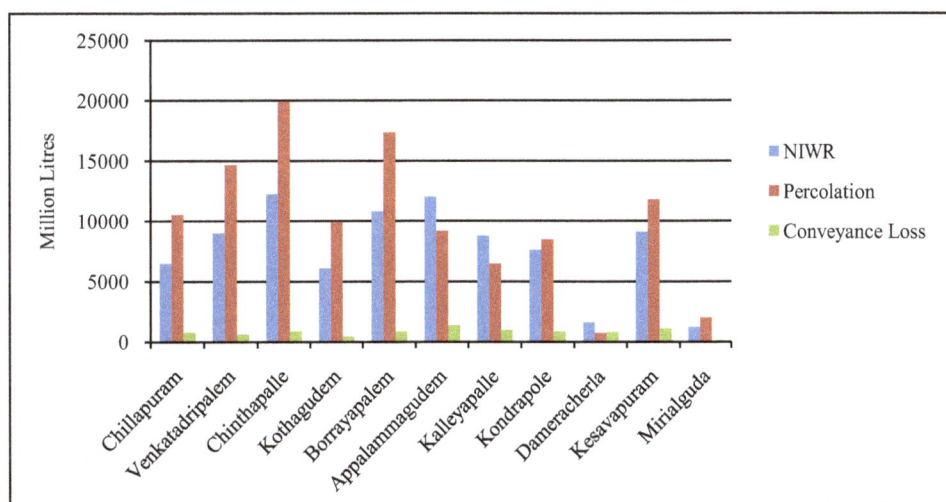

Figure 5. WUA wise irrigation demand in the year 2009.

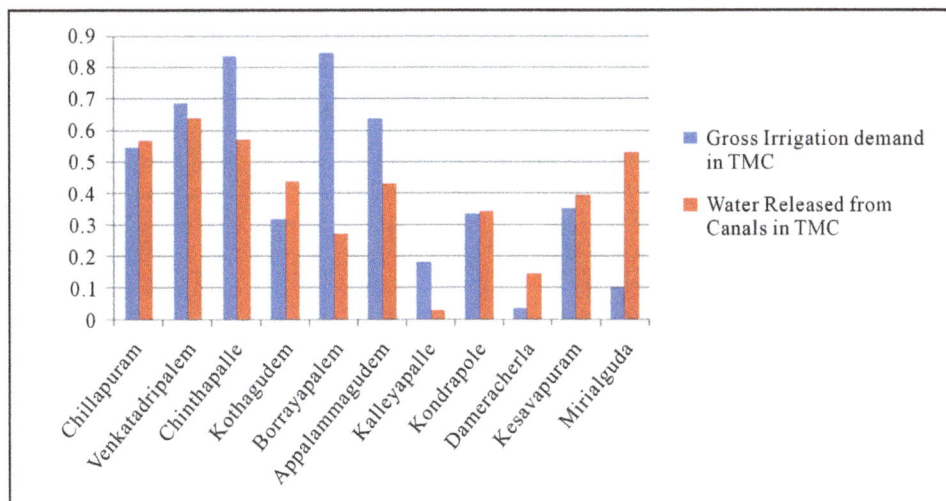

Figure 6. Irrigation demand vs. supply (year 2007).

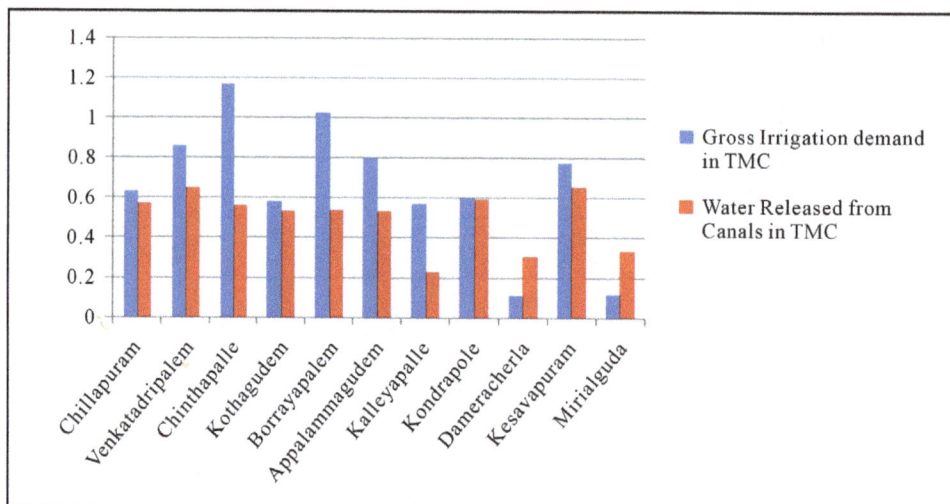

Figure 7. Irrigation demand vs. supply (year 2009).

4. Conclusions

From the study, 11 WUAs, 12 canal blocks and 212 canal sub-blocks were delineated under Wazirabad command area in GIS environment. The total command area was calculated in GIS platform as 26,725 hectares (ha). The gross command area was found maximum in Kalleyapalle WUA with 6482 ha and minimum area was found in Mirialguda WUA with 279 ha. In blocks, the maximum gross command area was found in WL5 canal block with 7721 ha and minimum area was in WL1 canal block with 213 ha. Utilizing GIS network model, the total length of the canal system was identified as 155.846 km. The clay soils were predominant and found 49% in the study area whereas sandy and loamy soils occupied 29% and 22% respectively.

The maximum rice crop was identified in Appalammagudem WUA with 1442 ha in the year 2007 and 1884 ha in the year 2009 in a gross command area of 3254 ha. Whereas in Dameracherla WUA, the rice crop was occupied very less area as 20 ha in the year 2007 and 257 ha in the year 2009 out of 3324 ha of gross command area which was situated in tail end of the command area. The increase of area under rice cultivation demands higher water, as a result greater losses were encountered in the command area.

Due to lengthy canal system coupled with low agriculture area (in Dameracherla WUA), the conveyance losses were higher and this created increase in irrigation demand. Significant difference was observed between the sandy and clayey command areas. From the study, it was observed that for WUAs of Chillapuram, Venkatadripalem, Kothagudem, Mirialguda and Chinthapalle, the percolation losses accounted 10 million litres per ha. Whereas for WUAs of Appalammagudem and Kalleyapalle, the percolation losses were 5 million litres per ha. For Kondrapole and Kesavapuram WUAs, the percolation losses accounted 7 to 8 million litres per ha.

References

[1]   Brouwer, C., Hoevenaars, J.P.M. and Van Bosch, B.E., (1992) Scheme Irrigation Water Needs and Supply. FAO, Irrigation Water Management: Training Manual No. 6, 7-43.

[2]   El-Magd, I.A. and Tanton, T.W. (2005) Remote Sensing and GIS for Estimation of Irrigation Crop Water Demand. *International Journal of Remote Sensing*, **26**, 2359-2370. http://dx.doi.org/10.1080/0143116042000298261

[3]   Palacios, E.V. and Day, J.C. (2012) A New Approach for Estimating Irrigation Conveyance Losses and Their Economic Evaluation. *Journal of the American Water Resources Association*, **13**, 709-720. http://dx.doi.org/10.1080/0143116042000298261

[4]   Akkuzu, E., Ünal, H.B. and Karata, B.S. (2002) Determination of Water Conveyance Loss in the Menemen Open Canal Irrigation Network. *Turkish Journal of Agriculture*, **31**, 11-22.

[5]   Dadhwal, V.K. (1999) Remote Sensing and GIS for Agricultural Crop Acreage and Yield Estimation. *Proceedings of WG VII/2 Workshop on Application of RS and GIS for Sustainable Development, Inter-national Archives in Photogrammetry and Remote Sensing*, **7**, 58-67.

[6]   Jayasekera and Walker (1990) Remotely Sensed Data and Geographic Information System for Management and Ap-

praisal of Large Scale Irrigation Projects in the Developing Countries. Advances in Planning, Design and Management of Irrigation Systems as Related to Sustainable Land Use, 1453-1461.

[7]   Navalgund, R.R. and Ray, S.S. (2000) Geomatics in Natural Resources Management. *Proceedings of Geomatics*-2000, Pune, NR-1-14.

[8]   EPTRI (2008) ISEA Study Report for the Proposed Nagarjuna Sagar Modernization Project. 5-9.

[9]   Krishna Rao, M.V., Hebbar, R. and Venkatartnam, L. (1997) Evaluation of IRS-1C Data for Discrimination and Acreage Estimation of Crops Grown under Multiple Cropping Situation. Remote Sensing for Natural Resources, Joint Publication of ISRS & NNRMS, 205-211.

[10]   Dong, Q.H., Herman, E. and Chen, Z.X. (2008) Crop Area Assessment Using Remote Sensing on the North China Plain. *Proceedings in ISPRS*, 957-962.

[11]   Allen, E.A. (1998) Crop Evapotranspiration: Guidelines for Computing Crop Water Requirements. FAO Irrigation and Drainage Paper 56, Rome, 300.

[12]   Clarke, D. (1998) CropWat for Windows: User Guide. Version 4.2. University of Southampton, UK.

[13]   Giridhar, M.V.S.S. and Viswanadh, G.K. (2005) Estimating of Reference Evapotranspiration for an Irrigation Project Site in Andhra Pradesh, India. *Proceedings of International Conference on Environmental Management*, JNTU, Hyderabad, 216-223.

[14]   Hassan (2005) Estimation of Rice Evapotranspiration in Paddy Fields Using Remote Sensing and Field Measurements. PhD Thesis, Universiti Putra Malaysia, 3-5.

[15]   Konex (1996) Mapping the Spatial Distribution of Irrigation Water Requirement for Main Crop Potatoes Using GIS. *Agricultural Water Management*, **31**, 1-15.

[16]   Rao, K.V.G.K., Bower, R.C., Gaur, A. and Visvanatha.(1998) Tertiary Level Irrigation System Management in the Chambal Command by Water User Associations. *Modernization of Irrigation System Operations*: *Proceedings of the 5th ITIS Network International Meeting*, 214-215.

12

Estimating and Plotting of Groundwater Quality Using WQI$_{UA}$ and GIS in Assiut Governorate, Egypt

Mohamed R. El Tahlawi, Mohamed Abo-El Kassem, Gamal. Y. Baghdadi, Hussein A. Saleem*

Mining and Metallurgical Engineering Department, Faculty of Engineering, Assiut University, Assiut, Egypt
Email: *hseleem2002@yahoo.com

Abstract

This paper aims to turn complex groundwater data into comprehensible information by indexing the different factors numerically comparative to the standards of World Health Organization (WHO) to produce Water Quality Index (WQI). Water Quality Index (WQI) has been used to assess groundwater quality and Geographic Information Systems (GIS) has been used to create maps representing the spatial distribution of groundwater categories in Assiut governorate, Egypt. Water Quality Index has been computed by Un-weighted Arithmetic Water Quality Index (WQI$_{UA}$) method and applied on 796 wells over eight years from 2006 to 2013. The results showed that WQI$_{UA}$ values for drinking purposes were high and most of them reached higher or close to 100, which indicated that the groundwater was polluted and unsafe for drinking. On the other hand, the quality index of groundwater for irrigation purposes in most of the study area ranges between 55.78 and 78.38 (poor and very poor category); this means that groundwater is moderately polluted and rather suitable for irrigation.

Keywords

Un-Weighted Arithmetic Water Quality Index (WQI$_{UA}$), Groundwater Pollution, Geographic Information Systems (GIS)

1. Introduction

Groundwater quality assessment is considered as a significant topic to make sure possible safe use of this resource. As the population continues to increase, it is necessary to find additional sources of water such as groundwater. Groundwater is deemed one of the major resources for potable water in Assiut governorate espe-

*Corresponding author.

cially in rural areas [1]. Methods to incorporate some factors correlated to groundwater quality in a particular index are gradually more desired in local and global scenarios [2]. Using of indicators and indices to monitor the environmental condition can abstract a huge number of characteristics, thus decision makers can be provided by an integrated and more instructive summary [3]. Geographic Information System (GIS) is able to become a strong tool to improve solutions for groundwater resource problems, for assessing groundwater quality, determining groundwater availability, and managing groundwater resources on a national or international range especially when groundwater has a geographical status [4].

This paper aims to assess groundwater quality using Un-weighted Arithmetic Water Quality Index (WQI_{UA}), and display spatial distribution of groundwater quality using GIS in Assiut governorate. In view of the previous studies conducted to assess the quality of groundwater it has been shown that the few number of such studies was interested with groundwater pollution and have been applied on limited regions in Assiut governorate. In 1988, Sobih et al. [5] talked about the chemical and bacterial pollution of groundwater by assessing 25 water wells established at Assiut city. Schlumberger geoelectrical depth soundings and horizontal geoelectrical profiling was used to determine the distribution of the contaminated and uncontaminated zones of groundwater in (El-Madabegh) area, northwest of Assiut city [6]. Sebaq et al. (2003) [7] used surface geoelectrical methods for delineation of groundwater pollution in Beni Ghalib area, northwest of Assiut city.

Geographic Information System (GIS) has come into view as "a powerful tool for storing, analyzing, and displaying spatial data and using these data for decision making in several areas including engineering and environmental fields" [8]. GIS has been applied to create groundwater quality classification map depending on the relationship between total dissolved solids (TDS) values and some characteristics of aquifer [9] or based on the correlation between TDS and land use [10]. Some studies developed a GIS-based methodology to design a groundwater monitoring system and to estimate distributed groundwater contamination risk [11]-[13]. GIS technology has been assigned to study groundwater nitrate contamination resulted from agrochemical fertilizers [14] [15].

Ahmed and Ali (2009) [16], combined between geochemical modeling techniques and GIS to study an incorporated role of three different processes to evolve groundwater composition, and their effect on quality of groundwater in Sohag governorate, Egypt. In 2009, Dawoud and Ewea [17] developed a numerical model using GIS based widespread database to simulate the stable and temporary flow in saturated and unsaturated zones in Wadi Al Assiuti Area. El-Alfy (2010) [18] applied multivariate statistics and GIS to examine the effects on groundwater resources and the possible contamination sources. Latha and Rao (2010) [19] used Water Quality Index (WQI) and GIS to assess and map groundwater quality. Ratnakanth et al. (2011) [20], Balakrishnan et al. (2011) [21] and Dar et al. (2012) [22] used GIS techniques to get groundwater quality map.

2. Study Area

Study area is considered as a part of the Nile Valley, Egypt. It extends south-north between Sohag and El-Minia Governorates from latitude 27°37'N to 26°47'N and extends west-east between New Valley and Red Sea Governorates from longitudes 30°37' to 31°34'E, as shown in **Figure 1**. The total Assiut governorate area is 25,926 km². The extension of the River Nile along study area is about 125 km, and the valley width is between 16 and 60 km [1]. River Nile divides the study area into a western and an eastern part. The Eastern part expands between El-Badary in the South and Manfalut in the North, while the Western part of the study area expands between El-Ghanayim in the South, and Dairut in the North. The sharp declination is the major feature of the fringes of the study area because of the limestone plateau which limits the area from the east and the west, except of the Northwestern part that has a moderate slope [17]. There are some wadies joined with the study area like: Wadi El-Assiuti and Wadi El Ibrahimi in the central east, Wadi Abu Shih in the south east.

3. Methodology

3.1. Computing Water Quality Index (WQI)

Un-weighted Arithmetic Water Quality Index (WQI_{UA}) is calculated to diminish the huge data to one numerical value that expresses the overall water quality [24]. In this research, Un-weighted Arithmetic Water Quality Index (WQI_{UA}) was computed to evaluate suitability of groundwater for drinking and irrigation purposes. To assess the groundwater quality for drinking, twelve parameters were included which are PH, Na^+, K^+, Ca^{2+}, Mg^{2+}, Cl^-, SO_4^{2-}, HCO_3^-, TDS, TH, Fe and Mn. These parameters were synthesized by indexing the different water

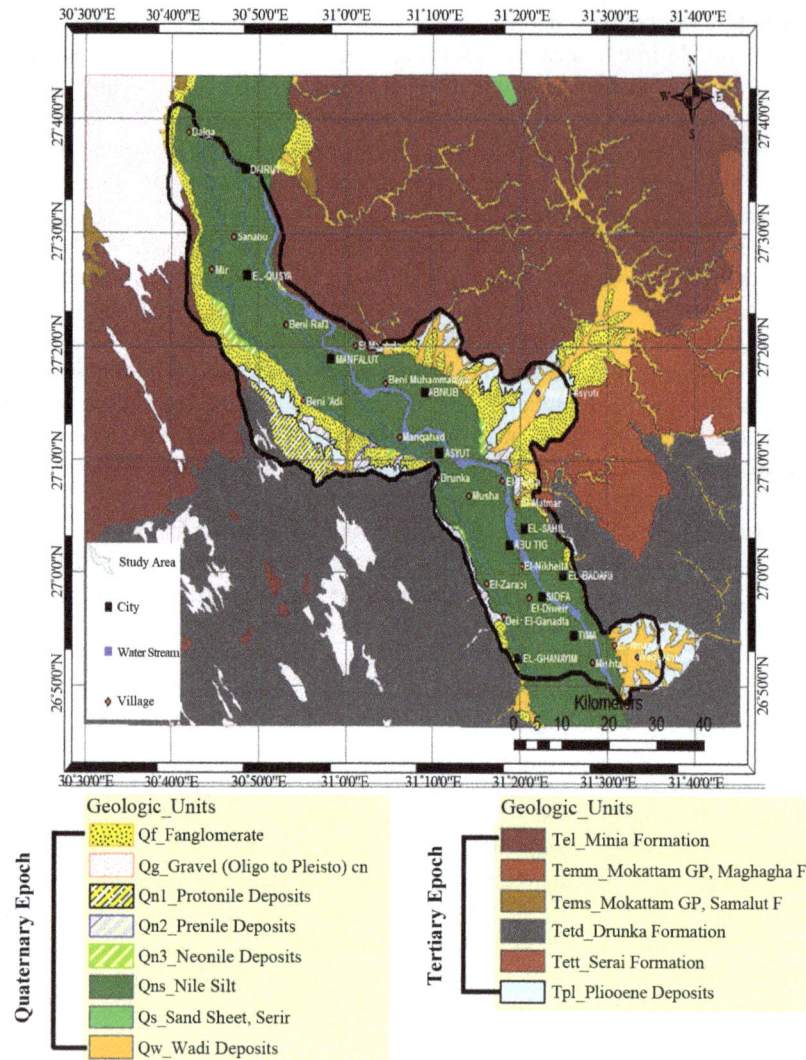

Figure 1. Geology map of Assiut region [23].

quality parameters numerically comparative to the World Health Organization (WHO) standards (1996) [25] and Egyptian Standards (1995) [26] as shown in **Table 1** to produce Un-weighted Arithmetic Water Quality Index (WQI$_{UA}$).

While SAR, Na%, PI, MR, and TDS were used to calculate Un-weighted Arithmetic Water Quality Index (WQI$_{UA}$) of groundwater if used for irrigation purposes. **Table 2** illustrates groundwater classification according to indexing categories of the five parameter used for assessing suitability of groundwater for irrigation.

The un-weighted arithmetic water quality index (WQI$_{UA}$) or quality rating (q) of ith parameter apart is defined as given in equation (1) [32] as follows,

$$q_i = 100 \frac{V_i}{S_i} \tag{1}$$

where:

q_i: the quality rating for the ith parameter;
V_i: the observed value of the ith parameter;
S_i: water quality standard value.

Un-weighted Arithmetic Water Quality Index (WQI$_{UA}$) or the average water quality index (Q) for the studied parameters is

Table 1. WHO guidelines and Egypt standards for drinking water quality [25] [26].

Parameter	Unit	WHO Guidelines	Egypt Standards
Sodium (Na)		200	200
Potassium (K)		12	-
Calcium (Ca)		200	200
Magnesium (Mg)		125	150
Chloride (Cl)		250	500
Sulphate (SO_4)	Mg/l	250	250
Bicarbonate (HCO_3)		350	-
Iron (Fe)		0.3	0.3
TDS		1000	1200
TH		-	500
Manganese (Mn)		0.4	0.4
pH		6.5 - 8.5	6.5 - 8.5

Table 2. Classification of groundwater to evaluate its suitability for irrigation.

Classification pattern	Ranges	Categories
Total Dissolved Solids (TDS) [27]	<500	Desirable for drinking
	500 - 1000	Permissible for Drinking
	1000 - 3000	Useful for irrigation
	>3000	Unfit for drinking and irrigation
Percent Sodium (% Na) [28]	0 - 20	Excellent
	20 - 40	Good
	40 - 60	Permissible
	60 - 80	Doubtful
	>80	Unsuitable
Sodium Absorption Ratio (SAR) [29]	0 - 10	Excellent
	10 - 18	Good
	18 - 26	Doubtful
	>26	Unsuitable
Permeability Index (PI) [30]	>75	Class-I
	25 - 75	Class-II
Magnesium Hazard (MH) [31]	>50%	Unsuitable
	<50%	Suitable

$$\mathrm{WQI_{UA}} \text{ or } Q = \frac{1}{n} * \sum_{i=1}^{n} q_i \tag{2}$$

where:

n: number of parameters.

$\mathrm{WQI_{UA}}$ or Q classifies groundwater into five categories Based on its value as follows in **Table 3** [10]. If the value of $\mathrm{WQI_{UA}}$ or Q is approaching 100; it points to that the groundwater contains high percentage of the considered parameters; thus, groundwater is in very poor category. While if $\mathrm{WQI_{UA}}$ or Q value ranges between 50

Table 3. Water Quality Index Categories [10].

Water quality index	Description
0 - 25	Excellent
26 - 50	Good
51 - 75	Poor
76 - 100	Very poor
>100	Unfit or unsuitable

and 75, it represents that the groundwater is in the poor category. On the other hand, if the average quality index is more than 100, it means that water falls under the fifth category, where groundwater is polluted and not suitable to use [10] [16]. In each of the previous cases (poor, very poor, and unfit) the groundwater is undesirable to use where the higher the value of water quality index, the lower the suitability of groundwater for use.

3.2. Mapping Spatial Distribution of Groundwater Quality

Interpolation can be defined as "a function used to generate a continuous surface from sampled point values" [33]. Interpolation system is based mainly on the Deriving of cell values from a limited number of data. The essential assumption of this technique is that the wells that are nearby together have similar properties more than those is far from each other. In ArcGIS 9.3 software three techniques of point interpolation were involved in spatial analyst, which are 1) Inverse Distance Weighted (IDW), 2) Spline, and 3) Kriging [34]. This paper is dealing with Inverse Distance Weighted (IDW) for interpolation of data.

The Inverse Distance Weighted (IDW) is used as interpolation method to create the spatial distribution map of groundwater quality, which infers the grid value for each cell by calculating the average of sample points. The calculated value depends on measured values of phenomenon in wells and the distance between wells and the calculated grid cell [35] [36]. The expected value is a weighted average of the neighboring groundwater wells in Inverse Distance Weighted method. Weights are calculated by taking the inverse of the distance from an observation's location to the location of the point being estimated [37]. This distance term is often raised to a power to manage the importance of locational separation in the estimation [29]. The distance between the points count, so the points of equal distance have equal weights [8]. The weight factor is determined as illustrated in "Equation (3)" [21]:

$$\lambda_i = \frac{D_i^{-\alpha}}{\sum_{i=1}^{n} D_i^{-\alpha}}$$

(3)

where:

λ_i: the weight of point;

D_i: the distance between point i and the unknown point;

α: the power ten of weight.

4. Results and Discussion

Firstly, the quality rating (q_i) was computed by "Equation (1)" for twelve parameters which used to determine validation of groundwater for drinking and purposes as shown in **Table 4**. It is worth mentioning that q_i isn't considered representative water quality index of all measured parameters in the year but it is just for individual parameter such as Na, K, and so on. It is clear from **Table 4** that all values of Ca and Mg are less than 50 overall the years from 2006 to 2013, this means that the groundwater does not contain dangerous levels of those elements. The quality rating for potassium K was more than 50 in 2006 and 2007 only, but the rest values were less than 50. It is indicated that groundwater does not suffer from increasing of potassium percentage.

The calculated values of q_i for such parameters have been plotted as illustrated in **Figure 2** during the period 2006-2013.

The sulfate quality rating (q_{SO4}) is similar with (q_K), it is also more than 50 in 2006, 2007 and 2008 and then became below 50 during the next period. It is noted that there is an increase above 50 in the quality rating of K,

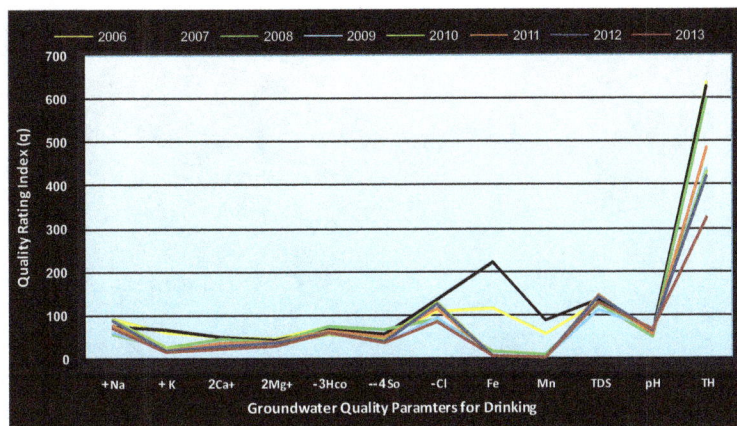

Figure 2. Quality rating index (q_i) of groundwater quality paramters for drinking.

Table 4. Calculating water quality rating (q_i) for drinking purposes.

Year	Quality Rating Index (q_i)											
	Na^+	K^+	Ca^{2+}	Mg^{2+}	HCO_3^-	SO_4^{2-}	Cl^-	Fe	Mn	TDS	pH	TH
2006	90.13	57.35	49.47	44.88	70.72	51.54	106.91	112.79	54.73	126.51	54.14	636.47
2007	75.26	67.80	49.63	43.12	68.70	58.51	136.67	220.00	87.13	135.84	58.66	625.46
2008	57.95	28.78	47.02	41.96	75.23	68.61	93.90	17.09	10.43	126.28	50.03	600.18
2009	57.90	22.64	31.59	33.29	63.00	34.44	91.70	3.58	0.348	109.87	65.21	438.13
2010	91.62	25.22	29.61	34.22	56.78	46.65	126.97	4.76	0.518	127.16	57.61	431.23
2011	76.97	16.49	33.85	37.90	63.40	48.09	114.15	3.31	0.25	143.75	53.66	484.62
2012	86.39	16.97	27.06	35.15	57.34	39.15	123.25	3.31	0.25	141.27	62.72	420.59
2013	70.67	14.99	19.74	28.07	58.41	34.36	84.42	3.46	0.256	126.97	65.81	323.40

So_4 and Mn in the years 2006, 2007, and 2008, but the ratios is changed to less than 50 in the period from 2009 until 2013.,

All quality rating indices (q) of Na and Hco_3 are limited to between 50 and 100, so sodium and bicarbonates ions have negative effect on groundwater quality. The quality rating index (q) for Cl is high throughout the observed years and it reached over 100, which represented that the water is polluted with high levels of Cl. In other case for Fe, the quality rating (q) rate was low in all observed years except 2006 and 2007. PH rate falls within the permissible limits in almost all wells and its quality rating was less than 100 over all years. According to the point that quality rating of TH reached higher than 100, it indicates that groundwater is classified as hard water. Although the quality rating values of Na, K, Ca, Mg, Hco_3, and So_4 were less than 100 but the quality rating for TDS reached over 100 in all years. This means the groundwater in Nile aquifer system along Assiut governorate is over saturated with total dissolved salts (TDS).

Therefore, quality rating (q) cannot be relied upon alone; all affected variables must be included to get one representative value of groundwater in each year which called the Un-weighted Arithmetic Water Quality Index (WQI_{AU}) or (Q) as demonstrated in Equation (2).

With regard to quality rating index (q) of the five used parameters to assess the suitability of groundwater for irrigation, **Table 5** summarizes the calculation of q_i, whereas, **Figure 3** illustrates quality rating index (q_i) of those parameters which are TDS, SAR, Na%, PI, and MR.

It is obvious from **Table 5** and **Figure 3** that all q_i values of TDS and SAR are below 50, this points to the groundwater has small percentage of these two parameters. While all q_i values of Na% and PI vary between 50 and 100, it means that those parameters have negative effect on groundwater quality. While most values of quality rating index (q_i) of MR exceed 100, it denotes that groundwater is oversaturated with magnesium ratio (MR).

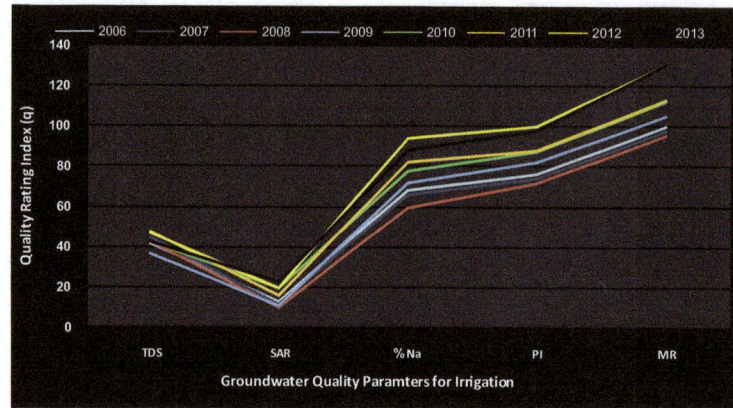

Figure 3. Quality rating index (q_i) of groundwater quality paramters for irrigation.

Table 5. Calculating water quality rating (q_i) for irrigation purposes.

Year	Quality Rating Index (q_i)				
	TDS	**SAR**	**Na%**	**PI**	**MR**
2006	42.17	13.32	68.26	76.14	99.81
2007	45.28	11.76	66.10	74.37	98.01
2008	42.09	9.82	59.14	72.19	95.68
2009	36.62	10.77	72.09	81.98	105.21
2010	42.39	19.45	77.90	87.74	112.51
2011	47.92	15.72	82.59	88.40	113.41
2012	47.09	20.04	94.22	99.78	130.76
2013	42.32	22.68	88.49	97.47	130.80

It is noted that it cannot be judged on the groundwater quality using a single factor alone, thus, it must be combined between five parameters to get one representative value which called Un-weighted Arthmetic Water Quality Index (WQI$_{AU}$) or (Q).

Data of 796 wells over the eight years have been used to estimate the Un-weighted Arithmetic Water Quality Index (WQI$_{UA}$) or (Q) yearly during the period from 2006 to 2013. **Table 6** shows the summary of data wells distributed over 8 years

The Un-weighted Arithmetic Water Quality Index (WQI$_{UA}$) or (Q) has been calculated yearly by Equation (2) as shown in **Table 7**. The Un-weighted Arithmetic Water Quality Index (WQI$_{UA}$) or (Q) results of groundwater whether it used for drinking or irrigation were plotted in two curves as cleared in **Figure 4**.

The tabulated results in **Table 7** show that WQI$_{UA}$ values of groundwater for drinking purposes are close to 100 or more than 100. In addition to, the curve of drinking water in **Figure 4** is located in two zones which are very poor category and unsuitable category. Hence, the groundwater is polluted and unsafe for drinking. On the other hand, the Un-weighted Arithmetic Water Quality Index (WQI$_{UA}$) or (Q) of groundwater for irrigation ranges between 55.78 and 78.38 (poor and very poor category); this means that groundwater is moderately polluted and hardly valid for irrigation.

Secondly, GIS has been used to plot the Un-weighted Arithmetic Water Quality Index (WQI$_{UA}$) results to create the spatial distribution map of WQI$_{UA}$ for drinking water and irrigation water as shown in **Figure 5** and **Figure 6** respectively.

The results revealed that the suitability of groundwater for drinking varies from location to another. Generally, the groundwater quality index for drinking around river Nile is less than the new reclaimed lands at the fringes of study area. The groundwater in most of the study area belongs to unsuitable and very poor categories, however good and poor categories occupy small areas. Accordingly, the majority of groundwater wells in study area

Table 6. Summary of data wells over eight years.

Year	No of Wells
2006	86
2007	101
2008	98
2009	102
2010	83
2011	113
2012	129
2013	84
Sum	**796 Wells**

Table 7. Summary of (WQI_{UA}) or (Q) of the groundwater.

Year	WQI_{AU} for drinking	WQI_{AU} for irrigation
2006	121.30	59.94
2007	135.57	59.10
2008	101.46	55.78
2009	79.31	61.33
2010	86.03	68.00
2011	89.70	69.61
2012	84.45	78.38
2013	69.21	76.35

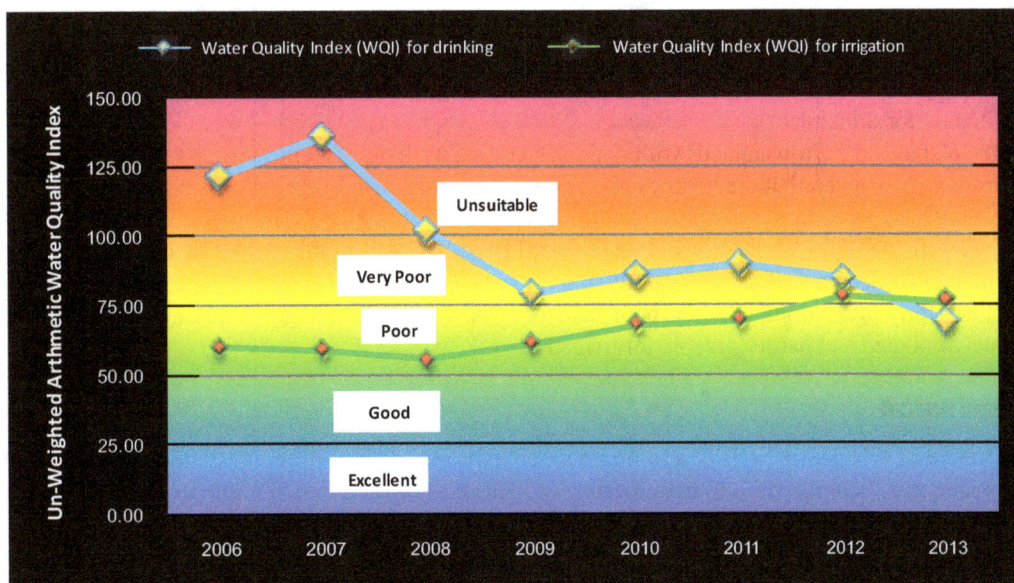

Figure 4. Average water quality index of the groundwater to evaluate its suitability for drinking and irrigation.

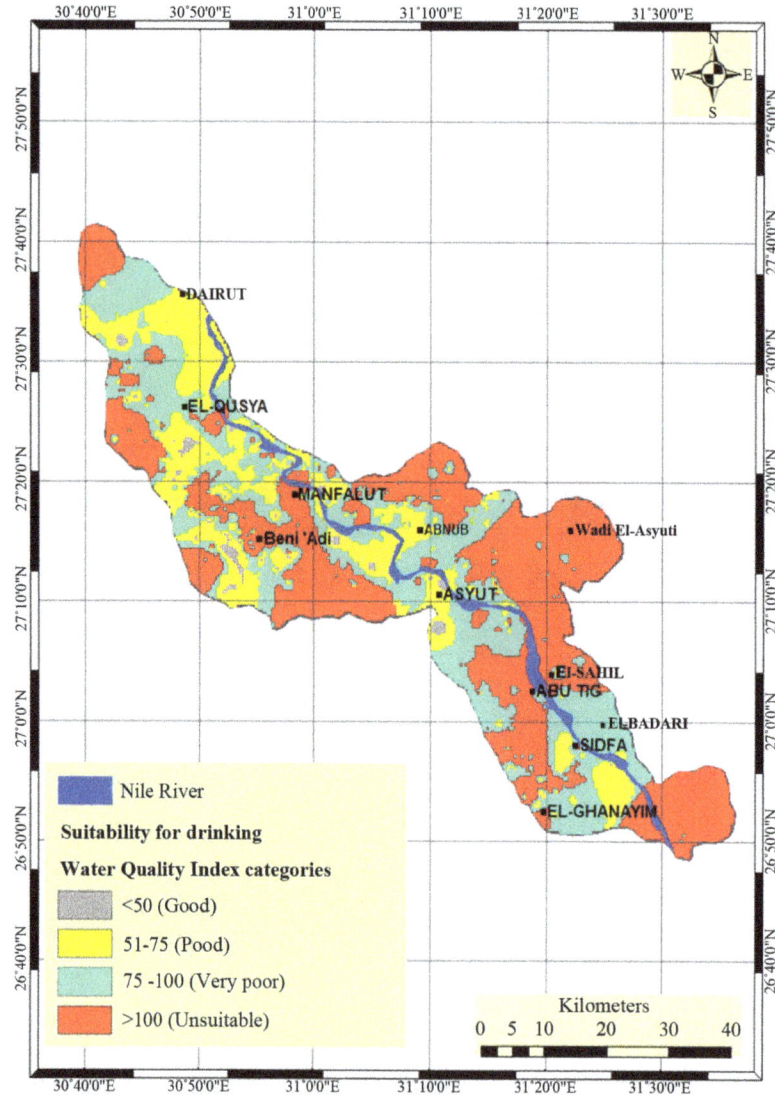

Figure 5. Spatial distribution map of the WQI_{AU} for the for drinking purposes.

are unsuitable for drinking.

Figure 6 shows the Un-weighted Arithmetic Water Quality Index (WQI_{UA}) results of groundwater for irrigation purposes. It is noted that the yellow color covers the most of study area which expresses the poor category. For the very poor category, it occupies small areas on the fringes of Assiut governorate, Wadi El-Asyuti and Abnub from the east side, Bani Adi and Al-Ghanaium on the west side district. The isolated areas in Wadi El-Asyuti and Abnub on the eastern side which have red color are unsuitable category. Consequently, the largest part of the study area varies between poor category and very poor category; this means that groundwater is moderately polluted and hardly suitable for irrigation.

5. Conclusion

The objective of calculating WQI_{UA} is to find out quality assessment of water by synthesizing different available groundwater data. Samples of 796 groundwater wells have been used to determine the Un-weighted Arithmetic Water Quality Index (WQI_{UA}) during the period from 2006 to 2013. The quality indices are worked out for PH, Na, K, Ca, Mg, HCO_3, SO_4, Cl, TDS, TH, Fe and Mn to evaluate the suitability of groundwater for drinking. While, SAR, Na%, PI, MR, and TDS were used to calculate WQI_{UA} for assessing the suitability of groundwater for irrigation. It is concluded that the groundwater of this study area is not suitable for drinking water but can be

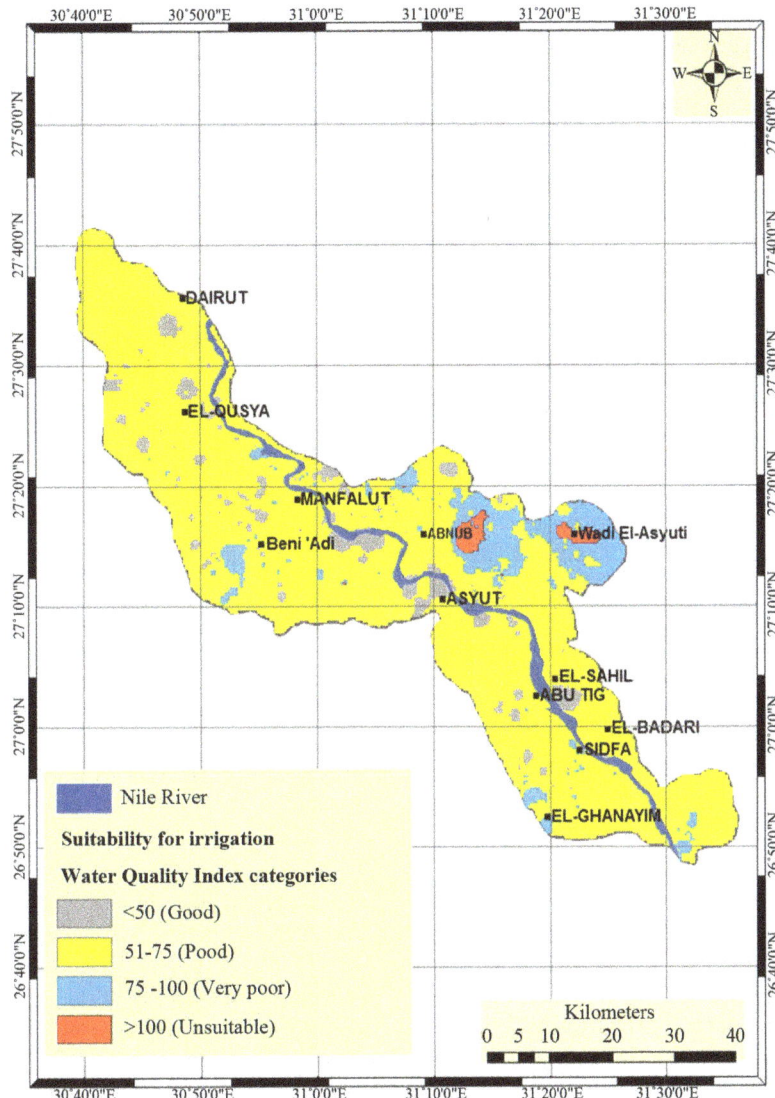

Figure 6. Spatial distribution map of the WQI$_{AU}$ for irrigation purposes.

hardly used for irrigation purposes. The results are also supported by the spatial distribution of the WQI$_{UA}$ map which was generated by ArcGIS 9.3. In case of using groundwater for drinking, the most study area belongs to unsuitable and very poor categories, whereas if groundwater is used for irrigation, the most study area belongs to poor and very poor categories.

References

[1]   ALDAR, CH2M Hill, Spain (2011) Study of Environmental and Social Impact Assessment Framework (ESIAF) of Assiut & Sohag Governorates. Executive Summary, Egypt, 82 p.

[2]   Lermontov, A., Yokoyama, L., Lermontov, M. and Machado, M.A.S. (2009) River Quality Analysis Using Fuzzy Water Quality Index: Ribeira do Iguape River Watershed, Brazil. *Ecological Indicators*, **9**, 1188-1197. http://dx.doi.org/10.1016/j.ecolind.2009.02.006

[3]   Smajgla, A., Larsona, S., Huga, B. and De Freitasb, D.M. (2010) Water Use Benefit Index as a Tool for Community-Based Monitoring of Water Related Trends in the Great Barrier Reef Region. *Journal of Hydrology*, **395**, 1-9. http://dx.doi.org/10.1016/j.jhydrol.2010.09.007

[4]   Tjandra, F.L., Kondhoh, A. and Mohammed, M.A. (2003) A Conceptual Database Design For Hydrology Using GIS. *Proceedings of Asia Pacific Association of Hydrology and Water Resources*, Kyoto, March 2003, 13-15.

[5]   Sobih, M.A., Reem, D., Kamel, Y.Y. and El-Gharably, G.A. (1988) Chemical and Bacteriological Evaluation of Drinking Ground Supplies in Assiut City. *Assiut Veterinary Medical Journal*, **38**, 99-104.

[6]   Ebrahim, M.O. (1997) Application of Surface Geoelecrtical Methods for the Detection of Groundwater Contamination in the Area Northwest of Assiut City (El Madabegh). Msc Thesis, Assiut University, Assiut.

[7]   Sebaq, A.S., El-Hussaini, A.H. and Ibrahim, H.A. (2003) Application of Electrical Resistivity and Self-Potential for Groundwater Exploration and Contamination Study in the Area Northwest of Assiut City, Egypt. *Journal of King Saud University*, **16**, 31-61.

[8]   Burrough, P.A. and McDonnell, R.A. (1998) Principles of Geographical Information Systems. Oxford University Press, New York, 333 p.

[9]   Butler, M., Wallace, J. and Lowe, M. (2002) Ground-Water Quality Classification Using GIS Contouring Methods for Cedar Valley, Iron County, Utah. Digital Mapping Techniques, Workshop Proceedings, US Geological Survey Open-File Report 02-370.

[10]   Asadi, S.S., Vuppala, P. and Reddy, M.A. (2007) Remote Sensing and GIS Techniques for Evaluation of Groundwater Quality in Municipal Corporation of Hyderabad (Zone-V), India. *International Journal of Environmental Research and Public Health*, **4**, 45-52. http://dx.doi.org/10.3390/ijerph2007010008

[11]   Dutta, D., Gupta, A.D. and Ramnarong, V. (1998) Design and Optimization of a Groundwater Monitoring System Using GIS and Multicriteria Decision Analysis. *Groundwater Monitoring & Remediation*, **18**, 139-147. http://dx.doi.org/10.1111/j.1745-6592.1998.tb00610.x

[12]   Giupponi, C. and Vladimirova, I. (2006) Ag-PIE: A GIS-Based Screening Model for Assessing Agricultural Pressures and Impacts on Water Quality on a European Scale. *Science of the Total Environment*, **359**, 57-75. http://dx.doi.org/10.1016/j.scitotenv.2005.07.013

[13]   Sinkevich, M.G., Walter, M.T., Lembo, A.G. and Richards, B.K. (2005) A GIS-Based Ground Water Contamination Risk Assessment Tool for Pesticides. *Ground Water Monitoring and Remediation*, **25**, 82-91. http://dx.doi.org/10.1111/j.1745-6592.2005.00055.x

[14]   Babiker, I.S., Mohamed, A.A., Terao, H., Kato, K. and Ohta, K. (2004) Assessment of Groundwater Contamination by Nitrate Leaching from Intensive Vegetable Cultivation Using Geographical Information System. *Environment International*, **29**, 1009-1017. http://dx.doi.org/10.1016/s0160-4120(03)00095-3

[15]   Nas, B. and Berktay, A. (2006) Groundwater Contamination by Nitrates in the City of Konya, (Turkey): A GIS Perspective. *Journal of Environmental Management*, **79**, 30-37. http://dx.doi.org/10.1016/j.jenvman.2005.05.010

[16]   Ahmed, A.A. and Ali, M.H. (2009) Hydrochemical Evolution and Variation of Groundwater and Its Environmental Impact at Sohag, Egypt. *Arabian Journal of Geosciences*, **4**, 339-352. http://dx.doi.org/10.1007/s12517-009-0055-z

[17]   Dawoud, M.A. and Ewea, H.A.R. (2009) Sustainable Development via Optimal Integration of Surface and Groundwater in Arid Environment: Nile River Quaternary Aquifer Case Study. *The International Conference on Water Conversation in Arid Regions*, Jeddah, 12-14 October 2009.

[18]   El Alfy, M. (2010) Integrated Geostatistics and GIS Techniques for Assessing Groundwater Contamination in Al-Arish Area, Sinai, Egypt. *Arabian Journal of Geosciences*, **5**, 197-215. http://www.springerlink.com/content/p7444736q3281k66/fulltext.pdf

[19]   Latha, S.P. and Rao, N.K. (2010) Assessment and Spatial Distribution of Quality of Groundwater in Zoneii and III, Greater Visakhapatnam, India using Water Quality Index (WQI) and GIS. *International Journal of Environmental Sciences*, **1**, 198-212.

[20]   Ratnakanth, M.J., Das, I.C., Jaisankar, G., Rao, E.N.D. and Kumar, P.A. (2011) Assessment of Groundwater Pollution in Parts of Guntur District using Remote Sensing & GIS. *International Journal of Earth Sciences and Engineering*, **4**, 1024-1030.

[21]   Balakrishnan, P., Saleem, A. and Mallikarjun, N.D. (2011) Groundwater Quality Mapping Using Geographic Information System (GIS): A Case Study of Gulbarga City, Karnataka, India. *African Journal of Environmental Science and Technology*, **5**, 1069-1084. http://dx.doi.org/10.5897/AJEST11.134

[22]   Dar, I.A., Sankar, K. and Dar, M.A. (2012) Groundwater Quality Evaluation of Mamundiyar River Basin, India Using ARCGIS 9.2 PLATFORM. *Wilolud Journals*, **1**, 1-24.

[23]   Conoco (1987) Geologic Map of Egypt. Egyptian General Authority for Petroleum (UNESCO Joint Map Project), 20 Sheets, Scale 1:500 000. Cairo.

[24]   Kolli, K. and Seshadri, R. (2013) Ground Water Quality Assessment Using Data Mining Techniques. *International Journal of Computer Applications*, **76**, 39-45. http://dx.doi.org/10.5120/13324-0885

[25]   World Health Organization (WHO) (1996) Guidelines for Drinking Water Quality. Vol. 2. Health Criteria and Other Supporting Information. WHO, Geneva, 940-949.

[26] Egyptian Higher Committee for Water (EHCW) (1995) Egyptian Standards for Drinking and Domestic Uses. EHCW, Cairo.

[27] Davis, S.N. and De Wiest, R.J.M. (1966) Hydrogeology. Vol. 463, Wiley, New York.

[28] Wilcox, L.V. (1955) Classification and Use of Irrigation Water. Agric Circular 969, USDA, Washington DC, 19 p.

[29] Richard, L.A. (1954) Diagnosis and Improvement of Saline and Alkali Soils. Agric Handbook 60, USDA, Washington DC, 160 p.

[30] Doneen, L.D. (1964) Water Quality for Agriculture. Department of Irrigation. University of Calfornia, Davis, 48 p.

[31] Szaboles, I. and Darab, C. (1964) The Influence of Irrigation Water of High Sodium Carbonate Content on Soils. In: Szabolics, I., Ed., *Proc 8th International Congress Soil Science Sodics Soils*, Research Institute for Soil Science and Agricultural Chemistry of the Hungarian Academy of Sciences, ISSS Trans II, 802-812.

[32] Farrag, A.A. (2005) The Hydraulic and Hydrogeological Impacts of the Nile System on the Groundwater in Upper Egypt. *Assiut University Bulletin for Environmental Researches*, **8**, 87-102.

[33] McCoy, J. and Johnston, K. (2001) Using Arcgis Spatial Analyst. Environmental Systems Research Institute, Inc., Redlands, 230 p.

[34] DeMers, M.N. (2000) Fundamentals of Geographic Information Systems. 2nd Edition, John Wiley & Sons, Inc., New York, 498 p.

[35] Buchanan, S. and Triantafilis, J. (2009) Mapping Water Table Depth Using Geophysical and Environmental Variables. *Ground Water*, **47**, 80-96. http://dx.doi.org/10.1111/j.1745-6584.2008.00490.x

[36] Arsalan, M.H. (2004) A GIS Appraisal of Heavy Metals Concentration in Soil. Technical Report for American Society of Civil Engineers, New York.

[37] Guan, W., Chamberlain, R.H., Sabol, B.M. and Doering, P.H. (1999) Mapping Submerged Aquatic Vegetation in the Caloosahatchee Estuary: Evaluation of Different Interpolation Methods. *Marine Geodesy*, **22**, 69-91. http://dx.doi.org/10.1080/014904199273506

13

GIS Based Inverse Distance Weighting Spatial Interpolation Technique for Fluoride Occurrence in Ground Water

M. Arif[1*], J. Hussain[2], I. Hussain[3], S. Kumar[4], G. Bhati[4]

[1]Department of Chemistry, Banasthali University, Niwai, District-Tonk, India
[2]National River Water Quality Laboratory, Central Water Commission, New Delhi, India
[3]Public Health Engineering Department (PHED) Laboratory, Bhilwara, India
[4]Department of Space, Regional Remote Sensing Centre (West), NRSC/ISRO, Jodhpur, India
Email: [*]dr.arifmohammed@gmail.com

Abstract

This study was conducted to evaluate fluoride contamination in ground water of the central part of Nagaur district where ground water is the main source of drinking as well as irrigation. Samples from hand pumps and tube wells of forty four stations were analyzed during the summer session with the help of standard methods of APHA. The analytical results show that eleven ground water samples are unfit for drinking purpose. A map has been prepared using an inverse distance weighting method which shows the fluoride concentration in the study area. The maximum concentration was recorded in the Sirsi village while the minimum was found in Sabalpura.

Keywords

Fluoride, Groundwater, Central Part, Nagaur, Rajasthan

Subject Areas: Environmental Sciences, Hydrology

1. Introduction

Water is very vital in nature and can be a limiting resource for men and other living beings. Without a well functioning water supply, it is difficult to imagine a productive human activity be it agriculture or livestock. The quality of water is of almost importance of quantity in any water supply planning. Water quality is influenced by natural and anthropogenic effects including local climate, geology and irrigation practices. In the world, around 200 million people from 25 nations have greater health risks, with high fluoride in the drinking water [1]. Vari-

[*]Corresponding author.

ous workers in our country have carried out extensive studies on water quality with reference to the fluoride concentration [2]-[8].

In Indian continent the higher concentration of F^- in groundwater is associated with igneous and metamorphic rocks. Some anthropogenic activities such as use of phosphatic fertilizers, pesticides and sewage and sludges, depletion of groundwater table etc., for agriculture has also been indicated to cause an increase in F^- concentration in groundwater [9] [10]. According to WHO (2006), the permissible limit for F^- in drinking water is 1.5 μg/ml. In India almost 60 - 65 million people drink fluoride contaminated groundwater and the number affected by fluorosis is estimated at 2.5 to 3 million in many States, especially, Andhra Pradesh, Bihar, Gujarat, Madhya Pradesh, Punjab, Rajasthan, Tamil Nadu and Uttar Pradesh [11]-[20]. The first case of endemic fluorosis in the country was reported as long as 1937 in the Prakasam District (erstwhile Guntur District), Andhra Pradesh [21]. WHO (2004) has laid down a limit of 1.5 mg/l fluoride in drinking water [22]. This maximum limit protects tooth decay and enhances proper bone growth. In India, safe limit of fluoride in potable water is between 0.6 and 1.2 mg/l [23]. Lower limit of fluoride (<0.6 mg/l) than that of the prescribed limit (0.6 mg/l) causes dental caries, while the higher limit of fluoride (>1.2 mg/l) than those of the recommended limit (1.2 mg/l) results in fluorosis. Rajasthan is the largest state in India is having 342,239 km^2 area with a relatively low population density, *i.e.*, 165 persons per square kilometer. According to physiographies divisions, the north and western part of the state is under the Great Plain of north India, while south and middle as well as eastern part is classified under the Peninsular Plateau. In the state, Thar Desert occupies about 61% of the total area. Groundwater is a major source for drinking and domestic and irrigation purpose [24]. Nagaur District is located at latitude 26°25' to 27°40' N and longitude 73°18' to 75°15' E. Its average elevation is about 300 m, ranging below 250 m in the south and 640 m in the north. There are 1396 habitations in the district. The main lithological units include gneisses, schists, granites, quartzites, phyllites, and limestones belonging to the Bhilwara and Delhi Supergroup of rocks of Archaean and Proterozoic ages, respectively. Although groundwater occurs mainly under water table condition in all the formations, the quaternary alluvium forms good aquifers in Nagaur District. In the hard rock terrain, the occurrence and movement of groundwater are controlled by secondary porosity such as fractures, fissures, joints, foliation, etc. A bibliographic survey has also shown that no studies have been undertaken in the district for assessment of the F^- concentration in its groundwater. This study was, therefore, undertaken to investigate the quality of underground drinking water of the central part of Nagaur district of Rajasthan, India in respect of F^- concentration.

These minerals are commonly associated with the country rocks through which the ground water percolates under variable temperature conditions. Besides these minerals, alkali rocks, hydrothermal solutions may also contribute to higher concentration of fluoride in groundwater. Robinson and Edington (1946) reported that the main source of fluoride in ordinary soil consists of clay minerals. The weathering and leaching process, mainly by moving and percolating water, play an important role in the incidence of fluoride in groundwater [25]. When fluoride rich minerals, which are present in rocks and soil, come in contact with water of high alkalinity; they release fluoride into groundwater through hydrolysis replacing hydroxyl (OH) ion. The degree of wreathing and leachable fluoride in a terrain is more important in deciding the fluoride bearing minerals in the bulk rocks or soil. Due to weathering of rocks the Ca-Mg/carbonate concentration which form in arid and semi arid areas appears to be good sink for the fluoride ion [26]. While fluoride is present in air, water and food, the most common way it enters the food chain is via drinking water.

2. Material and Method

Groundwater samples of 44 habitations located in the Nawa block of Nagaur District were collected in precleaned polyethylene bottles following standard sampling techniques. The fluoride concentration in water was determined electrochemically, using fluoride ion selective electrode [27]. This method is applicable to the measurement of fluoride in drinking water in the concentration range of 0.01 - 1000 mg/l. The electrode used to be an Orion fluoride electrode, coupled to an Orion electrometer. Standards fluoride solutions (0.1 - 10 mg/l) were prepared from a stock solution (100 mg/l) of sodium fluoride. As per experimental requirement, 1 ml of total ionic strength adjusting buffer grade III (TISAB III) was added in 10 ml of sample. The ion meter was calibrated for a slop of -59.2 ± 2. The composition of TISAB solution was 385.4 g ammonium acetate, 17.3 g of cyclohexylene diamine tetraacetic acid, and 234 ml of concentrate hydrochloric acid per liter. All the experiments were carried out in triplicate and the results were found reproducible with ±2% error.

2.1. Preparation of TISAB- III

Take 250 ml de-ionized water and add 234 ml of concentrated hydrochloric acid. Add 385.4 gm of Ammonium acetate. Add 17.3 gm of 1, 2-cyclohexylene-diaminetetraacetic (CDTA). Stir to dissolve and cool at room temperature. Make it up to 1000 ml.

2.2. Procedure

1) Calibration of instrument: Prepare a series of standards over the appropriate concentration range (0.1 mg/l, 1.0 mg/l and 10.0 mg/l) Calibrate the instrument to obtain −59.2 ±2 mv slop.

2) Take 10 ml sample add 1 ml of TISAB III and measure fluoride concentration.

2.3. Spatial Distribution of Fluoride Using Inverse Distance Weighted Method

Geographic Information System (GIS) is a computerized data based system for capture, storage, retrieval, analysis and display of spatial data. GIS is a general purpose technology for handling geographic data in digital form, and satisfying the following specific needs, among others. The Inverse distance weighted (IDW) interpolation explicitly implements the assumption that things that are close to one another are more alike than those that are farther apart. To predict a value for any unmeasured location, IDW uses the measured values surrounding the prediction location. The measured values closest to the prediction location have more influence on the predicted value than those farther away. IDW assumes that each measured point has a local influence that diminishes with distance. It gives greater weights to points closest to the prediction location, and the weights diminish as a function of distance, hence the name Inverse distance weighted

3. Result

Fluoride concentration in groundwater of 44 habitations of Nawa block was examined. All the habitations were categorized according to following concentration range (**Table 1**):

Category I—Fluoride concentration below 1.0 mg/l;

Category II—Fluoride concentration between 1.0 and 1.5 mg/l;

Category III—Fluoride concentration between 1.5 and 3.0 mg/l;

Category IV—Fluoride concentration between 3.0 and 5.0 mg/l; and

Category V—Fluoride concentration above 5.0 mg/l.

The distribution of fluoride in the groundwater of Nawa block is shown in **Figure 1**. **Figure 2** shows the no. of villages and their fluoride concentration range as per divided categorization. Fluoride concentration ranges from 0.4 to 5.9 mg/l. The minimum concentration was recorded in the Sablpura village while maximum concentration was recorded from Sirsi village (5.9 mg/l). It is clear from **Figure 2** that the maximum villages belong to the concentration of below 1.0 mg/L. The present investigation reveals that 19 habitations (43%) fall in category I .in which fluoride concentration is below 1.0 mg/l, a maximum desirable limit of standards for drinking water recommended by who. There is no possibility of fluorosis in these habitations because this concentration of fluoride is beneficial for calcification of dental enamel especially for children below 10 years of age. Once fluoride is incorporated into teeth, it reduces the solubility of the enamel under acidic conditions and thereby provides protection against dental carries. Out of 44 habitations of Nawa block, 14 habitations (32%) have fluoride concentration between 1.0 and 1.5 mg/l and fall in category II. The maximum permissible limit of fluoride in standard for drinking water is 1.5 mg/l who. In 32% population of these habitations, fluoride intake through drinking water is more than 4 mg/day in an individual. Therefore, an incidence of first and second degree dental fluorosis is possible in local residents of these habitations. About 17% of the population of 4 habitations (16%) consumes water with fluoride concentration between 1.5 and 3.0 mg/l, which is above the maximum permissible limit as recommended by who. Therefore, dental fluorosis is common in these habitations. At this concentration, the teeth lose their shiny appearance and chalky black, gray, or white patches develop known as mottled enamel [28]. In four habitations (9%), fluoride concentration in groundwater is above 3.0 mg/l and below 5.0 mg/l, and this fall in category IV. The intake of fluoride per day by an adult in these habitations is very high. About 6% population of these habitations may have all degrees of dental fluorosis (mild, moderate, moderately severe, and severe fluorosis) including skeletal fluorosis after 30 years of age. However, the probability of second stage skeletal fluorosis age may be more common after the age of 45 [29]. In the entire survey, Sirsi was the only vil-

Table 1. Fluoride categorization of villages of Nawa tehsil.

Category I (Below 1.0 mg/l)		Category II (Between 1.0 and 1.5 mg/l)		Category III (Between 1.5 and 3.0 mg/l)		Category IV (Between 3.0 and 5.0 mg/l)		Category V (Above 5.0 mg/l)	
Chawandiya	0.8	Lalas	1.2	Rooppura Torda	2.4	Piprali	4.2	Sirsi	5.9
Adaksar	0.9	Parewadi	1.4	Shyamgarh	2.4	Thikariya Khurd	4.4		
Todas	0.9	Kotra	1.2	Kooni	2.2	Prempura	3.0		
Charanwas	0.6	Jiliya	1.4	Mindha	1.5	Padampura	4.0		
Nalot	0.7	Rasal	1.3	Karkeri	2.4	Ghandi Gram	3.3		
Chitawa	0.8	Panchwa	1.1						
Sabalpura	0.4	Rewasa	1.4						
Aspura	0.8	Lichana	1.1						
Bhanwata	0.7	Panchota	1.4						
Kukanwali	0.9	Mandawara	1.2						
Nagwara	0.8	Kasari	1.1						
Rajliya	0.6	Govindi	1.2						
Nuwana	0.5	Bhatipura	1.1						
Gogor	0.7	Deoli	1.2						
Maroth	0.8								
Jeenwar	0.7								
Ulana	0.6								
Bawali Gurha	0.7								
Nanana	0.9								

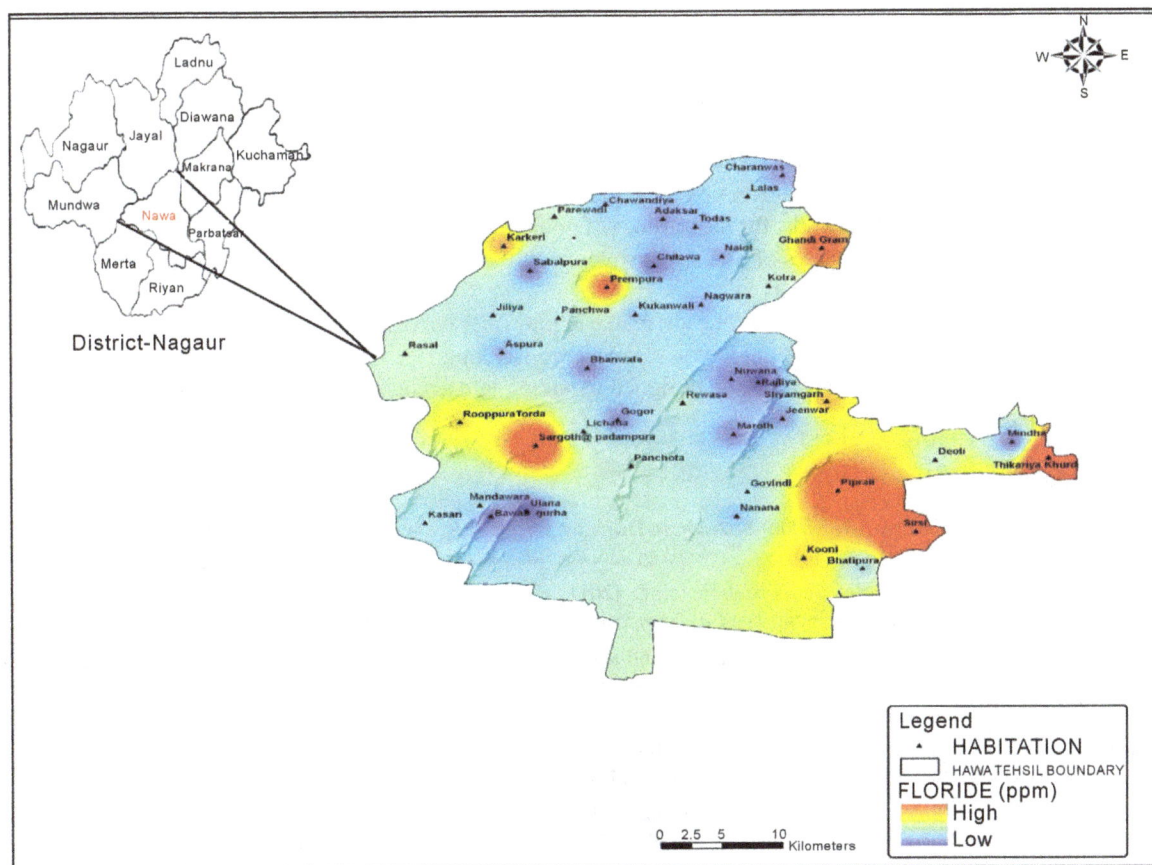

Figure 1. Fluoride distribution map of Nawa Tehsil using inverse distance weighting method.

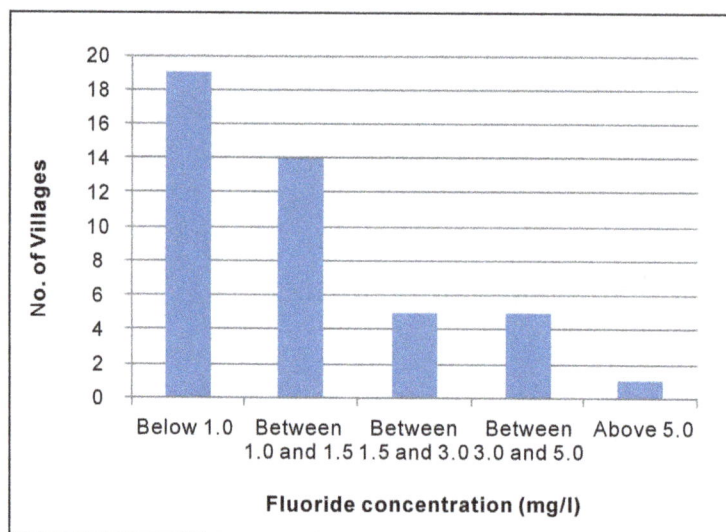

Figure 2. Categorization of villages as per their fluoride concentration range.

lage that falls in category V, which contributes 1% population of Nawa block. In this village, fluoride concentration is above 5.0 mg/l, which may result in all types of fluorosis among inhabitants. In the second clinical stage, the affected persons may have pain in bones, which causes further calcification in ligaments. A detailed survey of health hazards particularly fluorosis-induced symptoms and empirical data on affected population are required in Nawa.

4. Conclusions

The lack of resources and low-cost efficient technology acceptable to the affected populations restrict the development of an effective fluoride and fluorosis control and prevention program in developing countries.

- Provisions of alternative water sources such as the delivery of water from low fluoride sources have also been considered for fluorosis prevention strategies.
- Construction of piped networks for the purpose of potable water.
- It is suggested that the sources of municipal water supply must be established in a region where adequate levels of fluoride have been observed.
- One more way to avoid excessive fluoride intake is rainwater harvesting. The dilution of high fluoride water with rainwater will make small amount of rainwater to last long.
- A public health programme should be designed by taking into account all the fluoride sources, including dietary fluoride.

References

[1]   Ayoob, S. and Gupta, A.K. (2006) Fluoride in Drinking Water: A Review on the Status and Stress Effects. *Critical reviews in Environmental Science and Technology*, **36**, 433-487. http://dx.doi.org/10.1080/10643380600678112

[2]   Arif, M., Hussain, J., Hussain, I. and Neyol, S. (2011) Fluoride Contamination of Ground Water of Merta Block in Nagaur District, Rajasthan, India. *The Conference of Advance in Environmental Chemistry (AEC)*, Aizwal, 146-148.

[3]   Arif, M., Hussain, I., Hussain, J., Sharma, S. and Kumar, S. (2012) Fluoride in the Drinking Water of Nagaur Tehsil of Nagaur District, Rajasthan, India. *Bulletin of Environmental Contamination and Toxicology*, **88**, 870-875. http://dx.doi.org/10.1007/s00128-012-0572-4

[4]   Hussain, J., Sharma, K.C. and Hussain, I. (2005) Fluoride Distribution in Groundwater of Banera Tehsil in Bhilwara District, Rajasthan. *Asian Journal of Chemistry*, **17**, 457-461.

[5]   Hussain, J., Shrama, K.C., Arif, M. and Hussain, I. (2007) Fluoride Distribution and Modelling Using Best Subset Procedure in Nagour District of Central Rajasthan, India. *The 27th Conference of the International Society for Fluoride Research (ISFR XXVII)*, Beijing.

[6]   Hussain, I., Arif, M. and Hussain, J. (2011) Fluoride Contamination in Drinking Water in Rural Habitations of Central

Rajasthan, India. *Environmental Monitoring and Assessment*, **184**, 5151-5158.
http://dx.doi.org/10.1007/s10661-011-2329-7

[7]   Sharma, K.C., Arif, M., Hussain, I. and Hussain, J. (2007) Observation on Fluoride Contamination in Groundwater of District Bhilwara, Rajasthan and a Proposal for a Low Cost Defluoridation Technique. *The 27th Conference of the International Society for Fluoride Research* (*ISFR XXVII*), Beijing.

[8]   Choubisa S.L., Choubisa L. and Choubisa D.K. (2001) Endemic Fluorosis in Rajasthan. *Indian Journal of Environmental Health*, **43**, 177-189.

[9]   EPA (1997) Public Health Global for Fluoride in Drinking Water. Pesticide and Environmental Toxicology. Section Office of Environmental Health Hazard Assessment, California Environmental Protection Agency.

[10]   Ramanaiah, S.V., Venkatamohan, S., Rajkumar, B. and Sarma, P.N. (2006) Monitoring of Fluoride Concentration in Groundwater of Prakasham District in India: Correlation with Physico-Chemical Parameters. *Journal of Environmental Science and Engineering*, **48**, 129-134.

[11]   WHO (2006) Fluoride in Drinking Water. IWA Publishing, London, 144.

[12]   Athavale, R.N. and R.K. Das (1999) Beware! Fluorosis Is Zeroing in on You. *Down to Earth*, **8**, 24-25.

[13]   Susheela, A.K. (1999) Fluorosis Management Programme in India. *Current Science*, **77**, 1250-1256.

[14]   Chakraborti, D., Chanda, C.R., Samanta, G., Chowdhury, U.K., Mukherjel, S.C. and Pal, A.B. (2000) Fluorosis in Assam, India. *Current Science*, **78**, 1421-1423.

[15]   Muralidharan, D., Nair, A.P. and Satyanarayana, U. (2002) Fluoride in Shallow Aquifers in Rajgarh Tehsil of Churu District, Rajasthan—An Arid Environment. *Current Science*, **83**, 699-702.

[16]   Pillai, K.S. and Stanley, V.A. (2002) Implications of Fluoride—An Endless Uncertainty. *Journal of Environmental Biology*, **23**, 81-97.

[17]   Arif, M., Joshi, S. and Kumar, S. (2012) A Study of Fluoride Contaminated Ground Water in Uniara Tehsil, District-Tonk, Rajasthan, India. India Water Week.

[18]   Arif, M., Hussain, J., Hussain, I. and Kumar, S. (2013) An Assessment of Fluoride Concentration in Groundwater and Risk on Health of North Part of Nagaur District, Rajasthan, India. *World Applied Sciences Journal*, **24**, 146-153.

[19]   Arif, M., Hussain, J., Hussain, I. and Kumar, S. (2013) An Investigation of Fluoride Distribution in Ladnu Block of Nagaur District, Central Rajasthan. *World Applied Sciences Journal*, **26**, 1610-1616.

[20]   Arif, M., Yadav, B.S. and Garg, A. (2013) An Investigation of Fluoride Concentration in Drinking Water of Sanganer Tehsil, Jaipur District, Rajasthan, India and Defluoridation from Plant Material. *EQA-Environmental Quality*, **12**, 23-29.

[21]   Shortt, H.E., Mcrobert, G.R., Barnard, T.W. and Mannadinayer, A.S. (1937) Endemic Fluorosis in Madras Presidency, India. *Indian Journal of Medical Research*, **25**, 553-561.

[22]   WHO (2004) Guidelines for Drinking Water Quality. World Health Organization, Geneva, 515.

[23]   BIS (2003) Indian Standard Specifications for Drinking Water. Bureau of Indian Standards, New Delhi, India: IS: 10500, 10.

[24]   Census (2001) District Nagaur, Rajasthan, Government of Rajasthan.

[25]   Robinson, W.O. and Edington, G. (1946) Fluorine in Soils. *Soil Science*, **61**, 341-353.
http://dx.doi.org/10.1097/00010694-194605000-00001

[26]   Jacks, G., Bhattacharya, P., Chaudhary, V. and Singh, K.P. (2005) Controls on the Genesis of High-Fluoride Ground Waters in India. *Applied Geochemistry*, **20**, 221-228. http://dx.doi.org/10.1016/j.apgeochem.2004.07.002

[27]   APHA (1991) Standard Methods for the Examination of Water and Wastewater. 17th Edition, American Public Health Association, Washington DC.

[28]   Dean, H.T. (1942) The Investigation of Physiological Effects by the Epidemiological Method. *American Association for the Advancement of Science*, **19**, 23-33.

[29]   Olsson, B. (1979) Dental Finding in High Fluoride Areas. *Epidemiology*, **7**, 51-56.

14

Field Comparison of Soil Moisture Sensing Using Neutron Thermalization, Frequency Domain, Tensiometer, and Granular Matrix Sensor Devices: Relevance to Precision Irrigation

Clinton C. Shock[1*], André B. Pereira[2], Erik B. G. Feibert[3], Cedric A. Shock[4], Ali Ibrahim Akin[5], Levent Abdullah Unlenen[6]

[1]Malheur Experiment Station, Oregon State University, Ontario, USA
[2]Department of Soil Science and Agricultural Engineering, State University of Ponta Grossa, Ponta Grossa, Brazil
[3]Oregon State University, Malheur Experiment Station, Ontario, USA
[4]Oregon State University, Malheur Experiment Station, Ontario, USA
[5]Ankara Nuclear Agriculture and Animal Research Center, Ankara, Turkey
[6]General Directorate of Agricultural Researches, Potato Research Institute, Ministry of Agriculture and Rural Affairs, Nigde, Turkey
Email: *clinton.shock@oregonstate.edu

Abstract

The efficient use of irrigation water requires several kinds of information. One element of efficient irrigation scheduling is monitoring the soil moisture to assure that the crop irrigation goals are being met. Various soil moisture sensing devices were tested for irrigation scheduling in silt loam at the Malheur Agricultural Experiment Station, Oregon State University between 2001 and 2004. Neutron probes, frequency domain probes, tensiometers, granular matrix sensors, and Irrigas were compared as to their performance under field conditions at Ontario, Oregon, USA. Granular matrix sensors were tested as read automatically by a datalogger and read manually with a hand-held meter. Practical suggestions are provided to use soil moisture sensors to the benefit of crop production and water conservation.

---

*Corresponding author.

Keywords

Soil Water, Irrigation Scheduling, Responsiveness, Soil Moisture Sensor, Porous Ceramic

1. Introduction

Soil moisture monitoring is a well established method to govern irrigation scheduling [1]. Application of water at a criterion wetter than crop needs results in over-application of water and a potential loss of crop yield and quality [2] [3]. Irrigation that precisely matches crop soil water requirements can avoid unnecessary water losses and can optimize yield and quality [4]-[7].

Precise irrigation scheduling is necessary to optimize marketable yield of high value crops while conserving water and protecting surface water and groundwater quality. Irrigation scheduling is greatly facilitated by any soil moisture sensor which can provide timely and reliable information on soil water content or soil water potential. For a particular sensor to be useful for a particular crop and soil, it needs to respond rapidly and reliably to the range of variation of soil water status that is important for marketable yield and product quality. Sensor readings need to be meaningful to assist irrigation scheduling, avoiding excessively frequent or intensive irrigation. Several types of sensors were tested for their responsiveness and usefulness for irrigation scheduling in soils typical of the Treasure Valley of the Snake River Plain of Oregon and Idaho. Neutron probes [8], frequency domain sensors [9] [10], tensiometers [11], and granular matrix sensors [12] have been widely used for irrigation scheduling. Tensiometers and granular matrix sensors respond to soil water potential. The porcelain tip of the tensiometer takes up or releases water in the immediate vicinity of the tip in response to the dynamics of the soil water potential and the partial vacuum inside the tensiometer. The granular matrix sensor takes up or releases water to the soil in its immediate vicinity and responds with variable electrical resistance between its internal electrodes. The neutron probe and frequency domain sensors respond to volumetric soil water content but in very different ways. The neutron probe emits high speed neutrons into the soil and counts much lower speed neutrons that return following inelastic encounters with hydrogen atoms in the soil. The capacitance probes measure capacitance in various high frequencies in the immediate vicinity of the probe in the soil. In one experiment, sensors were tested that utilize air permeability of porous ceramics.

Air permeability of porous ceramics has been used to estimate soil water potential [13]. Air permeability of a specific porous ceramic is a function of its water content. As water dries from the ceramic, the pores allow the passage of air. The "initial bubbling pressure" (IBI) of a water saturated porous ceramic is the lowest applied pressure at which air permeability (bubbling) is observed. The IBI of a specific porous ceramic has been used to estimate whether a soil has reached a specific soil water potential, possibly an irrigation criterion. The National Center for Horticultural Research of EMBRAPA, Brasilia, Brazil developed "Irrigas" based on IBI of porous ceramics. Irrigas consists of a porous ceramic cup, a moveable container of water, a flexible tube, a transparent barrel, and a rigid thin plastic support. The porous ceramic cup is installed in the effective rooting zone of the crop and connected to a small transparent barrel by means of the flexible tube. The porous ceramic cup is designed to retard free air movement out of the cup until the soil and cup reach a predetermined water potential. To make a reading, the barrel is immersed in the container of water. The free air passage through the porous ceramic cup gets blocked whenever the soil water saturates the pores in the ceramic. When the soil dries, its moisture eventually drops below a critical tension value, and the porous cup becomes permeable to the passage of air. In dry soils when the barrel is immersed into the water, the meniscus (air-water boundary) rapidly moves upwards in the barrel to equalize it to the water level in the container. Whenever water enters the barrel, the soil is at least as dry as the calibration of the porous ceramic cup. The soil moisture is evaluated once a day to determine the moment to irrigate. In sandy soils the evaluation is made twice a day.

Various sensors were tested in four field trials over several years.

2. Materials and Methods

2.1. Experiment 1

Six soil moisture sensors were compared by their performance in response to wetting and drying in irrigated hybrid poplar (*Populus deltoides x P. nigra*) at the Malheur Experiment Station in Ontario, Oregon, 43°58'55"N 117°01'27"W, 683 m above mean seal level (amsl). The trees were planted in April 1997 on Nyssa-Malheur silt loam soil (coarse-silty, mixed, mesic Xerollic Durorthid) on a 4.27-m by 4.27-m spacing. The tree rows were oriented to the northwest. The trees were irrigated using a micro sprinkler system (R-5, Nelson Irrigation, Walla Walla, WA) with the risers placed between trees along the tree row at 4.27-m spacing. The sprinklers delivered water at the rate of 3.6 mm/hour at 176 kPa and a radius of 4.27 m. The area used for the sensor performance trial was managed to receive 50.8 mm of water whenever the soil water potential at 0.20-m depth reached −50 kPa.

Two frequency domain Aquaflex sensors (Streat Instruments, Christchurch, New Zealand) were installed on September 14, 2000. Each 3-m long sensor was installed at 0.20-m depth along the tree row and between two trees. The two Aquaflex sensors were connected to an Aquaflex datalogger (Streat Instruments). On July 23, 2001, six types of soil moisture sensors were added to the study and were read daily at 9 AM for 40 days. One sensor of each type was installed in four replicates: two replicates each were installed adjacent to each of the existing Aquaflex sensors. The order of the sensors was randomized within each replicate to achieve a randomized complete block statistical design. The sensors in each replicate were installed at 0.20-m depth in a line parallel to and 0.20 m from the Aquaflex sensors. The sensors in the four replicates were tensiometer (Irrometer SR, Irrometer Co. Inc, Riverside, CA, USA), granular matrix sensor (GMS, Watermark Soil Moisture Sensor model 200SS, Irrometer Co. Inc), neutron probe model 503 DR Hydroprobe (Boart Longyear, Martinez, CA), time domain reflectometry Moisture Point (E.S.I. Environmental Sensors Inc., Victoria, British Columbia, Canada), and two frequency domain devices, the Gro Point (E.S.I. Environmental Sensors Inc) and the Gopher (Cooroy, Queensland, Australia). The neutron probe and Moisture Point were capable of readings below the 20-cm depth, but only readings at 20 cm were used for comparisons. The four Gro Point sensors were connected to two Gro Point 3-channel dataloggers (E.S.I. Environmental Sensors Inc). The GMS were connected to an AM400 Soil Moisture Datalogger (M.K. Hansen Co., East Wenatchee, WA, USA). All other sensors were read manually at 9 AM from Monday through Friday.

The tensiometer and GMS required that a hole in the soil be made with a standard 22.2-mm (7/8-inch) diameter soil auger for installation. The tensiometers required regular resetting due to the column of water breaking suction around −60 to −70 kPa. The Gro Point sensor was relatively compact and was easy to bury. Both the neutron probe and the Gopher required the installation of PVC access tubes for each monitored location. The Moisture Point used a 0.9 m long probe permanently installed at each location to be monitored. The Moisture Point probe required a hole driven with a rectangular rod provided by Environmental Sensors Inc for installation. The neutron probe, Gopher, and Moisture Point allowed measurement of soil moisture at different depths at each location. The Aquaflex required a horizontal 3-m trench dug to the depth of installation, 0.20 m.

Both the neutron probe and Gopher required site specific calibration. One undisturbed core soil sample was taken in each instrument location during sensor installation. The soil samples were immediately placed in tin cans and weighed, then oven dried at 100°C for 48 hours and weighed again. Volumetric soil moisture content was calculated for the soil samples using the gravimetric method. After the sensors were installed, 50.8 mm of water was applied. On July 25, another set of soil samples was taken and volumetric soil moisture content was determined as before. The sensors were read at the same time as the soil samples were taken. The neutron probe was read as counts during 32 seconds. For calibration, the volumetric soil water content determined from the soil samples was regressed against the neutron probe and Gopher readings. The average soil moisture data from the neutron probe and from the tensiometers was compared using regression against the average soil moisture data for each of the other sensors. The GMS sensors were used to manage the irrigation. The entire area of the sensor comparison trial was irrigated when the GMS readings reached −50 kPa.

2.2. Experiment 2

Six types of soil moisture sensors were compared by their performance in response to wetting and drying in drip-irrigated potato (*Solanum tuberosum*) at the Malheur Experiment Station at Ontario, Oregon, 43°58'44"N 117°01'03"W, 676 m amsl. The sensors were Aquaflex, Gro Point, Moisture Point, neutron probe, tensiometer, and GMS as described above. The GMS were evaluated as read automatically by an AM400 datalogger and read manually with a hand-held meter, model 30 KTCD-NL (Irrometer Co. Inc.), as previously calibrated [15].

Potato seed of cultivar 'Mazama' was planted on April 26, 2002 in rows spaced 0.91 m apart. The potato seed pieces were spaced 0.23 m apart in the row. The soil was an Owyhee silt loam (coarse-silty, mixed, mesic Xerollic Camborthid) with a pH of 8.1 and 2 percent organic matter. Drip tape (T-tape, T-systems International, San Diego, CA) was laid at 0.10-m depth between two potato rows. The drip tape had emitters spaced 0.305 m apart and nominal emitter flow rate of 0.5 L/hr. The potato crop was irrigated daily to replace the previous day's evapotranspiration. Potato evapotranspiration (ET$_c$) was calculated with a modified Penman equation [14] using data collected at the Malheur Experiment Station by an AgriMet weather station (United States Bureau of Reclaimation, Boise, Idaho). In mid-June the sensor study was installed along one of the potato rows. Six types of sensors were installed between the drip tape and the potato row. The sensors were installed 0.20 m from the drip tape and 0.25 m from the potato row. The sensors were centered at 0.23-m depth. The experimental design was a randomized complete block design with four replicates. These instruments were installed, managed, and calibrated as in experiment 1 above.

From July 15 to July 25 and again from July 30 to August 7, the potato row containing the sensors was not irrigated so that sensor performance could be evaluated under variable soil moisture, during both wetting and drying conditions. Sensor readings were evaluated for 30 days near 9 AM.

2.3. Experiment 3

Four types of soil moisture sensors were evaluated in furrow-irrigated onion (*Allium cepa*) grown on Owyhee silt loam (coarse-silty, mixed, mesic, Xerollic Camborthid) over various wetting and drying cycles at the Malheur Experiment Station, 43°58'48"N 117°01'01"W, 675 m amsl. Onion seeds were direct-seeded on 17 March 2004 in double-rows on 0.56-in beds. The double onion rows were spaced 76 mm apart. The types of soil moisture sensors were tensiometers with pressure transducers (Irrometer Model RA, Irrometer Company, Inc.), ECH2O 10 dielectric aquameter (Decagon Devices, Pullman, Washington), GMS, and Irrigas (National Center for Horticultural Research of EMBRAPA, Brasilia, Brazil. On 15 July 2004 the sensitive parts of the sensors were installed at 0.20-m depth below double rows of onions. The statistical design was a randomized complete-block with 4 replicates and the replicates were placed 20 m apart down a 1.25-ha furrow-irrigated field.

Tensiometers, GMS, and ECH2O 10 dielectric aquameters were attached to three AM416 multiplexers (Campbell Scientific, Logan UT, USA) that in turn were wired to a CR10X datalogger (Campbell Scientific). The CR10X datalogger was programmed to take hourly readings. Two temperature sensors were installed at 0.20 m depth and the datalogger was programed to make temperature corrections of GMS readings [15]. Data were collected from the datalogger using a laptop computer from 15 July to 30 September 2004. Each replicate contained 2 tensiometers, 2 GMS, 1 ECH2O 10 dielectric aquameter, and 2 Irrigas. The Irrigas operated on the principle of air permeability of porous ceramics explained above. The Irrigas had a nominal calibration of −25 kPa and when we subjected Irrigas to progressive amounts of suction, the porous ceramic freely bubbled air at −25 kPa in the laboratory out of the soil. Irrigas readings were taken every day at 9 AM.

The ECH2O 10 dielectric aquameters were calibrated against volumetric soil water content by taking two soil samples near each probe centered at 0.20 m depth, once when the soil was relatively wet, and once when the soil was relatively dry, and also by preparing oven dry soil and placing the probes in the oven dry soil at the end of the trial. Gravimetric data were converted to volumetric water contents using the soil bulk density.

Prior to starting the sensor performance trial, the onions were irrigated at −25 kPa based on average GMS readings [3]. With the establishment of this experiment in the onion field, the onions in the entire sensor calibration trial were irrigated when the average GMS reading reached −25 kPa on July 17 and 22. Since the Irrigas had not provided positive readings at −25 kPa, the next five irrigations were delayed until at least half of the 8 Irrigas sensors indicated the need for irrigation.

2.4. Experiment 4

Automated reading of GMS was done in a furrow-irrigated Greenleaf silt loam (fine-silty, mixed, mesic Xerollic Haplargid) planted to onions at the Malheur Experiment Station, Ontario, Oregon, 43°58'04"N 117°01'17"W, 675 m amsl. Onion seeds were planted 14 March 2003 in double-rows on 0.56-in beds. The double onion rows were spaced 76 mm apart. The response of two independent sets of GMS to irrigation events and the termination of irrigation was read automatically using an AM400 datalogger and a Watermark Monitor (Irrometer Co. Inc.). Data from the last 41 days of the growing season were used for comparison purposes due to the larger range of soil water potential that occurred as the irrigation season ended and the soil dried.

The sensors were installed with their centers 0.20 m deep directly below the onion plants. The sensors were installed in the lower part of the 0.6 ha field where the furrow irrigations were less effective at wetting the soil. Six GMS and a temperature probe were connected to an AM400 datalogger which read the sensors three times a day. Data were recovered from the AM400 using a palm computer as previously described [16]. Seven GMS and a temperature probe were connected to a Watermark Monitor. A laptop computer and the WaterGraph program (Irrometer Co., Inc.) were used to set the sensor data collection frequency at 15 minutes. Data were recovered from the Irrometer Watermark Monitor using a laptop and the WaterGraph program.

2.5. Interpretation of Data

The soil moisture data from each type of sensor was graphed over time to determine if the data provided meaningful wetting and drying trends. In experiments 1, 2, and 3, soil moisture monitoring devices were compared using regression analyses over the range of wetting and drying observed in the particular trial (NCSS 97, Statistical System for Windows, [17]).

3. Results

3.1. Experiment 1

The coefficient of determination (r^2) for the regression equation for the neutron probe at 0.20-m depth was 0.93 at $P = 0.01$. The regression equation was used to transform the neutron probe readings to volumetric water content. A calibration for the Gopher sensor was not possible due to a lack of correlation between the Gopher readings and the volumetric soil water content determined from the soil samples. The tensiometer, Watermark, neutron probe, Gro Point, and Aquaflex responded to the wetting and drying cycles of the soil (**Figure 1**). The

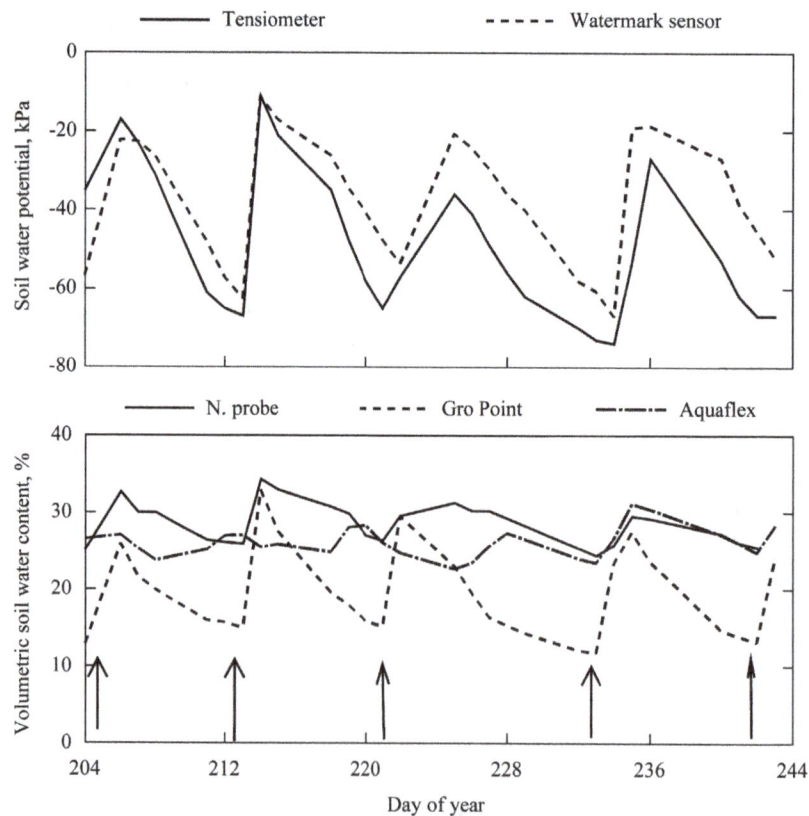

Figure 1. Soil moisture data over time for five types of soil moisture sensors in Experiment 1. Arrows denote irrigations with 50.8 mm of water applied. The Moisture Point sensor was not available during this time due to repairs being made. Malheur Experiment Station, Oregon State University, Ontario, OR, 2001.

Aquaflex sensors were less responsive to the soil drying between irrigations than the neutron probe and the Gro Point sensor was more responsive than the neutron probe. The lower responsiveness of the neutron probe than the Gro Point was not surprising since the response volume of the neutron probe is larger and integrates less rapidly changing soil moisture deeper in the profile. The magnitude of the drying indicated by the Gro Point was beyond what occurred. All sensors showed correlations ($r^2 > 0.7$) to the neutron probe ($P = 0.001$) and correlations ($r^2 > 0.5$) to the tensiometer ($P = 0.001$ to $P = 0.01$) except the Moisture Point sensor (**Figure 2** and **Figure 3**). Furthermore the Moisture Point estimates of soil water were substantially lower than the neutron probe data (**Figure 2** and **Figure 3**).

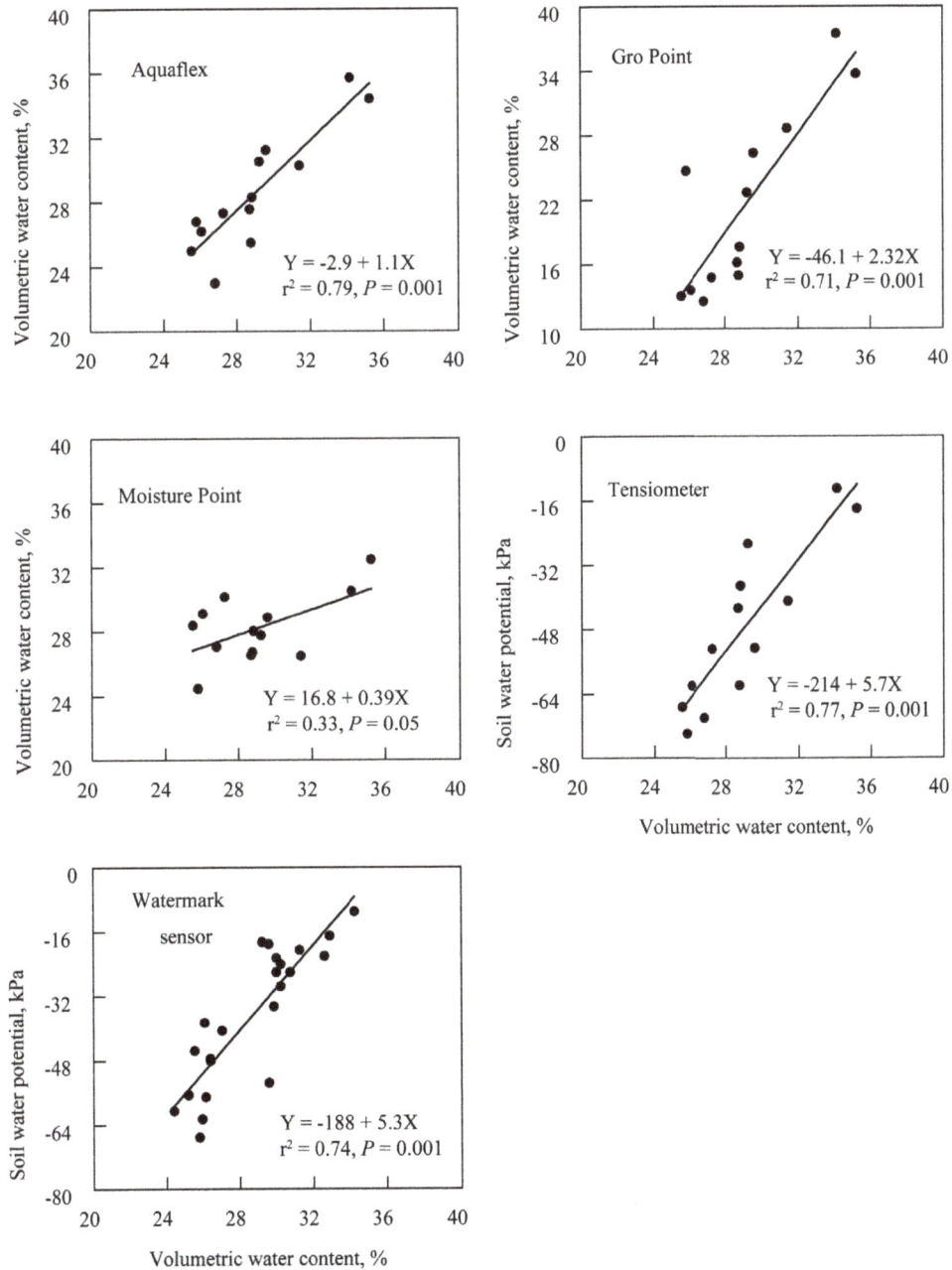

Figure 2. Volumetric soil water content measured in Experiment 1 by a neutron probe (X axis) regressed against soil moisture data (Y axis) measured by 5 types of soil moisture sensors. Data points for the Aquaflex sensor are the average of two sensors. Data points for the other sensors are the average of four sensors. Malheur Experiment Station, Oregon State University, Ontario, OR, 2001.

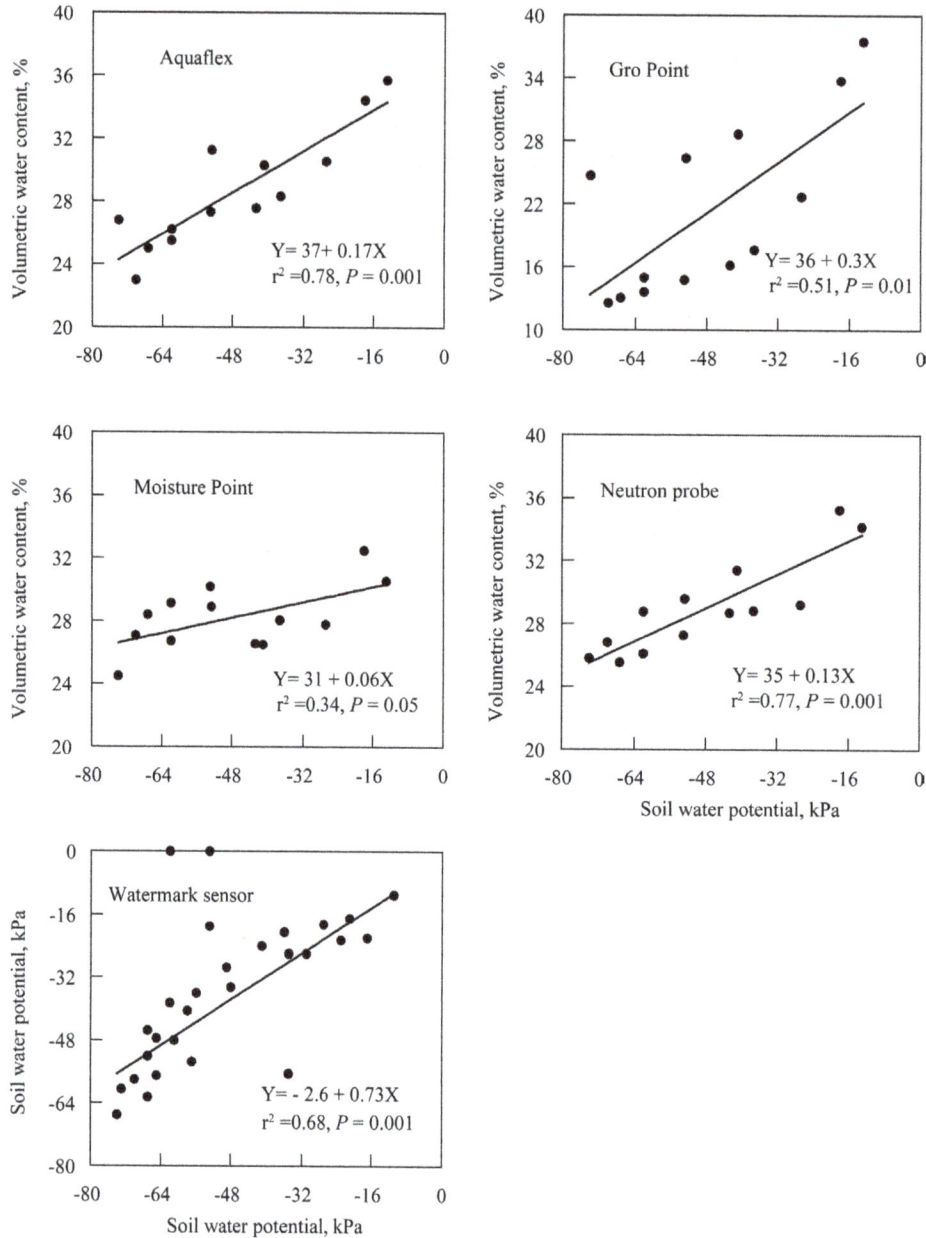

Figure 3. Soil water potential measured in Experiment 1 by tensiometers (X axis) regressed against soil moisture data (Y axis) measured by 5 types of soil moisture sensors. Data points for the Aquaflex sensor are the average of two sensors. Data points for the other sensors are the average of four sensors. Malheur Experiment Station, Oregon State University, Ontario, OR, 2001.

3.2. Experiment 2

The tensiometer, GMS, and neutron probe responded closely to the wetting and drying cycles of the soil (**Figure 4**). The Gro Point responded, but the amplitude of the response showed less fluctuation than that of the neutron probe. The Gro Point's variations could have been greater than the neutron probe, since the neutron probe measures soil water in a larger soil volume, including deeper soil that presumably would dry less. The Moisture Point was the least responsive to the wetting and drying cycles of the soil compared to the other sensors. For undetermined reasons, the Aquaflex datalogger only collected 3 days of data; this did not allow for conclusive results.

Readings of GMS by both the AM400 datalogger and the 30 KTCD-NL meter showed close correlations to tensiometer readings (**Figure 5**). The AM400 and the 30 KTCD-NL readings of different GMS were fairly

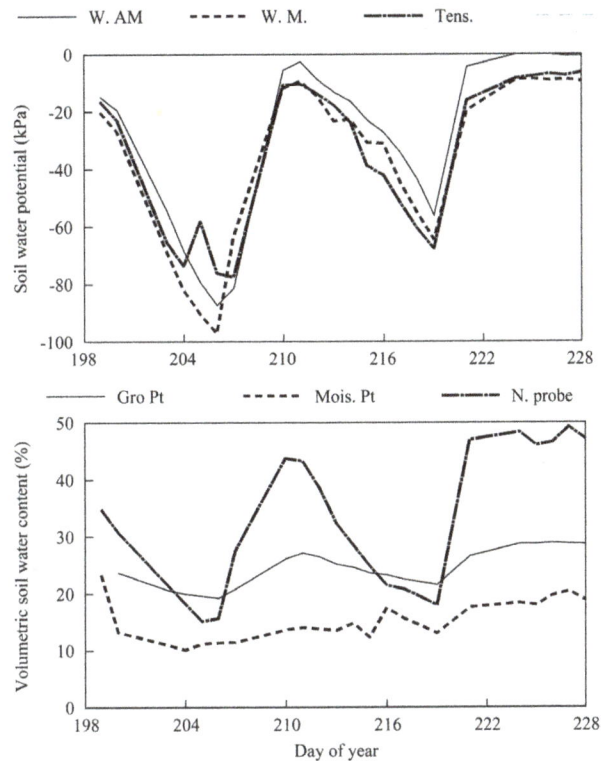

Figure 4. Soil moisture over time for five types of soil moisture sensors in Experiment 2, tensiometers (Tens.), Watermark soil moisture sensors read by a AM400 datalogger or manually (W. AM or W. M.), Gro Point (Gro Pt), Moisture Point (Mois. Pt) and neutron probe (N. probe). Malheur Experiment Station, Oregon State University, Ontario, OR, 2002.

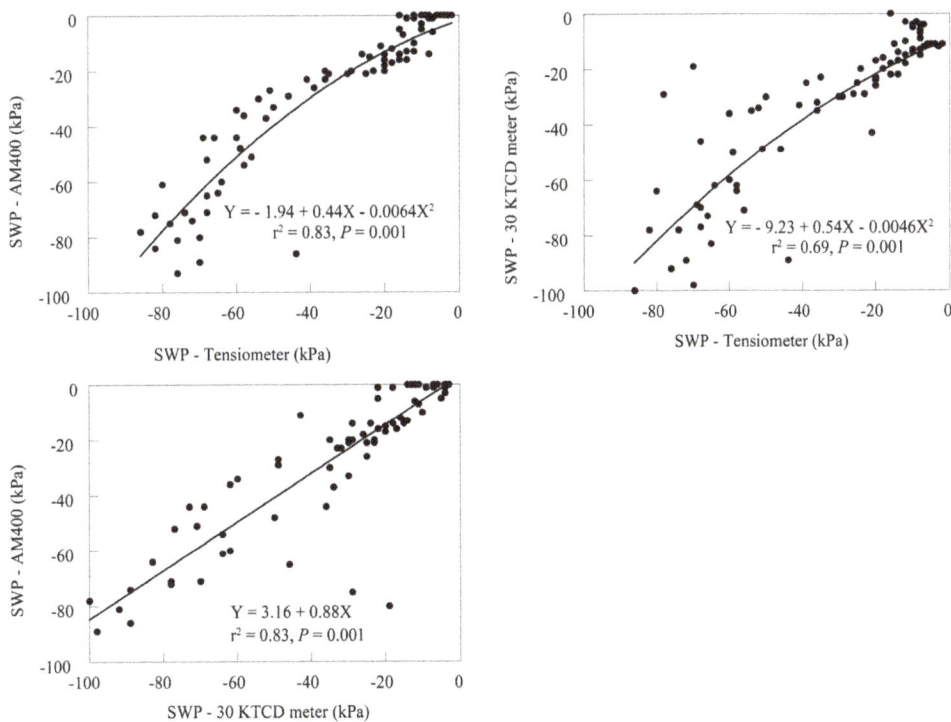

Figure 5. Regressions of soil water potential (SWP) measured in Experiment 2 by three instruments. Malheur Experiment Station, Oregon State University, Ontario, OR, 2002.

closely correlated to each other; both instruments used similar equations to convert Watermark sensor electrical resistance to soil water potential (Shock et al. 2001).

All sensors showed correlations ($r^2 > 0.6$) to the neutron probe ($P = 0.001$) except the Moisture Point sensor and Aquaflex (**Figure 6**). The Gro Point estimates of soil water were often lower than the neutron probe (**Figure 4** and **Figure 6**). The Moisture Point estimates of soil water were substantially lower than the neutron probe and Gro Point (**Figure 4** and **Figure 6**).

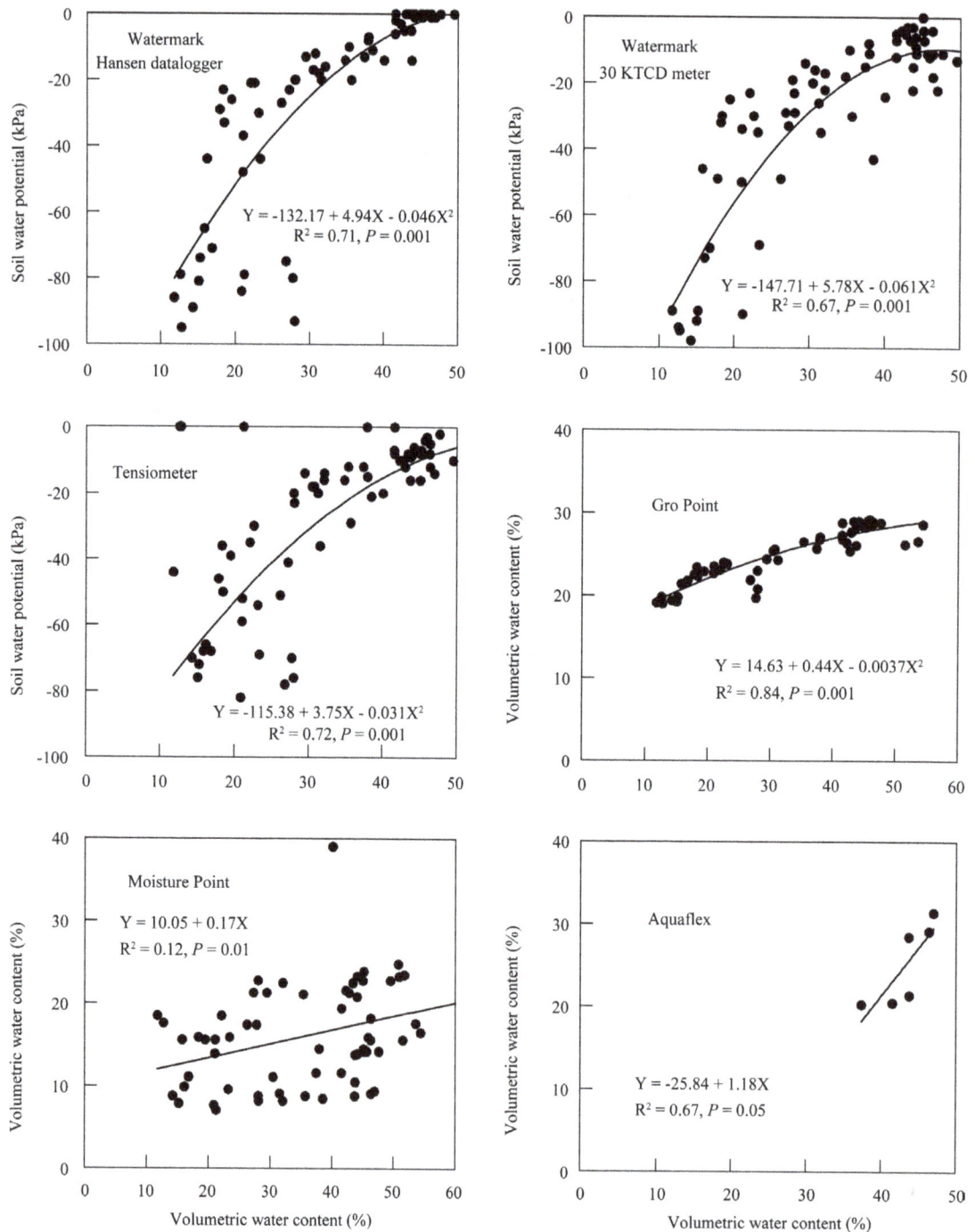

Figure 6. Volumetric soil water content measured in Experiment 2 by a neutron probe (X axis) regressed against soil moisture data (Y axis) measured by 6 types of soil moisture sensor. Malheur Experiment Station, Oregon State University, Ontario, OR, 2002.

3.3. Experiment 3

All the sensors used in this study had low unit cost and simple installation. There were two episodes of irrigation based on GMS readings at −25 kPa and five irrigation events based on the Irrigas criterion (**Figure 7**). Both tensiometers and GMS had similar responses to wetting and drying of the soil (**Figure 7**).

It took about 4 h for all the tensiometers and all GMS to indicate that the soil at 0.20 m had reached saturation after the onset of each furrow irrigation episode. The relative similarity in responsiveness between tensiometers with pressure transducers and granular matrix sensors (GMS) was confirmed by regression with a coefficient of determination of 0.92 ($P = 0.0001$) (**Figure 8**). The Irrigas had free air permeability close to −35 kPa for Owyhee silt loam in this trial (**Figure 7**).

Large changes in tensiometer readings from −10 to −40 kPa translated into small changes in water content readings for the ECH2O 10 dielectric aquameter (**Figure 9**). A comparison of the ECH2O 10 dielectric aquameter readings with soil volumetric water content from this field indicated that the readings were relatively flat and nonlinear in response to changes in volumetric soil water content (**Figure 10**). The relatively small changes in volumetric soil water content measured by the ECH2O 10 dielectric aquameter were unrealistic and the small changes in measurements occurred across the critical range of soil water potential for onion irrigation decisions, limiting the usefulness of the probe. The reasons for the low responsiveness of the ECH2O 10 dielectric aquameter to the soil water content were beyond the scope of this work.

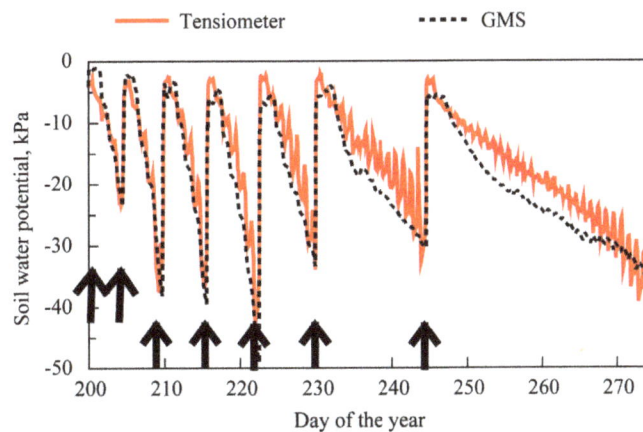

Figure 7. Soil water potential over time for tensiometers with transducers and granular matrix sensors in Experiment 3. Arrows denote furrow irrigations with 75 mm of water applied. Malheur Experiment Station, Oregon State University, Ontario, OR, 2004.

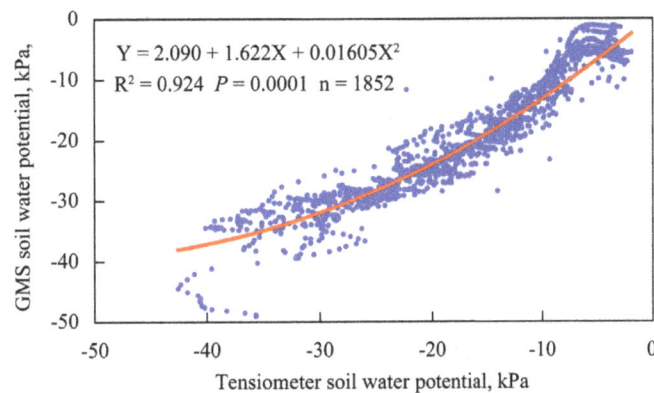

Figure 8. Soil water potential measured in Experiment 3 by a tensiometer with transducers (X axis) regressed against soil moisture suction measured by a granular matrix sensor (Y axis). Data points are the average of eight instruments. Malheur Experiment Station, Oregon State University, Ontario, OR, 2004.

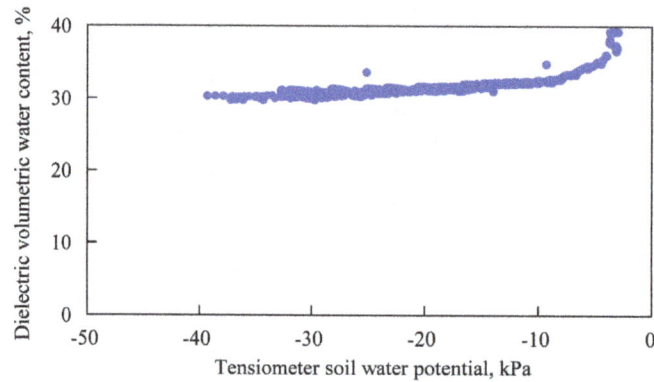

Figure 9. Soil water potential measured in Experiment 3 by a tensiometer with transducers (X axis) regressed against volumetric soil water content measured by an ECH2O 10 dielectric aquameter (Y axis). Data points for soil water potential are the average of eight tensiometers. Data points for the ECH2O 10 dielectric aquameter are the average of four sensors. Malheur Experiment Station, Oregon State University, Ontario, OR, 2004.

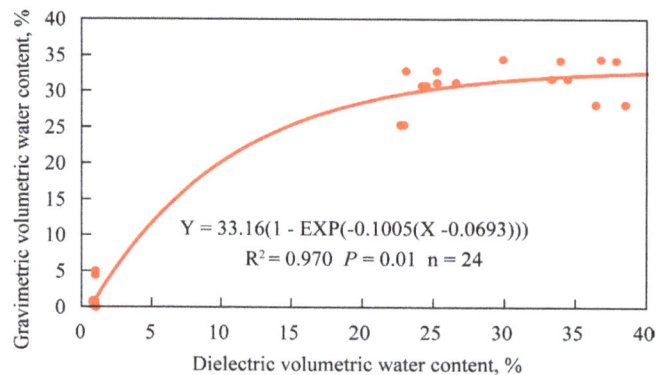

Figure 10. Regression of the volumetric soil water content measured by an ECH2O 10 dielectric aquameter (X axis) against the classical gravimetric method (Y axis). Data points from each of four ECH2O 10 dielectric aquameters were compared with two soil samples in each of three soil moisture ranges. Malheur Experiment Station, Oregon State University, Ontario, OR, 2004.

3.4. Experiment 4

The automated collection of GMS data by an AM400 datalogger and a Watermark Monitor provided similar interpretation of wetting and drying cycles (**Figure 11(a)**). The GMS started responding to irrigation within one hour of the irrigation onset. Small differences in calibration equations were noted (**Figure 11(b)**) and slight differences in the interpretation of soil water potential near saturation were evident (**Figure 11(a)**).

4. Discussion

4.1. Neutron Thermalization

The neutron probe was used in two of the four experiments. The neutron probe readings clearly indicated changes in soil water content. Due to the site specific calibrations, the changes in soil water content readings accurately followed the soil water content, as has been described by others [8]. Where the soil texture is variable, such as the silt loam soils of Ontario, Oregon, the practical use of neutron probe soil water content data requires field by field calibration. Calibrated data can be accurately interpreted to meet crop irrigation requirements and avoid over irrigation. Neutron probe use was awkard, time consuming, and required safety precautions. Readings can be downloaded automatically from the probe to a computer, but neutron probe readings cannot be automated in the sense of continuous readings.

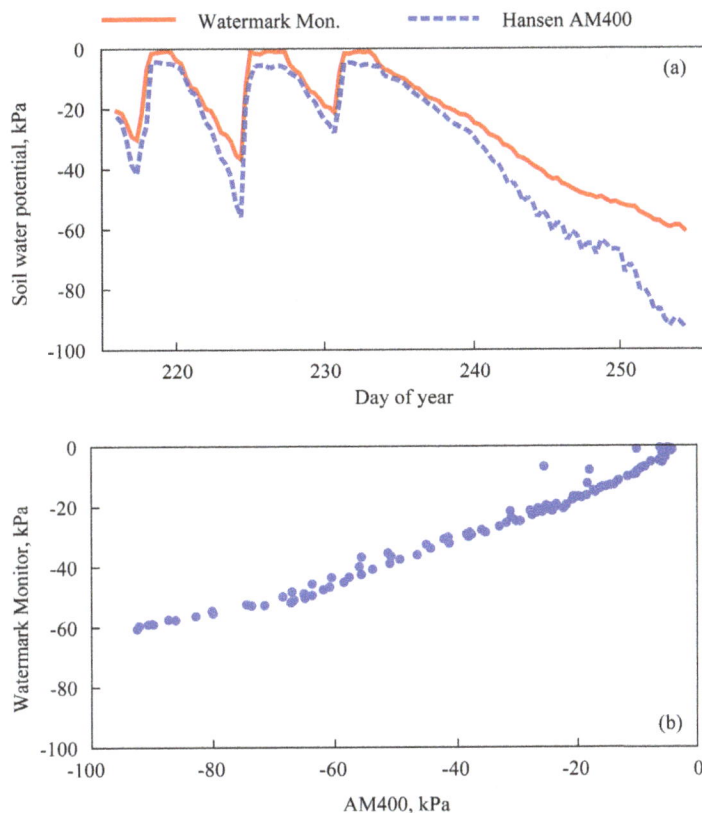

Figure 11. Response of Watermark soil moisture sensors to irrigation events and the termination of irrigation as measured by a Hansen AM400 datalogger and an Irrometer Watermark Monitor in Experiment 4. The average readings of an AM400 datalogger and a Watermark Monitor are compared over time (a) and over the measured range of soil water potential (b). Malheur Experiment Station, Ontario, OR, 2003.

4.2. Frequency Domain Sensors

The frequency domain sensors were more or less easy to automate. The ECH2O 10 dielectric aquameter was relatively easy to automate. The ECH2O 10 dielectric aquameter was used in only one experiment and the readings were relatively unresponsive to changes in soil water potential in the range of −10 to −40 kPa (**Figure 9**) and relatively unresponsive to changes to volumetric soil water content in the range of 23% to 38% (**Figure 10**). Limited ECH2O 10 dielectric aquameter responsiveness suggests the need for site specific calibrations prior to use. The ECH2O 10 is a 10 cm long, single panel capacitance soil moisture probe that uses low-frequency signals. Decagon Devices discontinued the ECH2O 10, replacing it with a high-frequency two pronged capacitance soil moisture probe (10HS), with much less susceptibility to the effects of soil salinity and soil texture.

Gro Point could be conveniently automated with proprietary dataloggers. The Gro Point readings followed wetting and drying cycles in the two experiments where they were used (**Figure 1** and **Figure 4**) with inconsistent results. The changes in soil water were overestimated in Experiment 1 (**Figure 2**) and underestimated in Experiment 2 (**Figure 6**). These results suggest that the Gro Point would benefit from site specific calibrations prior to use. The need for site specific calibrations noted here for both the Gro Point and ECH2O 10 dielectric aquameter is consistent with the work of Evett who tested a variety of capacitance probes in widely divergent soils and recommended site specific calibrations [9]. Gopher readings were not closely associated with changes in soil water. With the Aquaflex, the interpretation of soil wetting and drying trends were ambiguous in Experiment 1.

4.3. Tensiometers

Variations in soil water were clearly shown by tensiometers in the three experiments where they were used and the interpretations of the results were clear. The tensiometers with pressure transducers used in experiment 3 were easily automated. Due to variation in the tensiometer reading by time of day (**Figure 7**), visual readings at a consistent time of day as done here undoubtedly helped assure data quality in experiments 1 and 2. The tensiometers required servicing twice during the 76 days of the trial 3. More frequent servicing to replace lost water should be expected when soils are not maintained as wet as in these experiments.

4.4. Granular Matrix Sensors

Granular matrix sensors were used in all four experiments and were very responsive to wetting and drying in the soils used in these experiments (**Figure 1**, **Figure 4**, **Figure 7** and **Figure 11**). The GMS have limitations in reading soil water potential in soils wetter than -10 kPa (**Figure 5** and **Figure 8**), as has been described previously [15], and in responding in coarse textured soils [12].

The three methods for automated reading of GMS in several trials were all convenient. The AM400 was helpful for following and scheduling irrigation events in the field due to its graphic display. The Watermark Monitor provided convenient settings for datalogger reading frequency, easy retrieval, and automatic interpretation of the data. The Campbell Scientific dataloggers plus peripheral equipment had great flexibility for reading GMS, as has been shown by complex designs of GMS and their use in controlling irrigations [18].

4.5. Air Permeability of Porous Ceramics

The model of "Irrigas" tested here in only one comparison experiment, appeared to provide a clear signal for irrigation scheduling at -35 kPa in silt loam, and failed to signal the nominal specification of -25 kPa. Since previous research had shown furrow irrigations at a criterion of -27 kPa optimized long-day onion yield and grade on silt loam soil [3], a nominal signal at 25 kPa could have provided a useful irrigation scheduling tool. The Irrigas equipment configuration used had limited usefulness, since water stress sensitive vegetable crops have many different irrigation criteria, and the ideal criteria are complex since they also vary by climate, soil type, and irrigation system [7].

Soil particles in contact with a porous ceramic can interfere with air permeability [13]. From results of this trial, it is possible that the silt loam interfered with the air permeability. One might predict greater interference by fine textured soils and less interference with coarse textured soils. One might predict greater interference at relatively high (wetter) soil water potential and less interference at relatively low (drier) soil water potential. In the present experiment the soil texture was fine and the soil water potentials were wet.

5. Conclusion

The soil moisture measurement devices tested here generally provided data closely correlated with each other. Neither the neutron probe nor the Irrigas were convenient to automate. The frequency domain sensors were more or less easy to automate but were not adequately reliable without engineering improvements or soil specific calibrations. Both tensiometers and granular matrix sensors were relatively easy to calibrate and the readings were readily useful for irrigation scheduling.

Acknowledgements

Technical assistance of student employees Scott Jaderholm, Kendra Nelson, Autumn Tchida and Christie Linford is greatly appreciated. Funding for two sensor comparison trials was made available by the Agricultural Research Foundation at Oregon State University, Corvallis, Oregon, USA. Additional support was provided by Oregon State University and Hatch funds. Many thanks are also owed to the Conselho Nacional de Desenvolvimento Cientifico e Tecnologico for the provision of the Post-Doctoral scholarship, Brazil, as well as to the Fundação Araucária of Brazil for the concession of a productivity fellowship in research.

References

[1]   Charlesworth P. and Stirzaker, R.J. (2003) Irrigation Scheduling by Soil Water Status. In: Stewart, B.A. and Howell, T.A., Eds., *Encyclopedia of Water Science*, Marcel Dekker, Inc., New York, 528-531.

[2]   Eldredge, E.P., Holmes, Z.A., Mosley, A.R., Shock, C.C. and Stieber, T.D. (1996) Effects of Transitory Water Stress on Potato Tuber Stem-End Reducing Sugar and Fry Color. *American Potato Journal*, **73**, 517-530.

http://dx.doi.org/10.1007/BF02851697

[3]   Shock, C.C., Feibert, E.B.G. and Saunders, L.D. (1998) Onion Yield and Quality Affected by Soil Water Potential as Irrigation Threshold. *HortScience*, **33**, 1188-1191.

[4]   Shock, C.C., Feibert, E.B.G. and Saunders, L.D. (2000) Irrigation Criteria for Drip-Irrigated Onions. *HortScience*, **35**, 63-66.

[5]   Shock, C.C., Feibert, E.B.G., Seddigh, M. and Saunders, L.D. (2002) Water Requirements and Growth of Irrigated Hybrid Poplar in a Semi-Arid Environment in Eastern Oregon. *Western Journal of Applied Forestry*, **17**, 46-53.

[6]   Shock, C.C. and Feibert, E.B.G. (2002) Deficit Irrigation of Potato. In: Moutonnet, P., Ed., *Deficit Irrigation Practices*, Food and Agriculture Organization of the United Nations, Rome, Water Reports 22, 47-55. http://www.fao.org/docrep/004/Y3655E/Y3655E00.htm

[7]   Shock, C.C. and Wang F.-X. (2011) Soil Water Tension, a Powerful Measurement for Productivity and Stewardship. *HortScience*, **46**, 178-185.

[8]   Evett, S.R. (2003) Measuring Soil Water by Neutron Thermalization. In: Stewart, B.A. and Howell, T.A., Eds., *Encyclopedia of Water Science*, Marcel Dekker, Inc., New York, 889-893.

[9]   Evett, S.R., Laurent, J.P. Cepuder, P. and Hignett, C. (2002) Neutron Scattering, Capacitance, and TDR Soil Water Content Measurements Compared on Four Continents. *17th World Congress of Soil Science*, 14-21 August 2002, Bangkok, 1021-1. (CD-ROM)

[10]   Evett, S.R. and Steiner, J.L. (1995) Precision of Neutron Scattering and Capacitance Moisture Gauges Based on Field Calibration. *Soil Science Society of America Journal*, **59**, 961-968. http://dx.doi.org/10.2136/sssaj1995.03615995005900040001x

[11]   Hubbell, J.M. and Sisson, J. (2003) Soil Water Potential Measurement by Tensiometers. In: Stewart, B.A. and Howell, T.A., Eds., *The Encyclopedia of Water Science*, Marcel Dekker, New York, 904-907.

[12]   Shock, C.C. (2003) Soil Water Potential Measurement by Granular Matrix Sensors. In: Stewart, B.A. and Howell, T.A., Eds., *The Encyclopedia of Water Science*, Marcel Dekker, New York, 899-903.

[13]   Kemper, W.D. and Amemiya, M. (1958) Utilization of Air Permeability of Porous Ceramics a Measure of Hydraulic Stress in Soils. *Soil Science*, **85**, 117-124. http://dx.doi.org/10.1097/00010694-195803000-00001

[14]   Wright, J.L. (1982) New Evapotranspiration Crop Coefficients. *Journal of Irrigation and Drainage Division*, *ASCE*, **108**, 57-74.

[15]   Shock, C.C., Barnum, J.M. and Seddigh, M. (1998) Calibration of Watermark Soil Moisture Sensors for Irrigation Management. In: *Proceedings of the International Irrigation Show*, Irrigation Association, San Diego, 139-146.

[16]   Shock, C.C., Corn, A., Jaderholm, S., Jensen, L.B., and Shock, C.A. (2001) Evaluation of the AM400 Soil Moisture Datalogger to Aid Irrigation Scheduling. In *Proceedings of the International Irrigation Show*, Irrigation Association, San Diego, 111-116.

[17]   Hintze, J.L. (2000) NCSS 97 Statistical System for Windows, Number Cruncher Statistical Systems, Kaysville, Utah.

[18]   Shock, C.C. Feibert, E.B.G., Saunders, L.D. and Eldredge, E.P. (2002) Automation of Subsurface Drip Irrigation for Crop Research. *World Congress on Computers in Agriculture and Natural Resources*, American Society of Agricultural Engineers, Iguacu Falls, 809-816.

15

A Spatial Evapotranspiration Tool at Grid Scale

Sivarajah Mylevaganam*, Chittaranjan Ray

University of Nebraska-Lincoln, Lincoln, NE, USA
Email: *sivaloga@hushmail.com

Abstract

The drastic decline in groundwater table and many other detrimental effects in meeting irrigation demand, and the projected population growth have force to evaluate consumptive use or evapotranspiration (ET), the rate of liquid water transformation to vapor from open water, bare soil, and vegetation, which determines the irrigation demand. As underscored in the literature, Penman-Monteith method which is based on aerodynamic and energy balance method is widely used and accepted as the method of estimation of ET. However, the estimation of ET is oftentimes carried out using meteorological data from climate stations. Therefore, such estimation of ET may vary spatially and thus there exists a need to estimate ET spatially at different spatial or grid scales/resolutions. Thus, in this paper, a spatial tool that can geographically encompass all the best available climate datasets to produce ET at different spatial scales is developed. The spatial tool is developed as a Python toolbox in ArcGIS using Python, an open source programming language, and the ArcPy site-package of ArcGIS. The developed spatial tool is demonstrated using the meteorological data from Automated Weather Data Network in Nebraska in 2010.

Keywords

Evapotranspiration, Penman-Monteith Method, Aerodynamic Method, Energy Balance Method, Python, ArcPy, ArcGIS, Spatial Scale, Geoprocessing, Python Toolbox

1. Introduction

The world population is projected to grow from 6.9 billion in 2010 to 9.1 billion in 2050 [1]. With this projected population growth, it has been documented that by 2050 the food demand is expected to increase by 70% [2]. Having said this, as per the ongoing evaluation by [1], the current status of food scarcity is not at a level to

---

*Corresponding author.

appreciate or sustain, specifically in developing countries. Therefore, for the next few decades, the most striking focus would be on food security to meet the growing demand. With this statement in mind, to alleviate global hunger and to improve food security in the world, many initiatives and collaborations at global levels have been taken up to ensure availability of adequate food supply by means of efficient use of resources. One of the key limited resources that influences the definition of food security is the available freshwater [1] [2].

Although over 70% of Earth's surface is covered by water, the amount of freshwater available for appropriation is limited as 97.5% of all water on Earth is saline [3]. This limited freshwater is competed by many water users such as irrigation, household and municipal, and industrial. Among these water users, irrigation demand accounts for 87% of the total use globally [1] [2]. The consequences of such high demand on irrigation in addition to the demands from other water users have forced to pump water heavily from groundwater aquifers. This has arguably caused groundwater tables to decline in many regions, including in Afghanistan, China, India, Iran, Pakistan, and the US. The drastic decline in groundwater table and many other detrimental effects in meeting irrigation demand, and the projected population growth have force to evaluate the consumptive use or evapotranspiration (ET), the rate of liquid water transformation to vapor from open water, bare soil, and vegetation, which determines the irrigation demand.

As underscored in the literature [4]-[8], to date, there are many methods available to estimate ET. These methods are either empirical or climate data driven. Under empirical based estimation of ET, Blaney-Criddle method or its modified version is widely used in the arid western regions of the United States [4] [5]. However, this method doesn't account for humidity, wind speed, and other climate factors. On the other hand, using meteorological data from climate stations, the methods of estimation of ET include aerodynamic method, energy balance method, and combination methods such as Penman-Monteith method [4]-[6]. The aerodynamic method of determining evaporation considers the transport of water vapor by the turbulence of the wind blowing over a natural surface. The energy balance method considers all heat energy received and reflected/dissipated by a cropped area or a water body. Penman-Monteith method of evaporation is obtained by combining the evaporation computed by aerodynamic and energy balance method [6]. The weighting factors are applied in combining the methods (*i.e.*, aerodynamic and energy balance). The weighting factors account for aerodynamic resistance and surface resistance that accounts for movement of water vapor from the plant leaves to the atmosphere. As underscored in the literature [4]-[8], Penman-Monteith method is widely used and accepted as the method of estimation of ET. However, the application using Penman-Monteith is oftentimes evaluated using meteorological data from climate stations.

The current practices of water resources planning and management are at watershed scale and oftentimes at grid scale in parallel with the contemporary technology development. Having said this, as discussed previously, the estimation of ET is carried out using meteorological data from climate stations. However, such estimation of ET may vary spatially and thus there exists a need to estimate ET spatially. Moreover, in the absence of optimum spatial scale, and with a growing interest of watershed studies at different spatial scales [9], the need to evaluate the variation of ET at different spatial scales is also required. Thus, in this paper, a spatial tool that can geographically encompass all the best available climate datasets to produce ET at different spatial scales is developed using Python, an open source programming language supported by a growing user community for its extensive collection of standard and third-party libraries, and the ArcPy site-package of ArcGIS.

2. Estimation of Evapotranspiration

The need to manage the available freshwater wisely with ever increasing population and the demand from irrigation has brought ET as one of the critical subject areas to research in the field of hydrology. Over the years, with many research works, numerous methods have been developed to estimate ET. These methods mainly fall under these categories: 1) aerodynamic method, 2) energy balance method, and 3) combination of aerodynamic and energy balance methods.

2.1. Aerodynamic Method

This method of determining evaporation considers the transport of water vapor by the turbulence of the wind blowing over a natural surface. According to this method, the evaporation $\left(E_a\right)$, generally from lakes and reservoirs, is proportional to $\left(e_s - e_z\right)$. The mathematical expression of this method is given by Equation (1).

$$E_a = M * \left(e_s - e_z\right) * u_z \tag{1}$$

where M, e_s, e_z, and u_z are mass transfer coefficient, saturated vapor pressure at water temperature, vapor pressure at height z, and wind velocity at height z, respectively. The mass transfer coefficient is given by Equation (2).

$$M = 0.622 * \frac{\rho_a C_E}{\rho_w P} \tag{2}$$

where P, ρ_w, ρ_a, and C_E are atmospheric pressure at height z, density of water, density of air, and evaporation coefficient, respectively.

By substituting Equation (2) in Equation (1),

$$E_a = 0.622 * \frac{\rho_a C_E}{\rho_w P} * (e_s - e_z) * u_z \tag{3}$$

Considering aerodynamic resistance, $r_a = \dfrac{1}{C_E u_z}$, Equation (3) leads to

$$E_a = 0.622 * \frac{\rho_a}{\rho_w P r_a} * (e_s - e_z) \tag{4}$$

Since $\rho_a = \dfrac{3.486 P}{T + 273}$, Equation (4) becomes

$$E_a = 0.622 * \frac{3.486 P}{\rho_w P r_a * (T + 273)} * (e_s - e_z) \tag{5}$$

where T is the air temperature in degree Celsius.

$$E_a = \frac{2.17 (e_s - e_z)}{\rho_w r_a * (T + 273)} \tag{5'}$$

2.2. Energy Balance Method

As shown in **Figure 1**, this method considers all heat energy received and reflected/dissipated by a cropped area or a water body. The portion of energy that is used to warm the air in contact with the ground or water surface is known as sensible heat flux (H). The term G is the heat conduction from the water surface or soil to the layer of soil or water below. Since the energy required to evaporate a unit mass of water is called latent heat of vaporization (λ), the total energy absorbed per unit area to evaporate E_r is $\rho_w \lambda E_r$. Therefore, neglecting the other small energy terms that are dissipated/stored, the energy balance for the control volume shown in **Figure 1** is given by Equation (6).

$$R_n - H - G - \rho_w \lambda E_r = 0 \tag{6}$$

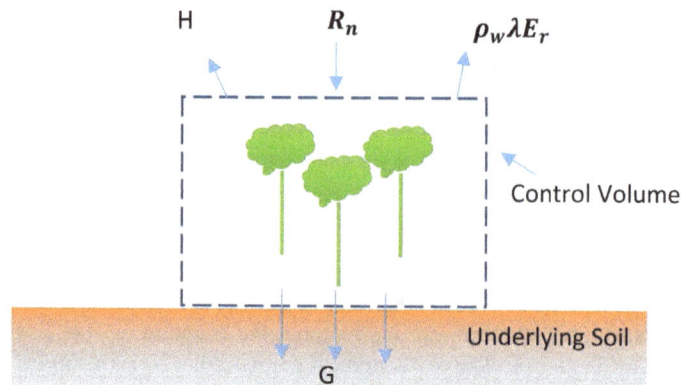

Figure 1. The energy flow diagram for a cropped area.

where R_n is the net radiation. The latent heat of vaporization is 2.45 MJ/kg at about 20 degree Celsius [6]. However, to account for temperature variation, the latent heat of vaporization is given by

$$\lambda = 2.501 - 0.002362T \qquad (6')$$

The sensible heat flux defined by Equation (7) is related to Bowen ratio, β

$$H = \beta * \rho_w \lambda E_r \qquad (7)$$

Bowen ratio, $\beta = \gamma \dfrac{T_2 - T_1}{e_2 - e_1}$ is derived from temperatures and vapor pressures at two heights above the water surface. By substituting Equation (7) in Equation (6),

$$E_r = \frac{R_n - G}{\rho_w \lambda (1 + \beta)} \qquad (8)$$

Since $G \approx 0$,

$$E_r = \frac{R_n}{\rho_w \lambda (1 + \beta)} \qquad (9)$$

As shown in Equation (10), the net radiation (R_n) is the sum of net long-wave radiation (L_n) and net short-wave radiation (S_n).

$$R_n = L_n + S_n \qquad (10)$$

The net short-wave radiation that is defined by Equation (11) is a function of total extraterrestrial radiation (S_0) and cloudiness fraction $\left(\dfrac{n}{N}\right)$.

$$S_n = (1 - \alpha)\left(0.25 + 0.5\frac{n}{N}\right)S_0 \qquad (11)$$

The net long-wave radiation which is in accord with Stefan-Boltzmann's law of black body radiation is given by Equation (12).

$$L_n = -\left(0.1 + 0.9\frac{n}{N}\right)\left(0.34 - 0.14\sqrt{e_d}\right)\sigma T^4 \qquad (12)$$

where e_d, σ, and T are vapor pressure at air temperature, Stefan-Boltzmann constant, and mean air temperature, respectively.

2.3. Combination Method of Penman

As shown in Equation (13), this method of evaporation is obtained by combining the evaporation computed by aerodynamic (E_a) and energy balance method (E_r). The weighting factors (i.e., $\dfrac{\Delta}{\Delta + \gamma}$ and $\dfrac{\gamma}{\Delta + \gamma}$) are applied in combining the methods (i.e., aerodynamic and energy balance). The weighting factors sum to unity.

$$E = \frac{\Delta}{\Delta + \gamma} E_r + \frac{\gamma}{\Delta + \gamma} E_a \qquad (13)$$

where γ is the psychrometric constant that is defined by Equation (14). The gradient of the saturated vapor pressure (Δ) is given by Equation (15).

$$\gamma = 0.0016286 \frac{P}{\lambda} \qquad (14)$$

$$\Delta = \frac{4098 e_s}{(273.3 + T)^2} \qquad (15)$$

2.4. Penman-Monteith Method

This method is same as the combination method of Penman. However, in this method, similar to r_a, another term called surface resistance (r_s) is introduced to account for resistance associated with movement of water vapor from the plant leaves to the air outside. This method is widely used to estimate evapotranspiration. The mathematical expression of this method is given by Equation (16).

$$E = \frac{\Delta}{\Delta + \gamma * \left(1 + \dfrac{r_s}{r_a}\right)} * \frac{R_n - G}{\rho_w \lambda} + \frac{\gamma}{\Delta + \gamma * \left(1 + \dfrac{r_s}{r_a}\right)} * \frac{2.17(e_s - e_z)}{\rho_w r_a * (T + 273)} \tag{16}$$

For grass reference crop, $r_s = 69 \text{ s/m}$ and $r_a = \dfrac{208}{u_2} \text{ s/m}$. Therefore, for grass reference crop

$$1 + \frac{r_s}{r_a} = 1 + 0.33 u_2 \,.$$

where u_2 is the wind speed at 2 m. When the wind speed is measured at different elevation, it can be adjusted from one level to another by using Equation (17).

$$u_2 = u_1 * \frac{\ln \dfrac{z_2}{z_0}}{\ln \dfrac{z_1}{z_0}} \tag{17}$$

where Z_1, Z_2 are measurement heights for levels 1 and 2, respectively. Z_0 is the reference height where velocity is zero. For open agricultural area, $Z_0 = 0.03$.

3. The Development of Spatial Evapotranspiration Tool (SET) at Grid Scale

ArcGIS provides easy-to-use platform to extend its desktop features by accessing geoprocessing functionalities through programming/scripting languages. Python, an open source programming language supported by a growing user community for its extensive collection of standard and third-party libraries, is one of the scripting languages supported by Environmental Systems Research Institute (Esri). The communication between ArcGIS and Python is through a site-package that is called ArcPy. The ArcPy site-package encompasses the modules, functions, and classes required to access the geoprocessing functionalities. The modules are the main gates to access the geoprocessing functionalities. The ArcPy site-package comes with a series of modules such as data access module, mapping module, and ArcGIS Spatial Analysis Extension module. To support the main modules, the ArcPy site-package also has some classes that are oftentimes used as shortcuts to complete geoprocessing parameters. Using the ArcPy site-package, the customization of desktop features could be in three ways: desktop add-in, standard toolbox, and Python toolbox.

3.1. Python Desktop Add-In

To extend desktop functionality, in ArcGIS 10, a new desktop add-in model that are authored using .NET or JAVA programming languages was introduced. The extension of functionality could be to make a customization that performs an action in response to an event such as dragging a rectangle over a geographical map to define an area of interest. This new add-in model is further enhanced by introducing Python to the list of supported programming languages, and the Python add-in wizard to reduce the development effort.

3.2. Standard and Python Toolboxes

ArcGIS tools that are boxed within toolboxes are a chunk of codes used to perform small, but essential tasks on geographical data. ArcGIS with its installation comes with a set of tools that are known as system tools. Often-times, geographic information system (GIS) professionals are required to repeat a task again and again by using one or more of the system tools. To facilitate this, a script is written using Python scripting language and the ArcPy site-package. This script is then attached to a newly created toolbox. The newly created toolbox could be either a standard or Python toolbox. In case of Python toolbox, which is an ASCII-based file, the toolbox is created entirely in Python.

Since the development of SET at grid scale does not involve an event such as dragging a rectangle over a geographical map to define an area of interest, Python toolbox is used to develop the SET.

3.3. The Spatial Evapotranspiration Tool (SET) at Grid Scale

The skeleton of ArcGIS Python toolbox is basically a class in Python. A toolbox can have more than one tool. Each tool is defined by a Python class. The tools are associated with the toolbox class by setting the "tools" property of the toolbox within the constructor or the class initialization method of the toolbox class. For example, in the below shown code, a tool named "SpatialET" is associated with the toolbox. The "label" and the "alias" are the properties of the toolbox, which are used in calling the tool in geoprocessing tasks.

```
class Toolbox(object):
def __init__(self):
self.label = "Spatial Evapotranspiration"
self.alias = "grid"
self.tools = [SpatialET]
class SpatialET(object):
def __init__(self):
    ...
def getParameterInfo(self):
    ...
defisLicensed(self):
defupdateParameters(self, parameters):
defupdateMessages(self, parameters):
def execute(self, parameters, messages):
    ...
```

The constructor or the class initialization method of Spatial ET class is used to set the properties of the tool. The "label" and "description" are the most important properties. The "canRunInBackground" property is used to let the tool to run in the background. As shown below, For SET, this property is set to "False".

```
class SpatialET(object):
def __init__(self):
self.label = "SpatialET"
self.description = "This tool is about Evapotranspiration at grid scale."
self.canRunInBackground = False
```

Within the Spatial ET class, the method named getParameterInfo() is used to collect the user specified inputs such as the location of raster data of temperature, relative humidity, wind speed, and solar radiation. These inputs are used to calculate ET using Penman-Monteith at grid scale. Within the getParameterInfo(), the parameter() object is used to collect each input from the user. For example, as shown in the below code, the variable named "param0" is used to collect the location of raster data of temperature. The parameter() object has few properties such as name, datatype, parametertype, and direction. Since the user input for the temperature is a raster, the datatype of param0 is set to "DERasterDataset". Furthermore, the temperature raster dataset is an input. Therefore, the direction property of "param0" is set to "Input". This same chunk of code is repeated for the other inputs (e.g., solar radiation, relative humidity, and wind speed) collected from the user through the tool.

```
def getParameterInfo(self):
param0 = arcpy.Parameter(
displayName="Temperature",
name="Temperature_raster",
```

```
datatype="DERasterDataset",
parameterType="Required",
direction="Input")
```

After collecting the parameters, they are stored within a list as shown in the below code, where param0, param1, param2, param3, param4, and param5 are the variables used to store the temperature raster, relative humidity raster, wind speed raster, location of tool generated ET raster, wind speed measurement height, and solar radiation raster, respectively. The default wind speed measurement height is also set to 3 m. This list is returned from getParameterInfo() method. The graphical user interface of the above outlined code is shown in **Figure 2**.

```
params = [param0,param1,param2, param3, param4, param5]
params[4].value=3
return params
```

Since the developed tool is in need of spatial analyst extension, the isLicensed()methodis used to check the existence of required licenses. The updateMessages() and updateParameters() are some of the other methods that could be used to customize the tool further. In SET, these methods are not modified as it doesn't require customization using these methods.

```
def isLicensed(self):
return True
```

After having stored the information on user specified inputs, the execute() method of the tool is used to setup the Penman-Monteith method as discussed in section 2.0. One of the arguments of the execute() method is "parameters" that has all the input data obtained through getParameterInfo() method. This argument is a list. Therefore, the individual inputs are retrieved using the indices of the list and the valueAsText() method of the parameter object. For example, as shown below, to retrieve the temperature raster, the zero[th] index of the list is called. This procedure is followed to retrieve the other inputs as well. Since the user specified temperature raster is in °F, a temporary raster of temperature in °C is created and stored in the Python variable named "inRasterTempC".

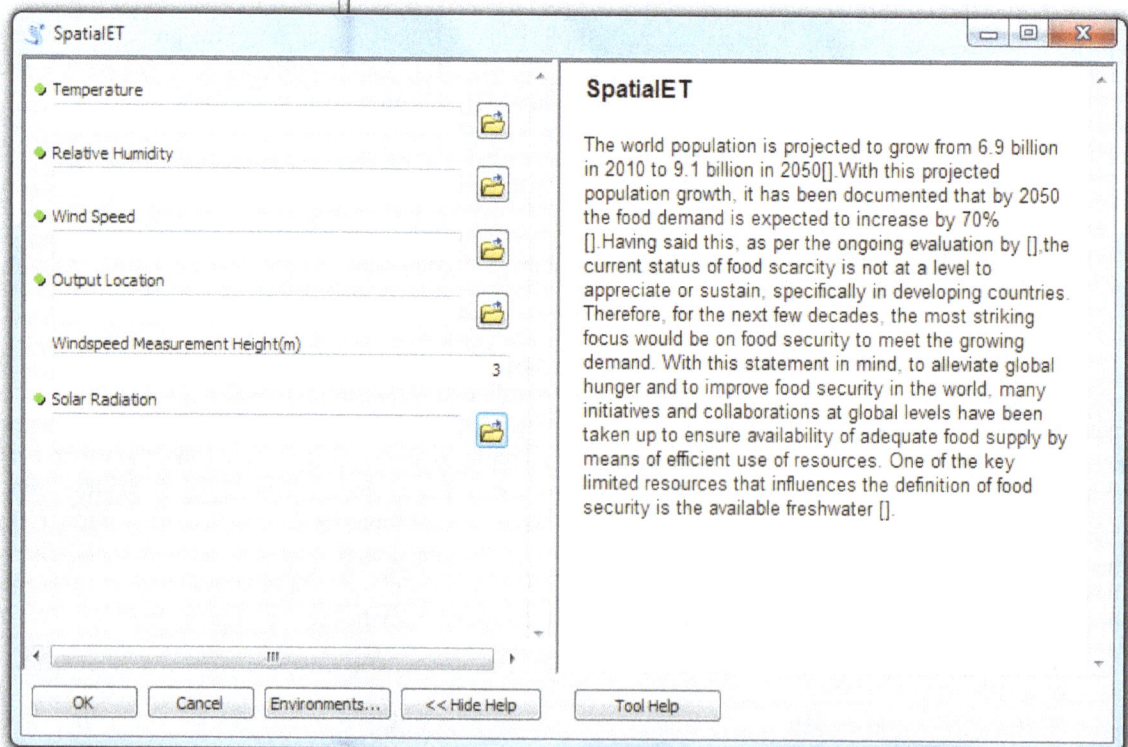

Figure 2. The graphical user interface of SET.

```
#User inputs for the tool
inRasterTempF=parameters[0].valueAsText #Temp
inRasterRH=parameters[1].valueAsText #RH
inRasterW=parameters[2].valueAsText #Wind speed
outRaster1=parameters[3].valueAsText # The name of the ET Raster
inWindHeight=parameters[4].value # Anero height
inSolarRadiation=parameters[5].valueAsText # So
inRasterTempC=(Raster(inRasterTempF)-32)/1.8
```

Within the execute() method, raster operations are carried out using the syntax of Map Algebra. As shown below, the raster of saturated vapor pressure is developed using the user specified temperature raster. The developed raster of saturated vapor pressure is multiplied by the raster of relative humidity to get the raster of vapor pressure. This is a raster operation. In other words, ArcGIS generates values for each grid based on the respective raster values of saturated vapor pressure and the relative humidity. This can be further explained as shown in **Figure 3**.

```
outSaturated=0.6108*Exp((inRasterTempC*17.27)/(237.3+inRasterTempC)) #Saturated vapor pressure
outSaturatedZ=outSaturated*Raster(inRasterRH)/100 #Vapor pressure
```

The developed rasters are temporary. In other words, they are not physically stored in a specified location. These temporary rasters are used in the subsequent raster operations to develop the other temporary rasters associated with Penman-Monteith method. The detailed raster operations in computing ET using Penman-Monteith method is outlined in **Figure 4**.

To calculate the ET using energy balance method, two inputs at grid levels are required: the latent heat of vaporization and the net radiation which is the sum of net long wave and net short wave radiation. As shown in the below code, the latent heat of vaporization is developed at grid level using equation (6'). The net radiation raster that is named "NetRadiation" is also developed to produce the raster of ET using energy balance method at grid scale.

```
outLamda=2.501-0.002362*inRasterTempC #Latent heat of vaporization
NetSRadiation=(1-0.23)*(0.25+0.5*0.25)*Raster(inSolarRadiation)#Net SW Radiaition
NetLRadiation=-1*(0.1+0.9*0.25)*(0.34-0.14*SquareRoot(outSaturatedZ))*(4.903/1000000000)*((273.2+
inRasterTempC)**4)
NetRadiation=NetLRadiation+NetSRadiation
EnergyBal=NetRadiation/1000/outLamda #Using energy balance method
```

Similarly, as shown in the below code, using raster operations and the temporary rasters, the ET using aerodynamic method is also developed. These raster operations are based on the equations discussed in section 2.0. The comments provided at each line explain the meaning of the variables.

```
Windspeed=Raster(inRasterW)
Windspeed2=0.44704*Windspeed*(Ln(2/0.03)/Ln(inWindHeight/0.03)) #Windspeed at 2m height
AerodynamicR=208/Windspeed2 #Aerodynamic resistance
inRasterP=1.225*(275+inRasterTempC)/3.486 #Pressure
PsychrConst=0.0016286*inRasterP/outLamda #Psychrometric constant
DeltaR=4098*outSaturated/((237.3+inRasterTempC)**2) #Vapor pressure gradient
AeroBal=2.17*(outSaturated-outSaturatedZ)/1000/AerodynamicR/(inRasterTempC+273)
#Using aerodynamic method
```

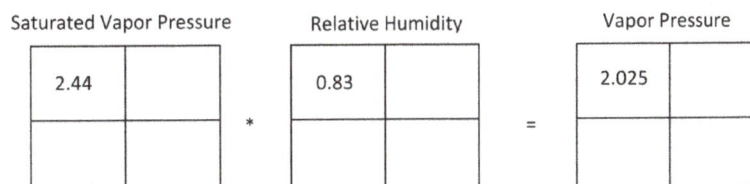

Figure 3. The computation of vapor pressure using raster operations at grid scale.

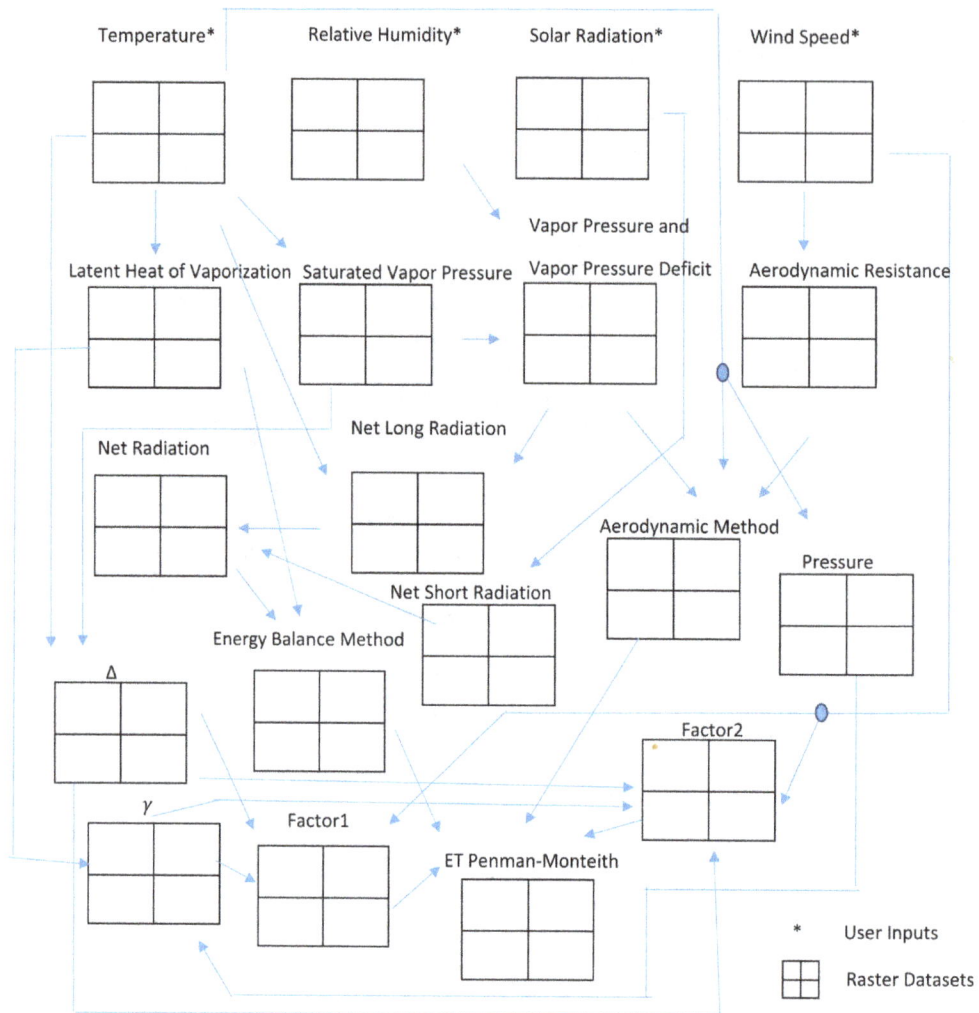

Figure 4. The raster operations in computing ET using Penman-Monteith.

The weights associated with each method (*i.e.*, aerodynamic and energy balance method) are generated at grid scale. The weights are generated as per Equation (13). These weights are then applied to the computed ET using aerodynamic and energy balance to develop the raster of ET using the Penman-Monteith method. The developed raster is stored using the save() method of the raster, as per the user specified output name and the location.

Factor1=DeltaR/(DeltaR+PsychrConst*(1+0.33*Windspeed2)) #Factor for energy balance method
Factor2=PsychrConst/(DeltaR+PsychrConst*(1+0.33*Windspeed2)) #Factor for aerodynamic method
PenmanMon=(Factor1*EnergyBal+Factor2*AeroBal)*1000 #Evapotranspiration using Penman-Mon in mm/day
PenmanMon.save(outRaster1)

4. The Application of SET

The state of Nebraska that lies in both the Great Plains and the Midwestern United States has a total geographical area of 200,520 km^2. The total population of the state is 1.8 Million [10]. Both the surface and groundwater are used to meet the demand for wide range of purposes. Around 94.8% of the estimated total groundwater withdrawals is used to meet the irrigation demand [11]. As of November 2014, around 95,000 irrigation wells are registered in the state. The developed tool is demonstrated for the state of Nebraska using the meteorological data from Automated Weather Data Network (AWDN) that gathers climatological observational data and the information for High Plains Region and provides its shareholders in fields such as agriculture. **Figure 5** shows the spatial locations of the climate stations in AWDN.

The AWDN has around 63 stations to cover the state of Nebraska. The data is available on hourly, daily, and sub-daily basis since 1985. To demonstrate the tool, the daily data in 2010 was downloaded from the online services provided by AWDN. As discussed in Section 3, the developed tool requires raster datasets on temperature, relative humidity, wind speed, and solar radiation. Therefore, at first, the average daily data in 2010 was developed based on the daily data of temperature, relative humidity, and wind speed. Using ArcGIS and the spatial locations of the climate stations, the tabular datasets of average daily data in 2010 were transformed to geographical data. Subsequently, the Spatial Analyst Extension of ArcGIS was used to develop the grid level values of temperature, relative humidity, and wind speed at a resolution of 1 km. The Kriging spatial interpolation technique packaged with ArcGIS was used to develop the rasters shown in **Figure 6**. To ensure that the Krigged data covers the whole state, the extent of the interpolation was set using the state map of Nebraska. The research work carried out by [4] was used to develop the solar radiation raster.

As depicted in **Figure 6**, in 2010, the temperature varies from 47°F to 53°F. The maximum temperature is observed in the Eastern part of Nebraska. The relative humidity that determines the vapor pressure is very high in the Eastern part of Nebraska and tends to decrease towards the Western part of the state. It is also worth to note that the trends of temperature and the wind speed are opposite. In other words, the wind speed tends to increase from Eastern part of Nebraska to Western part of Nebraska in the direction of South East to North West.

The estimated ET using Penman-Monteith method is shown in **Figure 7(a)**. The **Figure 7(b)** shows the categorized version of **Figure 7(a)**. For the state of Nebraska in 2010, the estimated ET using Penman-Monteith method varies from 0.77 to 1.04 mm/day. In other words, the maximum spatial variation of estimated ET using Penman-Monteith method is $= \dfrac{1.04 \text{ mm/day} - 0.77 \text{ mm/day}}{1.04 \text{ mm/day}} \times 100\% = 26\%$. Moreover, the highest estimated ET using Penman-Monteith method is registered in the Eastern part of Nebraska. However, the geographical extent of such high ET is limited to a very small area in contrast to the geographical area with estimated ET using Penman-Monteith method in the range of 0.77 - 0.96, as shown in **Figure 7(b)**. Furthermore, an increasing trend of spatial variation is observed from Western part of Nebraska to Eastern part of Nebraska in the direction of North West to South East.

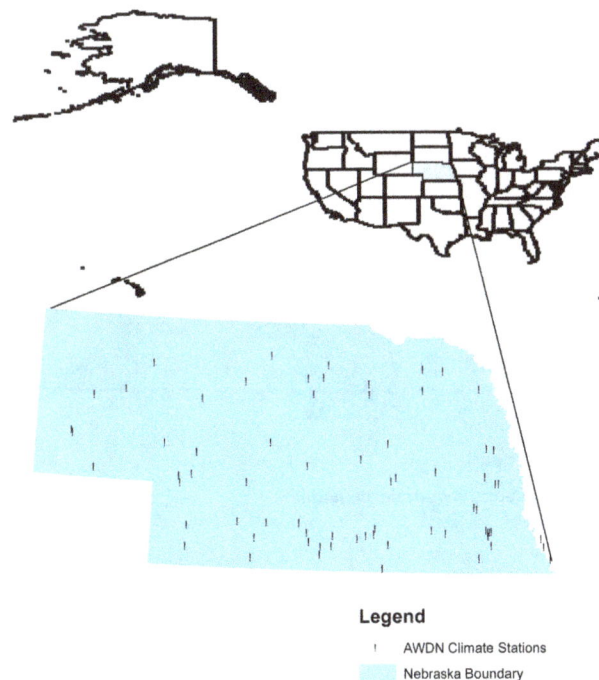

Figure 5. The spatial locations of the climate stations from AWDN Network in Nebraska.

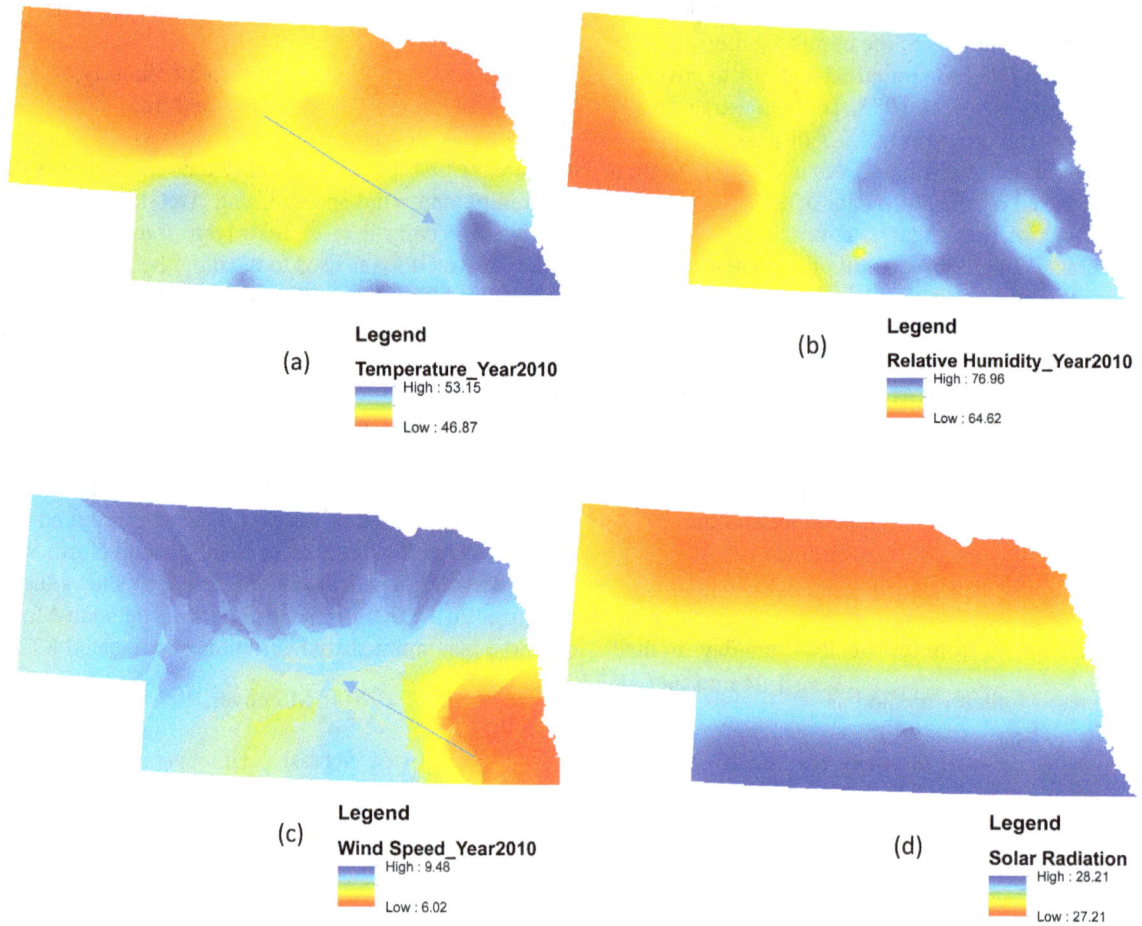

Figure 6. The input rasters to SET: (a) The average daily temperature in 2010 in °F; (b) The average daily relative humidity in 2010 in %; (c) The average daily wind speed in 2010 in MPH; (d) The average daily solar radiation in $MJm^{-2} \cdot d^{-1}$.

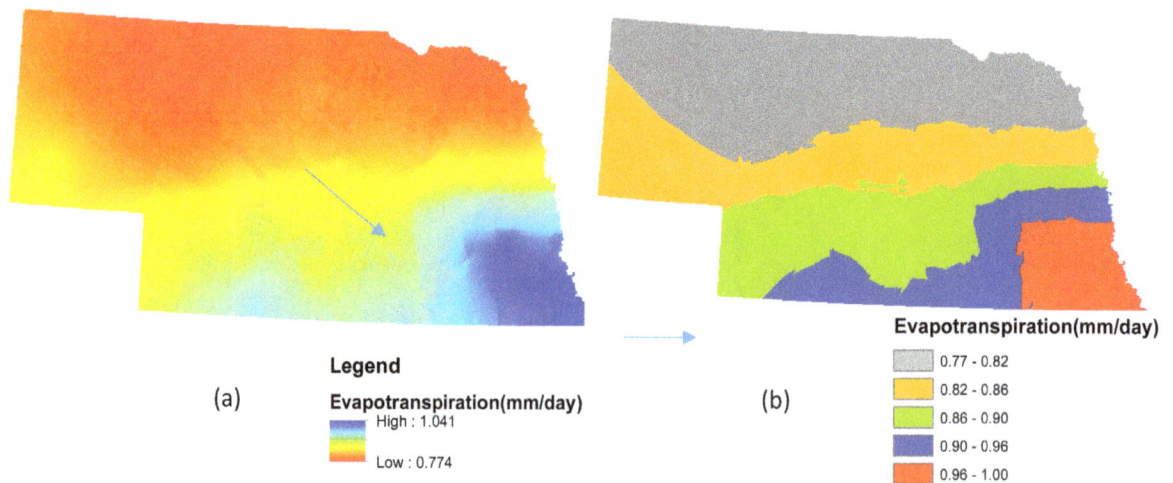

Figure 7. The Estimation of ET (mm/day) using Penman-Monteith method.

Figure 8(a) and **Figure 8(b)** show the spatial variation of evapotranspiration estimated using energy balance method and the aerodynamic method, respectively. These figures reveal that the contribution of aerodynamic method for the estimation of ET using Penman-Monteith method is not that siginificant compared to that of energy balance method. In other words, the estimation of ET using Penman-Moneith is totally defined by the energy balance method. The lowest and the highest estimated ET using energy balance method is 2.3 mm/day and 2.44 mm/day, respectively. Therefore, the maximum spatial variation of estimated ET using energy balance method is $= \dfrac{2.44 \text{ mm/day} - 2.3 \text{ mm/day}}{2.44 \text{ mm/day}} \times 100\% = 6\%$. Although the spatial variation of estimated ET using energy balance method is not significant, an increasing trend of spatial variation is observed from Western part of Nebraska to Eastern part of Nebraska in the direction of North West to South East. Moreover, it is also important to note that even though the estimation of ET using Penman-Moneith is totally defined by the energy balance method, the spatial variation of estimated ET using Penman-Monteith method is not influenced by the energy balance method, but bythe weights used to determine the contribution of energy balance method.

On the other hand, The lowest and the highest estimated ET using aerodynamic method is 2.88×10^{-5} mm/day and 5.33×10^{-5} mm/day, respectively. Therefore, the maximum spatial variation of estimated ET using aerodynamic balance method is $\dfrac{5.33 \times 10^{-5} \text{ mm/day} - 2.88 \times 10^{-5} \text{ mm/day}}{5.33 \times 10^{-5} \text{ mm/day}} \times 100\% = 46\%$. In other words, the variation of estimated ET using aerodynamic method is spatially significant compared to that of energy balance method. Moreover, in contrast to what is observed with the trend of estimated ET using energy balance method, the lowest estimated ET using aerodynamic method occurs in the Eastern part of Nebraska and increases towards the Western part of Nebraska, as shown in **Figure 8(b)**.

As shown in **Figure 9(a)**, the weighting factor that is used to determine the contribution of energy balance

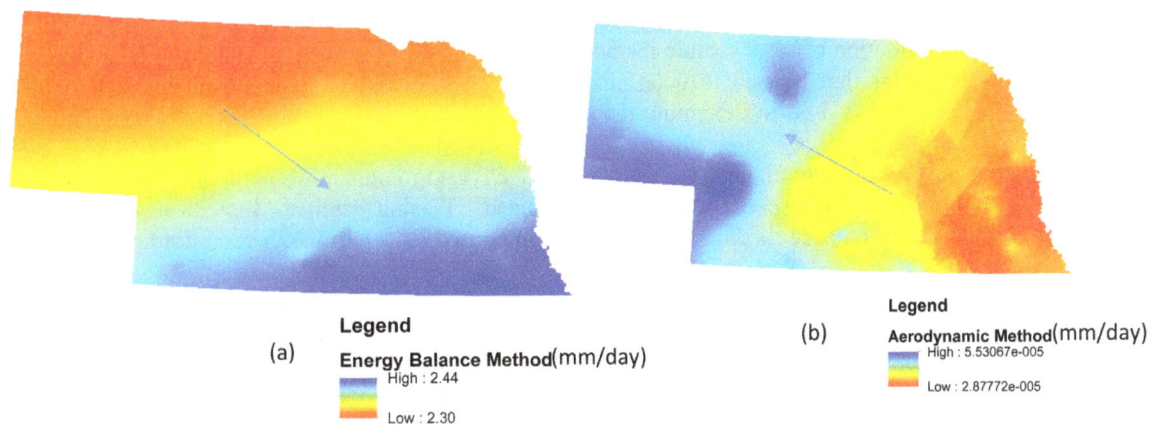

Figure 8. The estimation of ET (mm/day) using (a) Energy balance method and (b) Aerodynamic method.

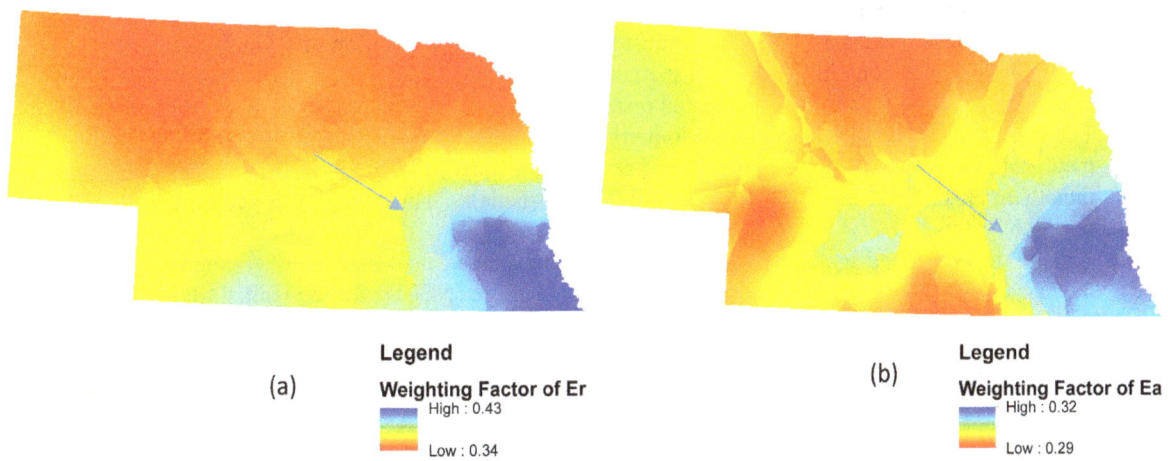

Figure 9. Weighting factor for (a) Energy balance method and (b) Aerodynamic method.

method on estimation of ET using Penman-Monteith ranges from 0.34 to 0.43. Similarly, as shown in **Figure 9(b)**, the weighting factor that is used to determine the contribution of aerodynamic method on estimation of ET using Penman-Monteith ranges from 0.29 to 0.32. In other words, in addition to the fact that the contribution of aerodynamic method on estimation of ET using Penman-Monteith is very small, the weighting factor that is used to determine the contribution of aerodynamic method is smaller than weighting factor that is used to determine the contribution of energy balance method. Furthermore, it is also noted that the weighting factor that is used to determine the contribution of energy balance method tends to follow the same trend as in **Figure 8(a)**. In other words, higher ET estimated using energy balance method is weighted more and lower ET estimated using energy balance method is weighted less. In contrast to this, in the case of aerodynamic method, higher ET estimated using aerodynamic method, although the value is very small, is weighted less and lower ET estimated using aerodynamic method is weighted more. This is notable from **Figure 8(b)** and **Figure 9(b)**.

5. Conclusion and Recommendations

In this paper, a spatial tool that can geographically encompass all the best available climate datasets to produce ET using Penman-Monteith method at different spatial scales is developed using Python, an open source programming language supported by a growing user community for its extensive collection of standard and third-party libraries, and the ArcPy site-package of ArcGIS. The developed spatial tool is demonstrated using the meteorological data from Automated Weather Data Network (AWDN) in Nebraska in 2010. Based on the development and demonstration of the tool, the following points are highlighted:

1) The contribution of aerodynamic method for the estimation of ET using Penman-Monteith method is not that significant compared to that of energy balance method. However, the variation of estimated ET using aerodynamic method is spatially significant compared to that of energy balance method. In addition to the fact that the contribution of aerodynamic method on estimation of ET using Penman-Monteith is very small, the weighting factor that is used to determine the contribution of aerodynamic method is smaller than weighting factor that is used to determine the contribution of energy balance method on estimation of ET using Penman-Monteith.

2) It is also observed that even though the estimation of ET using Penman-Moneith is totally defined by the energy balance method, the spatial variation of estimated ET using Penman-Monteith method is not influenced by the energy balance method, but by the weights used to determine the contribution of energy balance method.

3) The demonstration of the developed tool was based on solar radiation raster from the work done by [4]. However, the solar radiation toolbox that is distributed with the installation of ArcGIS is an option that could be coupled with the developed tool.

4) In this study, the Kriging spatial interpolation scheme was used to transform the observed climate data from AWDN. Though Kriging is considered as one of the best spatial interpolation schemes, it is also worth to research on the impact of other interpolation schemes on spatial ET at grid scales. This may also lead to identify the impact of different schemes on energy balance method and the aerodynamic method on estimation of ET using Penman-Monteith.

5) Though the developed spatial tool is demonstrated for a spatial resolution of 1 km, it can be used to estimate ET using Penman-Monteith at different spatial resolutions to decide the best scale that fits a particular region or area, and to decide the impact of spatial resolutions on ET using Penman-Monteith.

6) The sensitivity of ET using Penman-Monteith is oftentimes estimated using meteorological data from climate stations. However, such estimation of sensitive or most influential parameters may vary spatially and thus there exists a need to estimate sensitivity of ET spatially. Thus, the developed tool will be of useful to research on this.

7) Though the developed spatial tool does not include the landuse datasets, the integration of one of the existing landuse datasets, such as the National Land Cover Database (NLCD) for the United States of America, with the tool is also worth to research.

Acknowledgements

The authors would like to thank the Board of Regents, University of Nebraska-Lincoln, Lincoln, for providing the financial support to conduct this research. This research was conducted when author[*] was a researcher at University of Nebraska-Lincoln, Lincoln.

References

[1] WWAP (World Water Assessment Programme) (2015) United Nations World Water Development Report 2015: Water for a Sustainable World. United Nations Educational, Scientific and Cultural Organization, Paris.

[2] WWAP (World Water Assessment Programme) (2012) United Nations World Water Development Report 4: Managing Water under Uncertainty and Risk. United Nations Educational, Scientific and Cultural Organization, Paris.

[3] Postel, S.L., Daily, G.C. and Ehrlich, P.R. (1996) Human Appropriation of Renewable Fresh Water. *Science*, **271**, 785-788. http://dx.doi.org/10.1126/science.271.5250.785

[4] Jensen, M.E., Burman, R.D. and Allen, R.G. (1990) Evapotranspiration and Irrigation Water Requirements, Manuals and Reports on Engineering Practice. ASCE No. 70, New York.

[5] Richard, G.A. and William, O.P. (1986) Rational Use of the FAO Blaney-Criddle Formula. *Journal of Irrigation and Drainage Engineering*, **112**, 139-155. http://dx.doi.org/10.1061/(ASCE)0733-9437(1986)112:2(139)

[6] Allen, R.G., Pereira, L.S., Raes, D. and Smith, M. (1998) Crop Evapotranspiration-Guidelines for Computing Crop Water Requirements. FAO Irrigation and Drainage Paper 56, United Nations Food and Agriculture Organization, Rome.

[7] Gong, L.B., Xu, C.Y., Chen, D.L., Halldin, S. and Chen, Y.Q.D. (2006) Sensitivity of the Penman-Monteith Reference Evapotranspiration to Key Climatic Variables in the Changjiang (Yangtze River) Basin. *Journal of Hydrology*, **329**, 620-629. http://dx.doi.org/10.1016/j.jhydrol.2006.03.027

[8] Bakhtiari, B. and Liaghat, A.M. (2011) Seasonal Sensitivity Analysis for Climatic Variables of ASCE Penman-Monteith Model in a Semi-Arid Climate. *Journal of Agricultural Science and Technology*, **13**, 1135-1145.

[9] Kevin, F.D., Thomas, E.R. and William, L.C. (2015) Groundwater Availability in the United States: The Value of Quantitative Regional Assessments. *Hydrogeology Journal*, **23**, 1629-1632. http://dx.doi.org/10.1007/s10040-015-1307-5

[10] U.S. Census Bureau (2015) Annual Estimates of the Resident Population for the United States, Regions, States, and Puerto Rico: April 1, 2010 to July 1, 2014. U.S. Census Bureau, Suitland, Maryland.

[11] Nebraska Surface Water and Groundwater Use—2005. Nebraska Department of Natural Resources. http://www.dnr.ne.gov/nebraska-surface-water-and-groundwater-use-2005

16

Growth, Yield and Water Use Effeciency of Forage *Sorghum* as Affected by Npk Fertilizer and Deficit Irrigation

Mohamed M. Hussein[1], Ashok K. Alva[2]

[1]Water Relations and Irrigation Department, National Research Centre, Cairo, Egypt
[2]USDA-ARS, Prosser, WA, USA
Email: mmoursyhus@gmail.com, Ashok.Alva@ars.usda.gov

Abstract

Drought stress (DS) is an important limiting factor for crop growth and production in some regions of the world. Limitation in water availability precludes optimal irrigation in some production regions. Therefore, investigations on the interaction of other factors to mitigate the DS to varying degree are important. Two field experiments were conducted in the experimental farm of the National Research Centre, Shalakan, Kalubia Governorate, Egypt, during 2004 and 2005 summer seasons to evaluate the interactions between N, P, K rates and optimal vs. deficit irrigation regimes on biomass yield as well as water use efficiency (WUE) of forage *sorghum*. Omission of the 4th irrigation significantly decreased the biomass of *sorghum* c.v. Pioneer, as compared to that of the plants receiving optimal irrigation or subject to omission of the 2nd irrigation. The biomass yield increased with an increase in NPK fertilizer rates. Plant height and leaf area also decreased by omitting the 2nd irrigation as compared to that of the plants under optimal irrigation, and further declined with omission of the 4th irrigation. The biomass of the plants (dry weight basis) that received the high N, P, K rates was greater by 26%, 29%, and 35% as compared to that of the plants that received no N, P, K fertilizers, under optimal irrigation, omission of the 2nd, and omission of the 4th irrigation, respectively. The corresponding increases in water use efficiency (based on fresh weight yield) were 37%, 42%, and 55%.

Keywords

Sorghum-Forage-Omitting of Irrigation-NPK Fertilizer-Growth, Yield-Water Use Efficiency

1. Introduction

Drought stress (DS) is the most important limiting factor for crop production in arid and semi-arid regions of the world. Severe DS during vegetative growth stage and moderate DS during flowering stage of grain *sorghum* crop contributed to about 30% of reduction in grain yield, despite high water use efficiency [1]. New land used for cultivation of forage crops, including forage *sorghum* (*Sorghum bicolor* L.), are rather marginal in soil characteristics and productivity. Sustainable production can be achieved on these marginal soils only through use of cultivars tolerant to drought and salinity stress.

Forage *sorghum* is an important biomass crop for hay and silage [2]. *Sorghum* as a livestock feed is of significant importance, particularly in the tropical zone because of its adaptation to low fertility soils and other limiting factors, such as DS [3].

Mineral fertilizers play a vital role in improving crop yields but the major challenge is to ensure adequate balance between the different nutrients and support optimal yield. Current recommendations for producing optimal forage yields of *sorghum*-sudan grass hybrids suggest application of 50 to 100 $kg \cdot N \cdot ha^{-1}$, applied in two equal doses at planting and after the first cut [4]. Timing and placement of N application should be managed to avoid significant losses while ensuring availability of adequate N when needed by the crop. *Sorghum* sudan grass is usually managed with low N fertilizer inputs (≤ 80 $kg \cdot ha^{-1}$) since growth and yield responses to N rates have been reported only up to 80 $kg \cdot ha^{-1}$ [5].

The optimum P availability is important for improving mineral P concentrations and yields of most crops [6] [7]. Soil solution P, following the dissolution of P fertilizers applied to the soil, is either taken up by the plants, precipitated, or adsorbed on the exchange sites in the soil [8]. Pholsen and Somsungnoen [9] reported an increase in most growth parameters of forage *sorghum* plants with an increase in N and K rates from 450 to 650 and 50 to 100 $kg \cdot ha^{-1}$, respectively. Ogunlela and Yusuf [10] reported significantly greater K content in 3 cultivars of forage *sorghum* with 75 $kg \cdot ha^{-1}$ K as compared to that of the plants receiving 25 or 50 $kg \cdot ha^{-1}$ K. Sharma and Kumari [11] reported that plant height, leaf area index, leaf area duration, plant growth rate, total dry matter production, K concentration and grain yield increased with K application rate from 25 to 50 $kg \cdot ha^{-1}$. Adequate availability of soil water and nutrients is important to support optimal plant growth and production in the arid and semi-arid regions [12]. Best management of nutrients is a successful strategy to alleviate abiotic stresses [13].

The objective of this study was to investigate the interactions between different rates of N, P, K and DS on growth, yield, and water use efficiency of forage *sorghum*.

2. Material and Methods

Two field experiments were conducted at the National Research Centre, Shalakan, Kalubia Governorate, Egypt, in 2004 and 2005 to evaluate the effects different rates of N, P, K fertilization on mitigating the adverse effects of DS on forage *sorghum*. The experiment comprised of factorial combination of 3 irrigation (main) and 4 fertilizer rates (sub) as shown below with 6 replications.

Irrigation treatments: Optimal irrigation (OI; No water stress), vs. two deficit irrigation (DI) treatments, *i.e.* omitting 2nd (DI-1) or 4th (DI-2) irrigation. The water uses for OI and DI treatments were 7960 and 6960 $m^3 \cdot ha^{-1}$, respectively. Sub treatments were rates of N:P:K (in $kg \cdot ha^{-1}$); 0 (0:0:0), 1 (36:8.4:32), 2 (72:16.8:64), 3 (144:33.6:28). Plot size was 3×7 m.

The full rates of P (as calcium super phosphate, 6.8% P) and K (as potassium sulfate, 40.3% K), as per treatments, were broadcast and mixed with the soil during pre-plant tillage. Forage *Sorghum* (*Sorghum bicolor* (L.) Moench) cv. Pioneer was planted on July, 15th both years. The seed rate was 96 kg/ha^{-1} in both seasons. Nitrogen rates, as per treatments, were applied as ammonium sulfate (20.5% N) in two equal doses; 21 and 35 days after sowing. The standard production practices for forage *sorghum* followed in the province were adapted (Recommendations of Egyptian Ministry of Agriculture, unpublished). The amount of each irrigation was 750 $m^3 \cdot ha^{-1}$, except the irrigation after sowing was 1000 $m^3 \cdot ha^{-1}$.

The following measurements were made on two plants from every subplot before cutting (Cutting was done 70 days after planting): Plant height (cm); number of green leaves; leaf area; fresh and oven dry (70°C for three days) weights of stem and leaves.

Water Use Efficiency (WUE; $kg \cdot m^{-3}$) was calculated as marketable yield (kg) per unit water use (m^3). Fresh and dry yields of forage per plot were measured and yield per ha was calculated.

Statistical significance of the treatments effects was evaluated by analysis of variance (ANOVA) test as described by [14].

3. Results and Discussion

Water stress *i.e.* omitting either the 2nd or 4th irrigation, significantly influenced plant height, leaf area, fresh weight of plant tops, and stem dry weight (**Table 1**). The negative effects of water stress were greater by omitting the 4th irrigation as compared to those by omitting the 2nd irrigation. Our results concur with those of Carmier *et al.* [15] who reported that irrigation influenced plant height and dry matter.

Mohammadkhani and Heidri [16] subjected the six-day-old seedlings to different concentrations of poly ethylene glycol (PEG) 6000 to induce drought stress treatment. After 24 h treatment in PEG 6000 the electrolyte leakage increased. Under drought stress the activities of protective enzymes in roots and shoots increased sharply. Drought induced by 40% concentration of PEG, which induced water potential of 1.76 MPa, which affected soluble sugars and proline content. The soluble sugars play an important role in the production of other compounds, energy, and stabilization of membranes [17], act as regulators of gene expression [18] and signal molecules [19]. Proline is important in water adjustment through stomatal aperture which, in turn, affects transpiration and photosynthesis [20]. Boomsma and Vyn [21] reported that water stress influenced plant uptake of water as well as nutrients. Li *et al.* [22] demonstrated that water stress impaired the oxidative defense systems in plants.

Omitting the 4th irrigation significantly decreased the fresh and dry yield of *sorghum* as compared to those of the plants grown under optimal irrigation or omitting the 2nd irrigation (**Table 2**). Li *et al.* [22] reported that *sorghum* yield affected by spatial or temporal stress from drought. Akmal and Jansenes [23] concluded that water deficit affected growth and yield of ryegrass. Ferre and Faci [24] demonstrated that deficit irrigation or reduced frequency of irrigation during the grain filling stage did not significantly affect the corn yield. They concluded that flowering stage was the most sensitive to water deficit as evident from significant reduction in biomass yield and harvest index due to water stress during flowering stage. They reported a linear relationship between amount of irrigation and grain yield.

Our study also revealed that WUE was greater for the plants subjected to omission of the 2nd irrigation than those of plants grown under optimal irrigation or omission of 4th irrigation (**Table 2**). The latter treatment resulted in the least WUE, both based on fresh or dry biomass weight. Ferre and Faci [24] reported that the negative impact of water deficit on WUE was greater when subjected to deficit irrigation during flowering stage than that during any other growth stages.

Table 1. Effects of drought stress on growth of *sorghum* plants (per plant basis; mean across 2004 and 2005 seasons).

Irrigation	Plant height (cm)	No of Leaves	Leaf area (cm²)	Fresh weight (g)			Dry weight (g)		
				Stem	Leaves	Total	Stem	Leaves	Total
Optimal Irrigation	113	4.7	3653	69.8	30.3	100.1	37.1	12.5	49.6
Omit 2nd irrigation	92	4.4	2287	68.3	24.2	92.5	32.2	11.5	43.7
Omit 4th irrigation	82	3.9	1808	40.3	20.3	60.6	17.5	9.9	27.4
LSD ($P \leq 0.05$)	6	NS	453	23.3	3.3	32.1	16.2	NS	22.3

LSD = Least Significant Difference; NS = Non-Significant.

Table 2. Effects of drought stress on yield of *sorghum* and water use efficiency (WUE) (Mean across 2004 and 2005 seasons).

Irrigation	Fresh yield (Mg·ha⁻¹)	Dry yield (Mg·ha⁻¹)	WUE (kg·m⁻³)	
			Fresh wt basis	Dry wt basis
Optimal Irrigation	94.7	46.8	12.38	6.18
Omit 2nd irrigation	97.1	45.6	13.98	6.53
Omit 4th irrigation	81.7	36.8	11.68	5.28
LSD ($P \leq 0.05$)	1.70	4.3	ND	ND

LSD = Least Significant Difference; ND = LSD was not calculated.

The increased rates of N, P, K increased the plant growth and biomass (**Table 3**). This trend is in agreement with those reported by Bokhtiar and Sakurai [25] on sugar cane; Bayu *et al.* [26] on *sorghum*; and Barros *et al.* [27] on intercropped maize/cowpea. Nitrogen is an important component of major structural, genetic, and metabolic compounds in plant cells, including chlorophyll, amino acids, ATP (adenosine triphosphate), and nucleic acids such as DNA [28]. Phosphorus is a vital component of: DNA, RNA, and ATP. Thus phosphorus is essential for the general health and vigor of all plants. Adequate P nutrition is critical for root development, increased stalk and stem strength, increased flowering and seed production, uniform and early crop maturity, improved crop quality, and increased resistance to plant diseases [28]. Potassium is important for various regulatory functions in plants. It is essential in nearly all processes needed to sustain plant growth and reproduction. Potassium plays a vital role in photosynthesis, translocation of photosynthates, protein synthesis, ionic balance, regulation of plant stomata and water use, activation of plant enzymes, and many other processes [28]. The forage yield at the highest N, P, K rates was 27% and 61% greater than that of the plants that received no N, P, K on fresh and dry weight basis, respectively (**Table 4**). The corresponding increases for the medium N, P, K rates were 8% and 46%.

Shrotriya [29] reported that balanced application of NPK caused up to 122% increase in *sorghum* yield in India. Increased plant growth with optimal N, P, K application provides vegetative cover, thus enhancing moisture retention, nutrient use efficiency and soil productivity [30]. Pholsen and Somsungnoen [9] reported that an increase in N and K rates significantly increased most growth parameters of *sorghum* plants. The highest total dry weight and seed yield were obtained from plants receiving 650 and 100 kg·ha^{-1} N and K, respectively. Increased rates of N, P, K increased WUE (**Table 4**). Barros *et al.* [27] showed that WUE improved with increasing rate of nutrients.

The interaction between N, P, K rates and irrigation treatments was significant with respect to plant height, leaf area, total fresh weight and stem as well as total dry biomass weights (**Table 5**). The increases in leaf area at the highest N, P, K rates as compared to that of the plants receiving no N, P, K were 248%, 126%, and 37% re-

Table 3. Effects of NPK rates on growth parameters of *sorghum* plants (per plant basis; mean across 2004 and 2005 seasons).

Fertilizer rates	Plant height (cm)	No of Leaves	Leaf area (cm^2)	Fresh weight (g)			Dry weight (g)		
				Stem	Leaves	Total	Stem	Leaves	Total
0	74	4.0	1512	42.3	18.0	60.3	21.8	8.8	30.6
1	87	3.9	2191	58.3	23.5	81.8	26.8	10.8	37.7
2	107	4.2	3015	62.2	28.3	90.5	30.8	12.4	33.2
3	115	5.0	3616	75.5	31.4	106.8	36.3	13.1	49.4
LSD (P ≤ 0.05)	5	1.0	328	5.6	3.2	6.2	4.3	2.9	4.6

LSD = Least Significant Difference; N:P:K Rates (kg·ha^{-1}); 0 = None; 1 = 36:8.4:32; 2 = 72:16.8:64; 3 = 144:33.6:128.

Table 4. Effects of NPK rates on yield of *sorghum* and water use efficiency (WUE) (mean across 2004 and 2005 seasons).

Fertilizer rates	Fresh yield (Mg·ha^{-1})	Dry yield (Mg·ha^{-1})	WUE (kg·m^{-3})	
			Fresh wt basis	Dry wt basis
0	86	30.9	10.50	5.30
1	87	38.7	12.07	5.50
2	93	45.2	13.00	6.30
3	109	49.7	15.13	6.87
LSD (P ≤ 0.05)	0.50	0.53	ND	ND

LSD = Least Significant Difference; ND = LSD was not calculated; N:P:K Rates (kg·ha^{-1}). 0 = None; 1 = 36:8.4:32; 2= 72:16.8:64; 3 = 144:33.6:128.

spectively in optimal irrigation, omission of the 2nd, and omission of the 4th irrigation treatments, respectively. The corresponding increases in total fresh weight were 98%, 67%, and 67%.

The interaction of effects of irrigation and NPK fertilizer on yield was significant only on dry biomass yield (**Table 6**). The increases in biomass yield at the highest N, P, K rates as compared to that of plants receiving no N, P, K were 26%, 29%, and 35%, respectively, at optimal irrigation, omission of the 2nd, and omission of the 4th

Table 5. Effects of drought stress and N, P, K rates on growth of forage *sorghum* plants (per plant basis; average of two seasons).

Irrigation	Fertilizer rates	Plant height (cm)	No of Leaves	Leaf Area (cm^2)	Fresh weight (g)			Dry weight (g)		
					Stem	Leaves	Total	Stem	Leaves	Total
Optimal Irrigation	0	80	4.2	1537	46.3	19.0	65.3	25.0	9.5	34.5
	1	108	4.4	3201	69.8	29.8	99.6	37.3	11.7	49.0
	2	127	4.7	4500	72.3	34.5	106.8	38.8	13.7	52.5
	3	137	5.3	5354	90.7	38.0	128.7	47.1	15.0	62.1
Omit 2nd irrigation	0	73	4.1	1479	50.3	18.8	69.1	25.6	10.0	35.6
	1	88	3.9	1796	67.3	21.5	88.8	27.0	10.8	37.8
	2	103	4.2	2613	72.2	25.3	97.5	34.5	12.5	47.0
	3	110	5.3	3348	83.5	31.0	114.5	41.5	12.5	54.0
Omit 4th irrigation	0	68	3.8	1569	20.2	16.3	46.5	14.6	7.0	21.6
	1	72	3.4	1576	37.8	19.3	57.1	16.2	10.0	26.1
	2	91	3.8	1943	42.0	21.2	63.2	19.1	11.0	30.1
	3	98	4.5	2145	51.2	25.2	76.4	20.1	11.7	31.8
LSD (P ≤ 0.05)		8	NS	567	NS	NS	10.8	7.7	NS	8.0

LSD = Least Significant Difference; NS = Non-Significant; N:P:K Rates (kg·ha^{-1}). 0 = None; 1 = 36:8.4:32; 2 = 72:16.8:64; 3 = 144:33.6:128.

Table 6. Effects of drought stress and N, P, K, rates on yield of forage *sorghum* and water use efficiency (WUE).

Irrigation	Fertilizer	Fresh yield (Mg·ha^{-1})	Dry yield (Mg·ha^{-1})	WUE (kg·m^{-3})	
				Fresh wt basis	Dry wt basis
Optimal irrigation	0	80	42	10.5	5.5
	1	91	45	11.9	5.9
	2	97	48	12.7	6.3
	3	110	53	14.4	7.0
Omit 2nd irrigation	0	80	41	11.5	5.9
	1	93	39	13.4	5.6
	2	102	49	14.7	7.0
	3	113	53	16.3	7.6
Omit 4th irrigation	0	66	31	9.5	4.5
	1	76	35	10.9	5.0
	2	81	39	11.6	5.6
	3	102	42	14.7	6.0
LSD at 5%		NS	0.9	ND	ND

LSD = Least Significant Difference; ND = LSD was not calculated; NS = Non-Significant; N:P:K Rates (kg·ha^{-1}); 0 = None; 1 = 36:8.4:32; 2 = 72:16.8:64; 3 = 144:33.6:128.

irrigations. The results support that the beneficial effects of optimal N, P, K fertilizer was greater under increased DS.

Soil nutrient availability induces large changes in plant functional attributes, which affect the water and carbon economy of plants [31]. In particular, nitrogen (N) can affect the cold and drought tolerance, yet there is no clear consensus on the magnitude or direction of its effect. Low tissue N concentration may hinder either cold or drought hardening [32]. Proper available nitrogen in soils enhanced the growth of plants and lowered the adverse effect on growth caused by water stress [33].

Water use efficiency also responded favorably to increased rates of N, P, K (**Table 6**). The WUE was greater for omission of the 2nd irrigation as compared to that of the plants under optimal irrigation or omission of the 4th irrigation. The WUE (fresh weight basis) for the plants which received high rates of N, P, K as compared to those of the plants received no N, P, K were greater by 55%, 42%, and 37%, respectively, for omission of the 4th irrigation, the 2nd irrigation, and optimal irrigation. The corresponding increases on dry weight basis were 33%, 29%, and 27%. Therefore, the WUE response to N, P, K fertilizer was greater with severity of water stress.

References

[1]   Khalili, I. Akbari, N. and Chaichi, M.R. (2008) Limited Lrrigation and Phosphorus Fertilizer Effects on Yield and Yield Components of Grain *Sorghum* (*Sorghum bicolor* L.var. Kimia). *American-Eurasian Journal of Agricultural & Environmental Science*, **3**, 697-702.

[2]   Skerman, P.J. and Rivers, F. (1999) Tropical Grasses. FAO. Plant Production and Protection, Series No. 23, FAO, Rome.

[3]   Pholsen, S. and Suksri. A. (2007) Effects of Phosphorus and Potassium on Growth, Yield and Fodder Quality of IS 23585 Forage *Sorghum* Cultivar (*Sorghum bicolor* L.). *Pakistan Journal of Biological Sciences*, **10**, 1604-1610. http://dx.doi.org/10.3923/pjbs.2007.1604.1610

[4]   Ontario Ministry of Agriculture and Food (OMAF) (2002) Annual Report.

[5]   Ram, S.N. and Singh, B. (2001) Effect of Nitrogen and Harvesting Time on Yield and Quality of *Sorghum* (*Sorghum bicolor*) Intercropped with Legumes. *Indian Journal of Agronomy*, **46**, 32-37.

[6]   Cisar, G.D., Synder, G.H. and Swanson, G.S. (1992) Nitrogen, P and K Fertilization for Histosols Grown St. Augustine Grass Sod. *Agronomy Journal*, **84**, 475-479. http://dx.doi.org/10.2134/agronj1992.00021962008400030023x

[7]   Halliday, D.T. and Trenkel, M.E. (1992) World Fertilizer Use Manual. International Fertilizer Industry Association, Paris, 31-32.

[8]   Rashid, M., Ranjha, M. and Rehim, A. (2007) Model Based P Fertilization to Improve Yield and Quality of *Sorghum* (*Sorghum bicolor* L.) Fodder on an Ustochrept Soil. *Pakistan Journal of Agricultural Sciences*, **44**, 221-227.

[9]   Pholsen, S. and Sormsungnoen, N. (2005) Effects of Nitrogen and Potassium Rates and Planting Distances on Growth, Yield and Fodder Quality of a Forage *Sorghum* (*Sorghum bicolor* L. Moench). *Pakistan Journal of Biological Sciences*, **7**, 1793-1800.

[10]   Ogunlela, B. and Yusuf, Y. (1988) Yield and Growth Response to Potassium of Grain *Sorghum* as Influenced by Variety in a Savanna Soil of Nigeria. *Fertilizer Research*, **16**, 217-226.

[11]   Sharma, P.S. and Kumari, T.S. (1996) Effect of Potassium under Water Stress on Growth and Yield of *Sorghum* in Vertisol. *J. Potash. Res.*, **12**, 319-325.

[12]   Richards, A.R. (2006) Physiological Traits Used in the Breeding of New Cultivars for Water Scarce Environments. *Agricultural Water Management*, **80**, 197-211. http://dx.doi.org/10.1016/j.agwat.2005.07.013

[13]   Aulakh, M.S. and Malhi, S.S. (2004) Interaction of Nitrogen with Other Nutrients and Water: Effects on Crop Yield and Quality, Nutrient Use Efficiency, Carbon Sequestration and Environmental Pollution. *Advances in Agronomy*, **86**, 341-409. http://dx.doi.org/10.1016/S0065-2113(05)86007-9

[14]   Snedecor, G.W. and Cochran, W.G. (1990) Statistical Methods. 8th Edition, Iowa State University, Iowa.

[15]   Carmier, A., Aharoni, Y., Edelstein, M., Umiel, N., Hagiladi, A., Yousef, E., Nikbachat, M., Zenou, A. and Miron, J. (2006) Effects of Irrigation and Plant Density on Yield, Composition and *in Vitro* Digestibility of a New Forage *Sorghum* Variety, TAL, at Two Maturity Stage. *Animal Feed Science and Technology*, **131**, 121-133. http://dx.doi.org/10.1016/j.anifeedsci.2006.02.005

[16]   Mohammadkhani, N. and Heidri, R. (2008) Effects of Drought Stress on Soluble Proteins in Two Maize Varieties. *Turkish Journal of Biology*, **32**, 23-30.

[17]   Hoekstra, F.A., Golovinia, E.A. and Butinik, J. (2001) Mechanisms of Plant Desiccation Tolerance. *Trends in Plant Science*, **6**, 431-438. http://dx.doi.org/10.1016/S1360-1385(01)02052-0

[18] Koch, K.E. (1996) Carbohydrate-Modulated Gene Expression in Plants. *Annual Review of Plant Physiology and Plant Molecular Biology*, **47**, 509-540. http://dx.doi.org/10.1146/annurev.arplant.47.1.509

[19] Smeekens, S. (2000) Sugar-Induced Signal Transduction in Plants. *Annual Review of Plant Physiology and Plant Molecular Biology*, **51**, 49-81. http://dx.doi.org/10.1146/annurev.arplant.51.1.49

[20] Larher, F., Leport, L., Petrelavisky, M. and Chapart, M. (1993) Effects of Osmo Induced Proline Response in Higher Plants. *Plant Physiology and Biochemistry*, **31**, 911-922.

[21] Boomsma, C.R. and Vyn, T.J. (2008) Maize Drought Tolerance: Potential Improvements through Arbuscular Mycorrhizal Symbiosis. *Field Crops Research*, **108**, 14-31. http://dx.doi.org/10.1016/j.fcr.2008.03.002

[22] Li, K.R., Wang, H.H., Han, G., Wang, Q.J. and Fan, J. (2008) Effects of Brassinolide on the Survival, Growth and Drought Resistance of *Robinia pseudoacacia* Seedlings under Water-Stress. *New Forests*, **35**, 255-266.

[23] Akmal, M. and Janssens, J.J. (2004) Productivity and Light Use Efficiency of Perennial Ryegrass with Contrasting Water and Nitrogen Supplies. *Field Crop Research*, **88**, 143-155. http://dx.doi.org/10.1016/j.fcr.2003.12.004

[24] Ferré, I. and Faci, J.M. (2009) Deficit Irrigation in Maize for Reducing Agricultural Water in a Mediterranean Environment. *Agricultural Water Management*, **96**, 383-394. http://dx.doi.org/10.1016/j.agwat.2008.07.002

[25] Bokhtiar, S.M. and Sakurai, K. (2005) Effect of Application of Inorganic and Organic Fertilizers on Growth, Yield and Quality of Sugarcane. *Sugar Tech*, **7**, 35-37. http://dx.doi.org/10.1007/BF02942415

[26] Bayu, W., Rethman, N.F.G., Hammes, P.S. and Alemu, G. (2006) Effects of Farmyard Manure and Inorganic Fertilizers on *Sorghum* Growth, Yield, and Nitrogen Use in a Semi-Arid Area of Ethiopia. *Journal of Plant Nutrition*, **29**, 391-407. http://dx.doi.org/10.1080/01904160500320962

[27] Barros, I., Gaiser, T., Lange, F.M. and Römheld, V. (2007) Mineral Nutrition and Water Use Patterns of a Maize/Cowpea Intercrop on a Highly Acidic Soil of the Tropic Semiarid. *Field Crops Research*, **101**, 26-36. http://dx.doi.org/10.1016/j.fcr.2006.09.005

[28] Marschner, H. (1995) Mineral Nutrition of Higher Plants. 2nd Edition, Acadmic Press, London.

[29] Shrotriya, G.C. (1998) Balanced Fertilizer—India Experience. *Proceedings of Symposium on Plant Nutrition Management for Sustainable Agricultural Growth*, NFDC, 8-10 December 1997, Islamabad.

[30] Bumb, B.I. and Bannante, C.A. (1996) The Use of Fertilizer in Sustaining Food Security and Protecting the Environment-2020. *Proceedings of Conference on Agriculture and Fertilizer Use by* 2010, NEDC, Islamabad, 35.

[31] Salifu, K.F. and Timmer, V.R. (2003) Nitrogen Retranslocation Response of Young *Picea mariana* to Nitrogen-15 Supply. *Soil Science Society of America Journal*, **67**, 309-317. http://dx.doi.org/10.2136/sssaj2003.0309

[32] Andivia, E., Fernández, M. and Vázquez-Piqué, J. (2011) Autumn Fertilization of *Quercus ilex* ssp. *ballota* (Desf.) Samp. Nursery Seedlings: Effects on Morpho-Physiology and Field Performance. *Annals of Forest Science*, **68**, 543-553. http://dx.doi.org/10.1007/s13595-011-0048-4

[33] Villar-Salvador, P., Peñuelas, J.L. and Jacobs, D.F. (2013) Nitrogen Nutrition and Drought Hardening Exert Opposite Effects on the Stress Tolerance of *Pinus pinea* L. Seedlings. *Oxford Journals Life Sciences Tree Physiology*, **33**, 221-232.

17

Sustainable Mangement of Drainage Water of Fish Farms in Agriculture as a New Source for Irrigation and Bio-Source for Fertilizing

Abdelraouf Ramadan Eid[1]*, Essam Mohamed Hoballah[2], Sahar E. A. Mosa[3]

[1]Water Relations & Field Irrigation Department, National Research Centre, Cairo, Egypt
[2]Agricultural Microbiology Department, National Research Centre, Cairo, Egypt
[3]Bio-Engineering Department, Agricultural Engineering Research Institute (AEnRI), Agricultural Research Center (ARC), Ministry of Agriculture, Giza, Egypt
Email: *abdelrouf2000@yahoo.com

Abstract

Two field experiments were carried out during growing seasons 2011 and 2012. It was executed in research farm of National Research Center in Nubaryia region, Egypt to study the effect of irrigation systems, fertigation rates by using the wastewater of fish farms "WWFF" in irrigation of potato. Study factors were irrigation systems (sprinkler irrigation system "SIS" and trickle irrigation system "TIS"), water quality (traditional irrigation water "TIW" and WWFF) and fertigation rates "FR" (20%, 40%, 60%, 80% and 100% NPK). The following parameters were studied to evaluate the effect of study factors: 1) Calculating the total amount of WWFF per season; 2) Chemical and biological description of WWFF; 3) Clogging ratio of emitters; 4) Yield of potato; 5) Irrigation water use efficiency of potato "$IWUE_{potato}$". Statistical analysis indicated that, maximum values were obtained of yield under SIS × $FR_{100\% NPK}$ × WWFF, also, there were no significant differences for yield values under the following conditions: SIS × $FR_{100\% NPK}$ × WWFF > SIS × $FR_{80\% NPK}$ × WWFF > SIS × $FR_{60\% NPK}$ × WWFF > TIS × $FR_{100\% NPK}$ × TIW. This means that, using WWFF in the irrigation can save at least 40% from mineral fertilizers and 100% from irrigation water under sprinkler irrigation system.

Keywords

Wastewater of Fish Farms, Potato, Arid Regions, Fertigation Rates, Irrigation Systems

---

*Corresponding author.

1. Introduction

Whenever good quality water is scarce, water of marginal quality will have to be considered for use in agriculture. Although there is no universal definition of "marginal quality" water, for all practical purposes it can be defined as water that possesses certain characteristics which have the potential to cause problems when it is used for an intended purpose. Many countries have included wastewater reuse as an important dimension of water resources planning. In the more arid areas of the world, wastewater is used in agriculture, releasing high quality water supplies for potable use.

This diverted attention to fish farming. However, recycling the drainage water (DW) of fish farming, rich with organic matter for agriculture use can improve soil quality and crops productivity [1], and reduce the total costs since it decreases the fertilizers use, whose demand became affected by the prices and the framer's education [2]. Meanwhile, organic matter content supports the cation exchange process in soils, which is important to the nutrition of plants [3]. Plants grow rapidly with dissolved nutrients that are excreted directly by fish or generated from the microbial breakdown of fish wastes. In closed recirculating systems with very little daily water exchange (less than 2 percent), dissolved nutrients accumulate in concentrations similar to those in hydroponic nutrient solutions. Dissolved nitrogen, in particular, can occur at very high levels in recirculating systems. Fish excrete waste nitrogen, in the form of ammonia, directly into the water through their gills. Bacteria convert ammonia to nitrite and then to nitrate. Aquaponic systems offer several benefits. Dissolved waste nutrients are recovered by the plants, reducing discharge to the environment and extending water use (*i.e.*, by removing dissolved nutrients through plant uptake, the water exchange rate can be reduced). Minimizing water exchange reduces the costs of operating aquaponic systems in arid climates and heated greenhouses where water or heated water is a significant expense. Having a secondary plant crop that receives most of its required nutrients at no cost improves a system's profit potential. The daily application of fish feed provides a steady supply of nutrients to plants and thereby eliminates the need to discharge and replaces depleted nutrient solutions or adjusts nutrient solutions as in hydroponics. The plants remove nutrients from the culture water and eliminate the need for separate and expensive biofilters. Directly absorbed and assimilated by plants, these compounds stimulate growth, enhance yields, increase vitamin and mineral content, improve fruit flavor and hinder the development of pathogens. The potato is the 5th most important crop in the world. It is nutritious and highly productive, and has a good value when sold, and is an effective cash crop for a developing country that has both local and export markets [4]. Quality of irrigation water also affects the degree of emitter clogging [5]. A high concentration of soluble salts in the water is the most important factor in clogging. When the concentrations of calcium, magnesium, bicarbonate and sulfate are high, the calcium carbonate, calcium sulfate and magnesium sulfate can occur. Calcium carbonate precipitation will also depend on the pH of the water. Precipitation of insoluble salts can also occur due to chemical reactions among the elements added as fertilizers in irrigation water [6]. Precipitated salts can easily clog emitters. Fertilizers injected into a microirrigation system may contribute to plugging [7]. The most important disadvantage of fertigation is precipitation of chemical materials and clogging of emitters [8]. Any fertilizer with calcium should not be used with sulfates together because they could form insoluble gypsum [7] [9] [10]. The objective of this study was maximizing utility from wastewater of fish farms in agriculture (potato cultivation) under arid regions conditions.

2. Materials and Methods

2.1. Site Description

Field experiments were conducted during two wheat seasons from Jan. to May of 2011-2012 at the experimental farm of National Research Center, El-Nubaria, Egypt (latitude 30°30'1.4"N, and longitude 30°19'10.9"E, and mean altitude 21 m above sea level). The experimental area has an arid climate with cool winters and hot dry summers prevailing in the experimental area. The monthly mean climatic data for the two growing seasons 2011 and 2012, for El-Nubaria city, are nearly the same. The data of maximum and minimum temperature, relative humidity, and wind speed were obtained from "Central Laboratory for Agricultural Climate (CLAC)". There was no rainfall that could be taken into consideration through the two seasons, because the amount was very little and the duration didn't exceed few minutes as shown in **Table 1**.

2.2. Estimation of the Seasonal Irrigation Water for Potato Plant

Seasonal irrigation water was estimated according to the meteorological data of the Central Laboratory for

Table 1. The monthly mean climatic data for the two growing seasons 2011 and 2012.

Items	Precipitation [mm]	Wind speed [m/sec]		HC Air temperature [°C]			HC Relative humidity [%]
	Sum	Aver.	Maxi.	Aver.	Mini.	Maxi.	Aver.
	0.3	0.5	2.1	14.1	9.6	19.1	78.9
	0.1	0.6	2.2	14.3	9.7	19.7	79.6
2011	0.0	0.6	2.3	17.3	11.6	24.0	75.6
	0.0	0.6	2.3	20.0	14.1	26.4	73.2
	0.0	0.5	2.1	14.1	9.6	19.1	78.9
	0.4	0.7	2.5	11.8	7.5	17.1	81.2
	0.0	0.6	2.1	11.7	7.7	16.4	79.7
2012	0.1	0.6	2.2	14.8	8.6	21.9	80.4
	0.0	0.6	2.2	18.8	12.1	27.1	74.4
	0.1	0.6	2.3	22.0	15.3	30.3	74.1

Agricultural Climate (CLAC) depending on Penman-Monteith equation shown in **Figure 1**. The volume of applied water increased with the growth of plant then declined at the end of the growth season. The seasonal irrigation water applied was found to be 2847 m³/fed/season for sprinkler irrigation system and 2476 m³/fed/season for trickle irrigation system.

2.3. Some Physical and Chemical Properties of Soil and Irrigation Water

Some Properties of soil and irrigation water for experimental site are presented in (**Table 2**, **Table 3** and **Table 4**). **Table 5** showed that, the determination of total bacteria, total fungi and some algal microorganisms and some physical and chemical determinations of wastewater of fish farm.

2.4. Potato Variety

Spunta Netherland production was used.

2.5. Experimental Design

Irrigation system components consisted of a control head and a pumping unit. It consisted of submersible pump with 45 m³/h discharge driven by electrical engine back flow prevention device, pressure regulator, pressure gauges, flow-meter and control valves. Main line was of PVC pipes with 110 mm in diameter (OD) to convey the water from the source to the main control points in the field. Sub-main lines were of PVC pipes with 75 mm diameter (OD) connected to the main line. Manifold lines: PE pipes of 63 mm in diameter (OD) were connected to the sub main line through control valve 2" and discharge gauge. Layouts of experiment design consisted of two irrigation systems. Sprinkler is a metal impact sprinkler 3/4" diameter with a discharge of 1.17 m³h⁻¹, wetted radius of 12 m, and working pressure of 250 kPa. Emitters, built in laterals tubes of PE with 16 mm diameter (OD) and 30 m in length (emitter discharge was 4 lph at 1.0 bar operating pressure and 30 cm spacing between emitters and all details about the experiment design and the source of wastewater of fish farm collected from 12 basin (5 m × 5 m × 2 m depth) are shown in **Figure 2**.

2.6. Methods

2.6.1. Sampling Site Description

Wastewater for fish farm samples were collected at the outlet of water basin used for fish breeding and production.

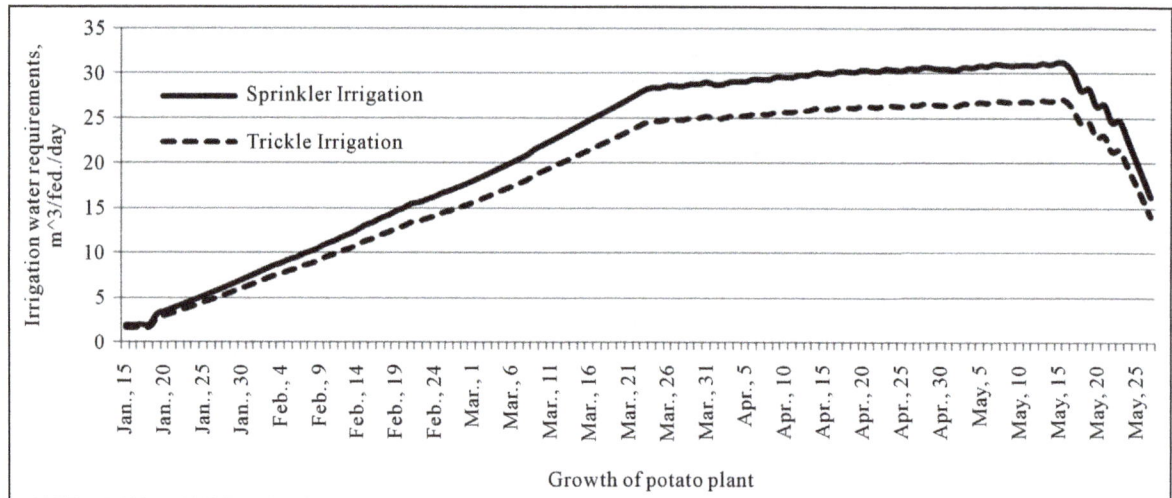

Figure 1. The relation between growth of potato plant and irrigation water requirements.

Table 2. Some chemical and mechanical analyses of soil.

| Depth | Chemical analysis | | | | Mechanical analysis, % | | | Texture |
	OM (%)	pH (1:2.5)	EC (dSm^{-1})	CaCO$_3$ %	Course sand	Fine sand	Clay + Silt	
0 - 20	0.65	8.7	0.35	7.02	47.76	49.75	2.49	
20 - 40	0.40	8.8	0.32	2.34	56.72	39.56	3.72	Sandy
40 - 60	0.25	9.3	0.44	4.68	36.76	59.40	3.84	

Table 3. Characteristics of soil.

Depth	SP (%)	F.C (%)	W.P (%)	A.W (%)	Hydraulic conductivity (cm/hr)
0 - 20	21.0	10.1	4.7	5.4	22.5
20 - 40	19.0	13.5	5.6	7.9	19.0
40 - 60	22.0	12.5	4.6	7.9	21.0

Table 4. Some chemical characteristics of irrigation water of open channel.

| pH | EC (dSm^{-1}) | Cations and anions (meq/L) | | | | | | | | SAR% |
| | | Cations | | | | Anions | | | | |
		Ca^{+2}	Mg^{+2}	Na$^+$	K$^+$	CO$_3^-$	HCO$_3^-$	Cl$^-$	SO$_2^-$	
7.35	0.41	1	0.5	2.4	0.2	--	0.1	2.7	1.3	2.8

2.6.2. Physico Chemical Characters of Wastewater for Fish Farm

The physicochemical characteristics were carried out according to [11]. pH, EC, N, P, K and potential toxic elements (Cu, Zn, Pb,… etc.)

2.6.3. Biological Parameters

1) Total Viable Count of Bacteria: TVCB was determined using the standard plate count method and nutrient agar culture medium according to [11]; 2) Total count of fungi: was determined using the standard plate count method and Rose-bengal agar culture medium according to [12]; 3) Faecal coliform bacteria were counted using MacConky broth [13] and most probable number method [14]; 4) Total counts of free N$_2$ fixers using Ashby's

Table 5. Some physical and chemical and biological determinations of drainage water of fish farm under search.

Physical Determinant		Value	Biological Determinant	Counts as CFU/ml
EC		1.82 dsm^{-1}	Total counts of bacteria	1.5×10^4
pH		7.02	Total count of faecal coliform	3×10^2
Chemical elements			Total counts of fungi	500
Chromium	Cr	0.0 ppm	Total counts of free N$_2$ fixers	600
Copper	Cu	0.33 ppm	Green algae	
Nickel	Ni	0.0 ppm	*Chlorella* sp. Count	400
Zinc	Zn	1.1 ppm	*Scenedesmus* sp. Count	150
Nitrogen	N	4.79 ppm	*Pediastrum* sp. Count	120
Phosphorus	P	10.2 ppm	Cyanobacteria	
Potassium	K	35 ppm	*Oscillatoria* sp. Count	100
Sodium	Na	205 ppm	*Nostoc* sp. Count	50

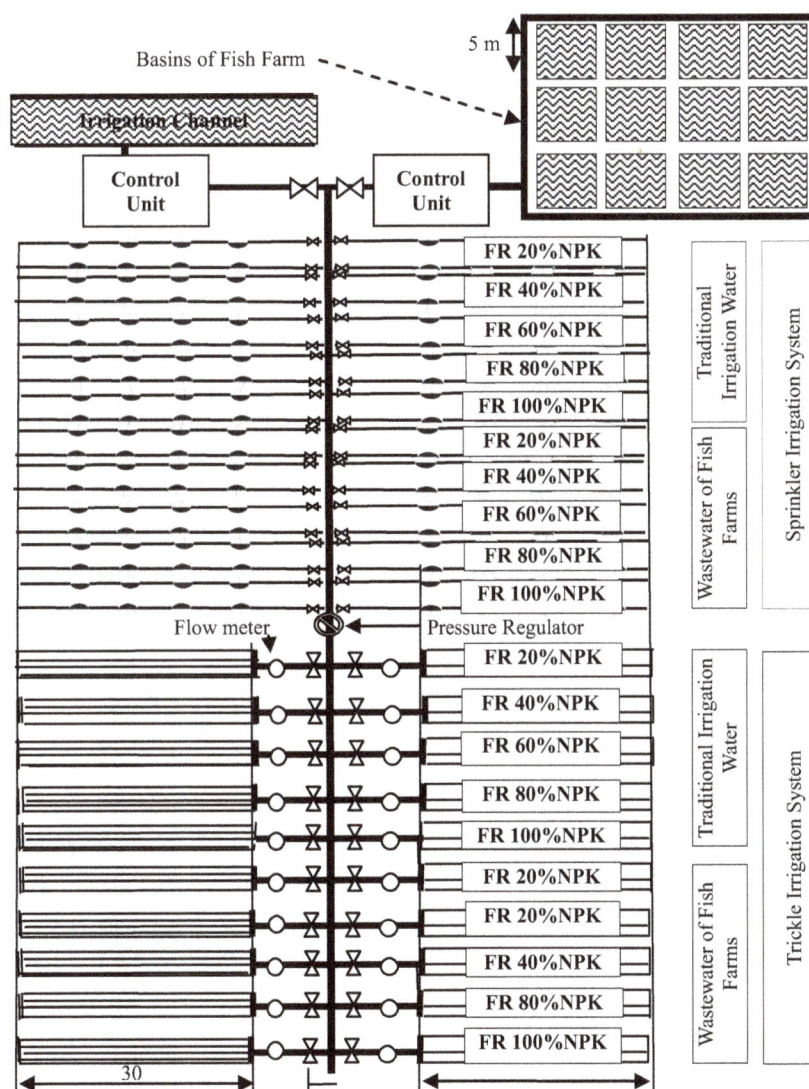

Figure 2. Layout of experiment design.

medium [15]; 5) Algae enumeration: The grouping of green algae and blue-green algae were accomplished and counted depending on morphological shape under light microscope using the Sedgwick-Rafter (S-R) cell count chamber according to [11], then calculated algae counts from the following equation:

$$\text{No./mL} = \left(C * 1000 \text{ mm}^3\right) / \left(L * D * W * S\right)$$

where: C = number of organisms counted, L = length of each strip (S-R cell length), mm, D = depth of a strip (S-R cell depth), mm, W = width of a strip (Whipple grid image width), mm, and S = number of strips counted.

2.6.4. Determination of Clogging Ratio

The flow cross section diameter of the long-path emitter was 0.7 mm; discharging 4 L/h with lateral length of 30 m. Distance between emitter along the lateral was 30 cm. The emitter is considered laminar-flow-type (Re < 2000) [16]. To estimate the emitter flow rate cans and a stopwatch were used. Nine emitters from each lateral had been chosen to be evaluated by calculating their clogging ratio at the beginning and at the end of the growing season for the two seasons. Three emitters at the beginning, three at middle and three at the end of the lateral were tested for the flow rate. Clogging ratio was calculated according to [17] using the following equations:

$$E = \left(qu/qn\right) \times 100$$

$$CR = \left(1 - E\right) \times 100$$

where: E = the emitter discharge efficiency (%), qu = emitter discharge at the end of the growing season (L/h), qn = emitter discharge, at the beginning of the growing season (L/h), CR = clogging ratio of emitters (%).

2.6.5. Determination Yield of Potato Crop

At the end of the growing season, potato yields were determined, Ton/Fadden for each treatment by the following steps; step 1 measuring the area to determine the yield, step 2 collecting the potato for each treatment on the buffer zone and step 3 weighing potato for each treatment.

2.6.6. Determination of Irrigation Water Use Efficiency of Potato Crop

Irrigation water use efficiency "IWUE" is an indicator of effectiveness use of irrigation unit for increasing crop yield. Water use efficiency of potato yield was calculated according to [16] as follows: IWUE_{potato} (kg/m^3) = Total yield (kg$_{tuber}$/fed)/Total applied irrigation water (m^3/fed/season)

2.6.7. Fertigation Method

The recommended doses of chemical fertilizer were added as fertigation *i.e.* nitrogen fertilizer was added at a rate of 120 kg/Fadden as ammonium sulfate (20.6% N), 150 kg calcium super phosphate/fed (15.5% P_2O_5) and 50 kg potassium sulfate (48% K_2O)) were added.

2.6.8. Statistical Analysis

The standard analysis of variance procedure of split-split plot design with three replications as described by [18] was used. All data were calculated from combined analysis for the two growing seasons 2011 and 2012. The treatments were compared according to L.S.D. test at 5% level of significance.

3. Results and Discussion

3.1. Calculating the Total Amount of Wastewater of Fish Farm per Season

To calculate the total amount of wastewater for fish farm in NUBARIA farm, the volume of water discharged per week must be calculated. There are 12 basin in the fish farm and the dimensions of the basin are 5 m × 5 m × 2 m, but the depth of the actual exchange is 1.5 m and therefore the size of the outgoing water per week = 5 × 5 × 1.5 × 12 basin = 450 m^3 of water. If we consider that potato cultivation needs 18 weeks, the total volume lost from this farm during the potato growing season = 18 × 450 = 8100 m^3/season of water as shown in **Figure 3**.

3.2. Chemical and Biological Description of Wastewater of Fish Farm

The data aforementioned in **Table 5** showed that, the EC was 1.82 ds/m, pH was 7.02. On the other hand, the

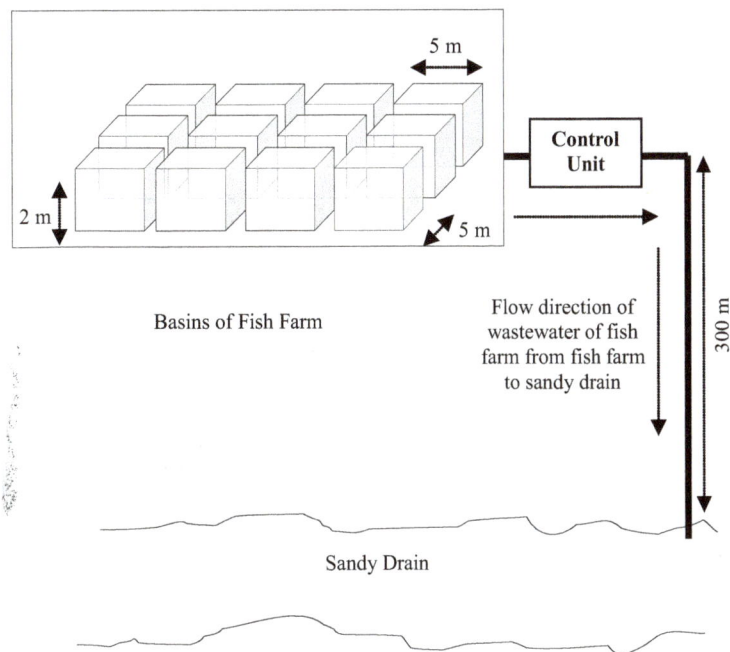

Figure 3. Loss of wastewater of fish farm.

results in **Table 5** showed that Chromium, Copper, Nickel, Zinc, total Nitrogen as N_2, Phophorus as P, Potassium and Sodium reached 0.0, 0.33, 0.0, 1.1, 4.79, 10.2, 35 and 205 ppm, respectively. The data mentioned above showed quantitative fertigation capacity of the wastewater of fish farm under study to be used as irrigation water. Wastewater of fish farm could supply seasonally the soil with 13.637 and 11.86 kg of nitrogen/Fed. from the whole quantities of irrigation water to sprinkler and trickle irrigation methods used, respectively, that are equivalent to 64.938 and 56.476 kg of ammonium sulphate fertilizer (21% N) to sprinkler and trickle irrigation methods used, respectively. Also, this water could supply seasonally the soil with 29.039 and 25.252 kg of phosphorus from the whole quantities of irrigation water to sprinkler and trickle irrigation methods used, respectively, that are equivalent to 351 and 306 kg of superphosphate fertilizer (8.25% P) to sprinkler and trickle irrigation methods used, respectively.

3.2.1. Quantitative Estimation of Bacteria and Fungi

The data aforementioned in **Table 4** showed that, the total counts of bacteria reached 1.5×10^4 CFU/ml; also total counts of free N_2 bacterial fixers determined by Ashby's medium [15] (Kizilkaya, 2009) were 600 CFU/ml however the total count of faecal coliform was 3×10^2 CFU/ml. On the other hand, total counts of fungi reached 500 CFU/ml. The results aforementioned before are partially in agreement with the findings stated by [19] in which the possible counts of total counts of bacteria in domestic wastewater reached between 10^3 to 10^5 CFU/ml and also, the Coliform group of bacteria comprises mainly species of the genera *Citrobacter*, *Enterobacter*, *Escherichia* and *Klebsiella* and includes Faecal Coliforms, of which *Escherichia coli* is the predominant species were 10^2. Several of the Coliforms are able to grow outside the intestine, especially in hot climates; hence their enumeration is unsuitable as a parameter for monitoring wastewater reuse systems. The Faecal Coliform test may also include some non-faecal organisms which can grow at 44°C, so the *E. coli* count is the most satisfactory indicator parameter for wastewater used in agriculture.

3.2.2. Quantitative Estimation of Phytoplankton

The morphological studies using a light microscope were done on the water samples under estimation. Water samples showed various phytoplankton structures belonging to two main groups, namely, Chlorophyceae (Green Algae) and Cyanophyceae (Blue-Green Algae). The general distribution of phytoplankton is demonstrated in **Table 4**. It may be important to note that genera, *chlorella*, *Pediastrum* and *Scenedesmus* as green algae were detected, whereas, *Oscillatoria* and *Nostoc* represented the most abundant genera of cyanobacteria in the investigated samples. The algae biomass contains nutrients such as C, N, P and k essential for microorganism development. The general microalgae biochemical structure has been successfully utilized as feedstock for digesters and as nutrient supplements in dairy farming. Algae biomass components such as protein, carbohydrates, poly-unsaturated fatty acids, are rich in nutrients vital for development of fish and shellfish consumption and other aquatic microorganisms as shown in **Figure 4**.

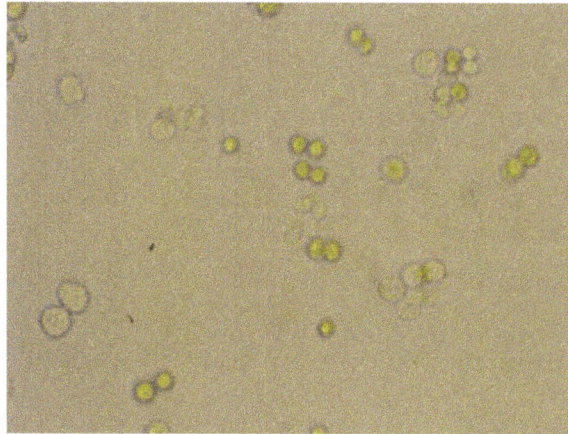
Chlorella sp.

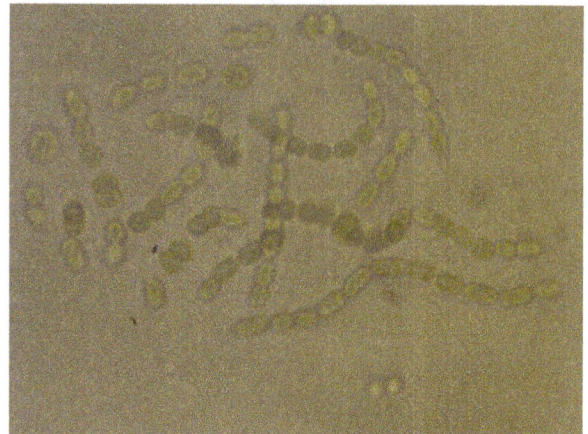
Nostoc sp.

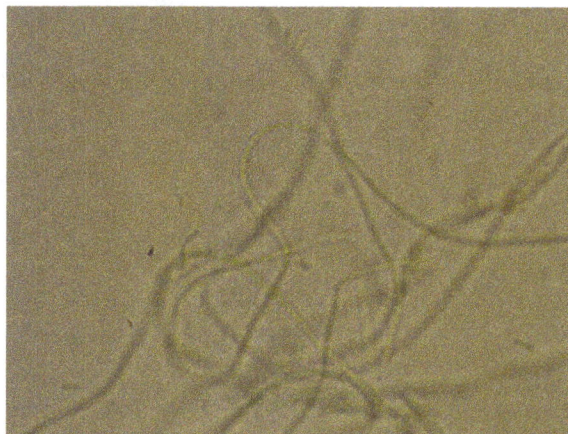
Oscillatoria sp.

Pediastrum sp.

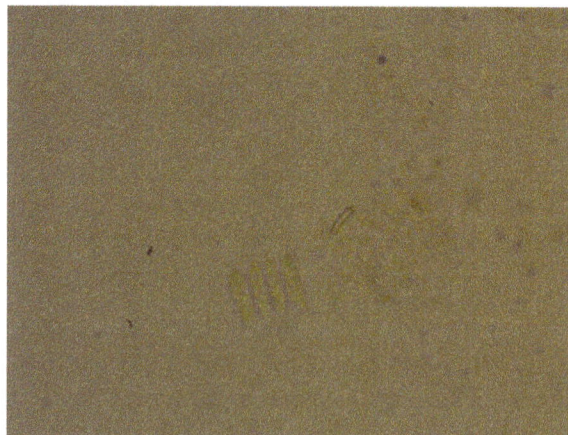
Scenedesmus sp.

Figure 4. Types of *Chlorella* sp., *Nostoc* sp., *Oscillatoria* sp., *Pediastrum* sp. and *Scenedesmus* sp. were found in the wastewater of fish farm.

3.3. Effect of Irrigation Systems, Wastewater of Fish Farms and Fertigation Rates on Clogging Ratio, Yield of Potato and Irrigation Water Use Efficiency of Potato

3.3.1. Effect of Irrigation Systems on Clogging Ratio, Yield of Potato and Irrigation Water Use Efficiency of Potato Crop

Table 6 showed that, the effect of irrigation systems on clogging ratio, yield of potato and irrigation water use efficiency of potato crop (Table 6). Clogging ratio was increased under trickle irrigation system more than sprinkler irrigation system this may be due to the increase in orifices diameter of sprinkler than dripper especially in the absence of a filtering system (Table 6). Yield of potato was decreased under trickle irrigation system more than sprinkler irrigation system this may be due to water stress under trickle irrigation system more than sprinkler irrigation system which comes from the increasing in clogging ratio (Table 6). Increasing of irrigation water use efficiency of potato under trickle irrigation system compared with sprinkler irrigation system this may be due to increasing of water requirements under sprinkler irrigation system.

3.3.2. Effect of Wastewater of Fish Farms on Clogging Ratio, Yield of Potato and Irrigation Water Use Efficiency of Potato

Table 6 showed that, the effect of wastewater of fish farms on clogging ratio, yield of potato and irrigation water use efficiency of potato crop (Table 6). Clogging ratio was increased under WWFF more than WIT this may be due to the increasing in increase the proportion of suspended materials such as organic material and algae in WWFF than WIT (Table 6). Yield of potato was decreased under WIT more than WWFF this may be due to increasing of bio-components in WWFF than in WIT. Table 6 indicated that increasing of irrigation water use efficiency of potato under WWFF and the difference between WWFF and WIT were nonsignificant.

3.3.3. Effect of Fertigation Rates on Clogging Ratio, Yield of Potato and Irrigation Water Use Efficiency of Potato

Table 6 and Figure 5 show the relation between fertigation rates and clogging ratio, yield of potato and irrigation water use efficiency of potato crop. Figure 5(a) shows that clogging ratio was increased by increasing the fertigation rates this may be due to increasing the amount and concentration of dissolved mineral fertilizers in irrigation water that lead to the increase in clogging ratio (Figure 5(b)). Yield of potato was increased by increasing fertigation rates this may be due to increasing the amount and concentration of mineral fertilizers in the root zone. Figure 5(c) indicated the increase of irrigation water use efficiency of potato by increasing the fertigation rates this may be due to increasing the yield of potato by increasing the fertigation rates.

3.4. Effect the Interaction between Irrigation Systems, Wastewater of Fish Farms and Fertigation Rates on Clogging Ratio, Yield of Potato and Irrigation Water Use Efficiency of Potato

Table 7 and Figure 6 show the effect of the interaction between irrigation systems, wastewater of fish farms

Table 6. Effect of irrigation systems, wastewater of fish farms and fertigation rates on clogging ratio, yield of potato and irrigation water use efficiency of potato (IWUE).

Study Factors	Clogging Ratio, %	Yield, Ton/Fed	IWUE, kg/m^3
SIS	1.3 b	10.0 a	2.9 a
TIS	33.2 a	5.7 b	2.9 a
WWFF	28.2 a	8.0 n.s	2.9 n.s
TIW	6.3 b	7.8 n.s	2.9 n.s
FR$_{20\% NPK}$	13.5 e	5.4 d	2.0 d
FR$_{40\% NPK}$	15.6 d	6.3 c	2.4 c
FR$_{60\% NPK}$	17.1 c	8.5 b	3.1 b
FR$_{80\% NPK}$	18.8 b	8.9 b	3.3 b
FR$_{100\% NPK}$	21.3 a	10.1 a	3.7 a

SIS: Sprinkler Irrigation System, TIS: trickle Irrigation System, WWFF: wastewater of fish farms, TIW: Traditional Irrigation Water, FR: Fertigation Rates. Letters a, b, c, d and e represent the significant between values.

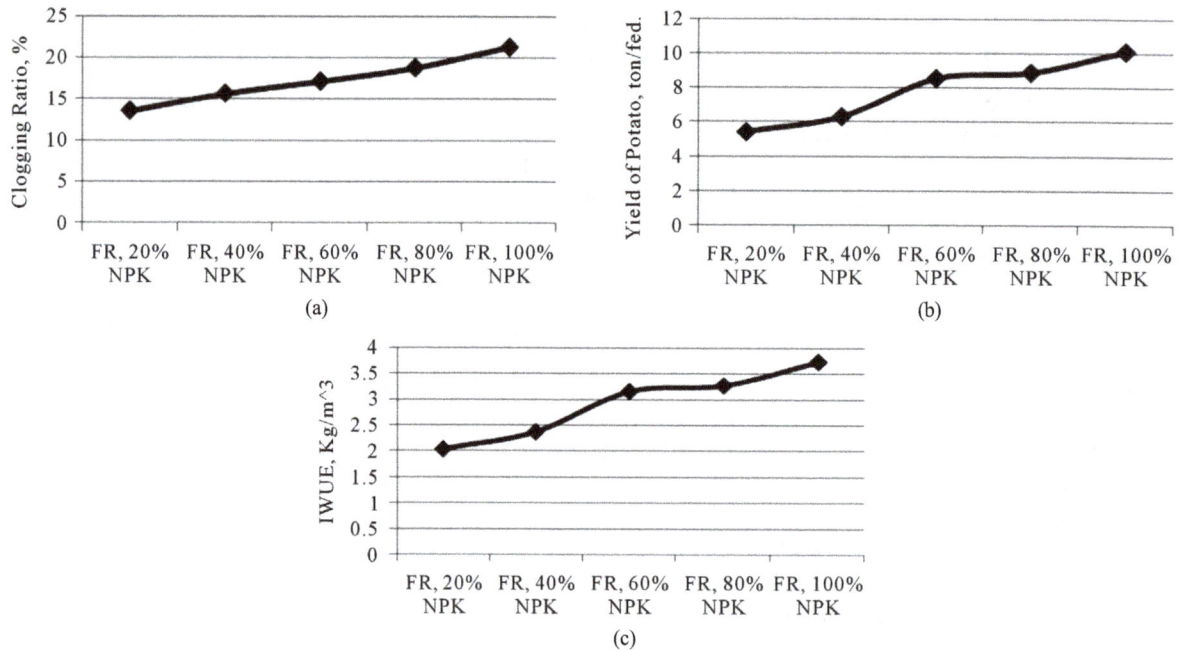

Figure 5. Effect of fertigation rates "FR" on (a) Clogging ratio, (b) Yield of potato and (c) Irrigation water use efficiency of potato "IWUE".

Table 7. Effect the interaction between irrigation systems, wastewater of fish farms and fertigation rates on clogging ratio of emitters, yield of potato and irrigation water use efficiency of potato crop.

Study Factors			Clogging ratio, %	Yield, Ton/Fed	IWUE, kg/m³
IS	Water quality	FR			
SIS	WWFF	FR, 20% NPK	1.0 jk	7.7 fg	2.7 g
		FR, 40% NPK	1.4 jk	8.3 f	2.9 fg
		FR, 60% NPK	2.2 jk	15.2 a	5.3 a
		FR, 80% NPK	2.5 j	14.9 ab	5.2 ab
		FR, 100% NPK	2.5 j	15.7 a	5.5 a
	TIW	FR, 20% NPK	0.4 k	5.0 i	1.8 ij
		FR, 40% NPK	0.5 k	6.0 f	2.1 h
		FR, 60% NPK	0.6 k	7.6 fg	2.7 g
		FR, 80% NPK	0.8 jk	9.3 e	3.3 e
		FR, 100% NPK	0.9 jk	10.5 d	3.7 d
TIS	WWFF	FR, 20% NPK	44.8 e	4.1 j	1.6 j
		FR, 40% NPK	50.8 d	4.0 j	1.6 j
		FR, 60% NPK	55.1 c	3.7 jk	1.5 jk
		FR, 80% NPK	59.3 b	3.2 kl	1.3 kl
		FR, 100% NPK	62.3 a	2.9 l	1.2 l
	TIW	FR, 20% NPK	7.9 i	5.0 i	2.0 hi
		FR, 40% NPK	9.8 h	7.0 g	2.8 g
		FR, 60% NPK	10.6 h	7.7 fg	3.1 ef
		FR, 80% NPK	12.4 g	8.2 f	3.3 e
		FR, 100% NPK	19.3 f	11.3 c	4.6 c

SIS: Sprinkler Irrigation System, TIS: trickle Irrigation System, WWFF: wastewater of fish farms, TIW: Traditional Irrigation Water, FR: Fertigation Rates. Letters a, b, c, d and e represent the significant between values.

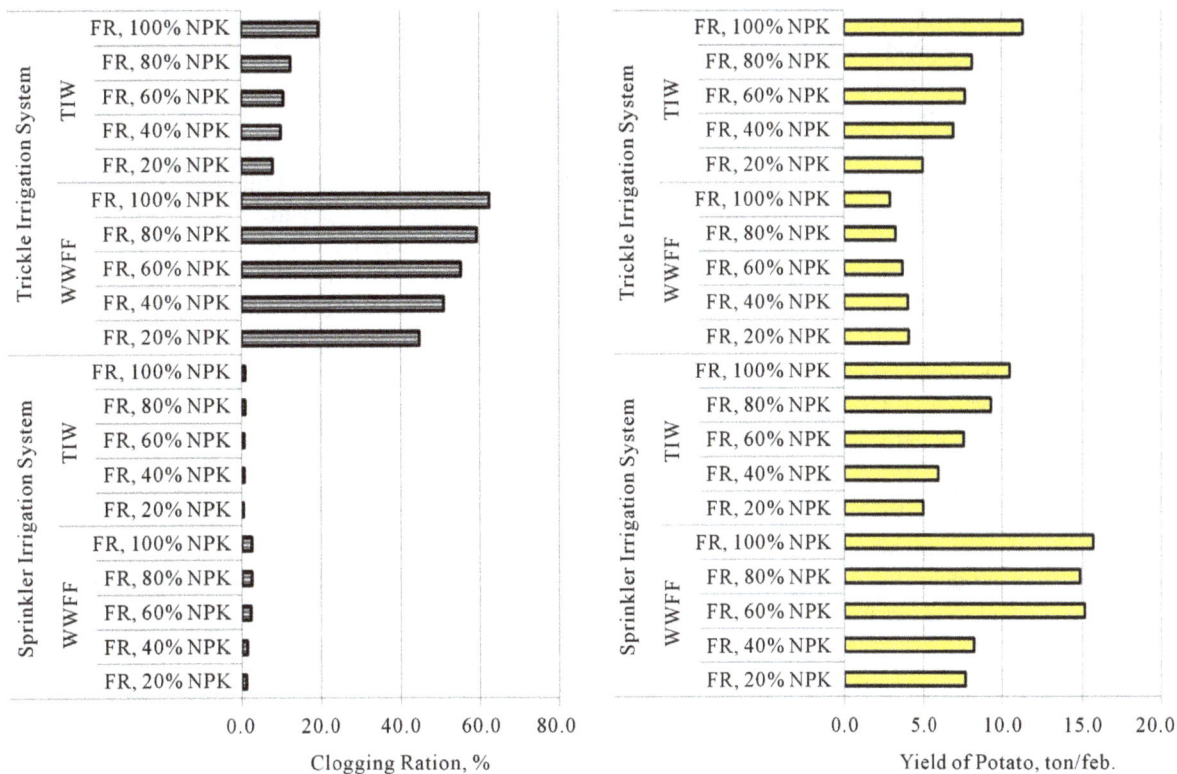

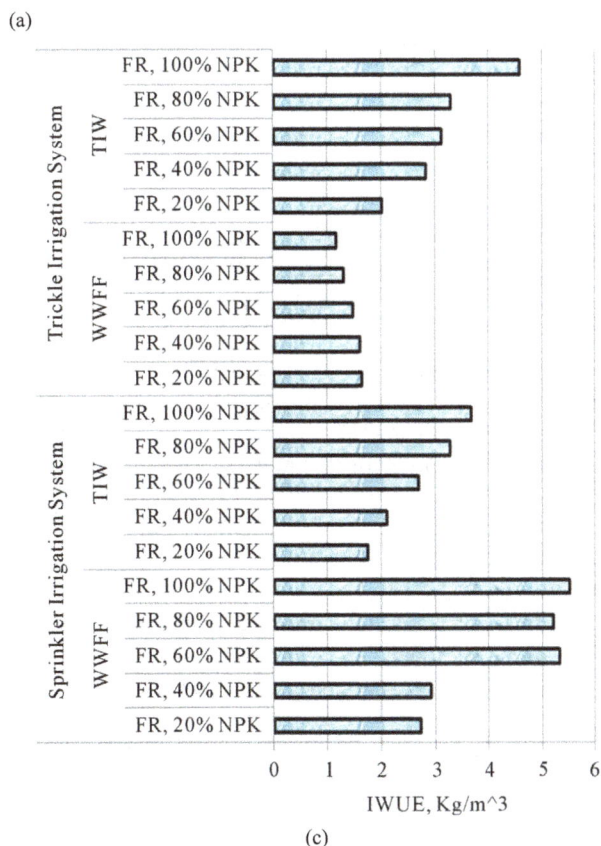

Figure 6. Effect of the interaction between irrigation systems, wastewater of fish farms "WWFF" and fertigation rates "FR" on (a) Clogging ratio, (b) Yield of potato and (c) Irrigation water use efficiency of potato crop.

"WWFF" and fertigation rates "FR" on clogging ratio, yield of potato and irrigation water use efficiency of potato crop. **Figure 6(a)** show the relation between study factors on clogging ratio. Maximum values of clogging ratio occurred under trickle irrigation system + WWFF + $FR_{100\% NPK} > FR_{80\% NPK} > FR_{60\% NPK} > FR_{40\% NPK} > FR_{20\% NPK}$ this may be due to the increase in orifices diameter of sprinkler than dripper and the increase in proportion of suspended materials such as organic material and algae in WWFF than WIT in addition to increasing the amount and concentration of dissolved mineral fertilizers in irrigation water. **Figure 6(a)** show the relation between study factors on yield of potato. Minimum values of clogging ratio occurred under sprinkler irrigation system + WWFF and WIT. **Figure 6(b)** show the relation between study factors on yield of potato. Maximum values of yield of potato occurred under sprinkler irrigation system + WWFF + $FR_{100\%, 80\%, 60\% NPK}$ this may be due to reduction in water stress resulting from reduction in clogging ratio under sprinkler irrigation system and increasing of bio-components in WWFF in addition to increasing the amount and concentration of mineral fertilizers in the root zone by increasing of FR. **Figure 6(c)** showed the relation between study factors on IWUE. Maximum values of IWUE occurred under sprinkler irrigation system + WWFF + $FR_{100\%, 80\%, 60\% NPK}$ this may be due to increasing the yield of potato.

4. Conclusion

Recycling the drainage water of fish farming, rich with organic matter for agriculture use can improve soil quality and crops productivity and reduce the total costs of fertilizers by adding minimum doses from minerals fertilizers and using sprinkler irrigation system.

References

[1]   Elnwishy, N., Salh, M. and Zalat, S. (2006) Combating Desertification through Fish Farming. The Future of Drylands. *Proceedings of the International Scientific Conference on Desertification and Drylands Research*, Tunisia 19-21 June 2006, UNESCO, 855.

[2]   Ebong, V. and Ebong, M. (2006) Demand for Fertilizer Technology by Smallholder Crop Farmers for Sustainable Agricultural Development in Akwa, Ibom State, Nigeria. *International Journal of Agriculture and Biology*, **8**, 728-733.

[3]   Altaf, U., Bhattihaq, N., Murtaz, G. and Ali, M. (2000) Effect of pH and Organic Matter on Monovalent-Divalent Cation Exchange Equilibria in Medium Textured Soils. *International Journal of Agriculture and Biology*, **2**, 1-2.

[4]   Dave, D. (2003) Egypt and the Potato Tuber Moth. Michigan State University, 280.
      http://potatobg.css.msu.edu/commercial_releases.shtml

[5]   Bucks, D.A., Nakayama, F.S. and Gilbert, R.G. (1979) Trickle Irrigation Water Quality and Preventive Maintenance. *Agricultural Water Management*, **2**, 149-162. http://dx.doi.org/10.1016/0378-3774(79)90028-3

[6]   Tuzel, I.H. and Anac, S. (1991) Emitter Clogging and Preventative Maintenance in Drip Irrigation Systems. *Journal of Agriculture Faculty*, **28**, 239-254.

[7]   Pitts, D.J., Haman, D.Z. and Smajstrla, A.G. (1990) Causes and Prevention of Emitter Plugging in Microirrgation Systems. Bulletin 258, Florida Cooperative Extension Service, Institute of Food and Agricultural Science, University of Florida, Gainesville. http://edis.ifas.ufl.edu/AE032

[8]   Papadopoulos, I. (1993) Agricultural and Environmental Aspects of Fertigation-Chemigation in Protected Agriculture under Mediterranean and Arid Climates. *Proceedings of the Symposium on Environmentally Sound Water Management of Protected Agriculture under Mediterranean and Arid Climates*, Bari, 16-18 July 1993, 1-33.

[9]   Burt, C.M., Connor, K.O. and Ruehr, T. (1995) Fertigation Irrigation Training and Research Center. California Polytechnic State University, San Luis Obispo, 295.

[10] Burt, C.M. (1998) Fertigation Basics. I.T.R.C. California Polytechnic State University san luis obispo ca 93407 Pacific Northwest Vegetable Association Convention Pasco. Washington.

[11] APHA (1998) Standard Methods for the Examination of Water and Wastewater. 20th Edition, American Public Health Association.
      http://www.worldcat.org/title/standard-methods-for-the-examination-of-water-and-wastewater/oclc/807586782?referer=di&ht=edition

[12] Tsao, P.H. (1970) Selective Media for Isolation of Pathogenic Fungi. *Annual Review of Phytopathology*, **8**, 157-186.
      http://dx.doi.org/10.1146/annurev.py.08.090170.001105

[13] Atlas, R. (2005) Handbook Media for Environmental Microbiology. CRC Press, Taylor & Francis Group 6000 Broken Sound Parkway NW Boca Raton, 33487-2742.

[14] Munoz, F.E. and Silverman, M.P. (1979) Rapid, Single-Step Most-Probable-Number Method for Enumerating Fecal Coliforms in Effluents from Sewage Treatment Plants. *Applied and Environmental Microbiology*, **37**, 527-530.

[15] Kizilkaya, R. (2009) Nitrogen Fixation Capacity of *Azotobacter* spp. Strains Isolated from Soils in Different Ecosystems and Relationship between Them and the Microbiological Properties of Soils. *Journal of Environmental Biology*, **30**, 73-82.

[16] James, L.G. (1988) Principles of Farm Irrigation System Design. John Willey & Sons. Inc., Washington State University, 73, 152-153, 350-351.

[17] El-Berry, A.M., Bakeer, G.A. and Al-Weshali, A.M. (2003) The Effect of Water Quality and Aperture Size on Clogging of Emitters.
http://afeid.montpellier.cemagref.fr/old/Mpl2003/AtelierTechno/AtelierTechno/AtelierEdite/Edite-Res-Atelier-oral/El-BerryResu-N48.pdf

[18] Snedecor, G.W. and Cochran, W.G. (1982) Statistical Methods. 7th Edition, Iowa State University Press, Towa, 511.

[19] Feachem, R.G., Bradley, D.J., Garelick, H. and Mara, D.D. (1983) Sanitation and Disease: Health Aspects of Excreta and Wastewater Management. World Bank Studies in Water Supply and Sanitation 3, Wiley, Chichester.

List of Abbreviations

Abbreviations	Definitions
WWFF	Wastewater of fish farms
SIS	Sprinkler irrigation system
TIS	Trickle irrigation system
TIW	Traditional irrigation water
FR	Fertigation rate
IWUE$_{potato}$	Irrigation water use efficiency of potato

18

Influence of Soil Moisture and Air Temperature on the Stability of Cytoplasmic Male Sterility (CMS) in Maize (*Zea mays* L.)

Heidrun Bueckmann, Katja Thiele, Joachim Schiemann

Institute for Biosafety in Plant Biotechnology, Julius Kühn-Institut, Federal Research Centre for Cultivated Plants (JKI), Quedlinburg, Germany
Email: katja.thiele@jki.bund.de

Abstract

Cytoplasmic male sterility (CMS) is a maternally inherited trait that suppresses the production of viable pollen. CMS is a useful biological tool for confinement strategies to facilitate coexistence of genetically modified (GM) and non-GM crops in case where it is required. The trait is reversible and can be restored to fertility in the presence of nuclear restorer genes (*Rf* genes) and by environmental impacts. The aim of this study was to investigate the influence of the level of irrigation on the stability of CMS maize hybrids under defined greenhouse conditions. Additionally the combination of irrigation and air temperature was studied. Three CMS maize hybrids were grown with different levels of irrigation and in different temperature regimes. Tassel characteristics, pollen production and fertility were assessed. The CMS stability was high in hot air temperatures and decreased in lower temperatures. The level of irrigation had no major effect on the level of sterility. The extent of these phenomena was depending on the genotype of CMS maize and should be known before using CMS for coexistence purposes.

Keywords

Soil Moisture, Air Temperature, Biological Confinement, Cytoplasmic Male Sterility (CMS), Genetically Modified (GM) Maize (*Zea mays* L.)

1. Introduction

Cytoplasmic male sterile (CMS) maize is proven as a reliable biological confinement method to prevent cross-pollination of pollen into neighbouring fields [1]. This might be necessary if genetically modified (GM) maize is grown in arable areas with a high level of non-GM crops or even organic farming. Maize plants produce a huge amount of pollen [2] [3], but 95% of this pollen does not contact the stigmatic surface of the maize silk and is deposited into the environment [4]. Several physical containment methods for maize such as isolation distances and border rows are already applicable. In regions with a wider sowing season, e.g. Spain, different sowing dates and—therefore—different blooming periods can avoid cross-pollination [5] [6].

CMS is a maternally inherited trait that prevents the development of functional pollen [7] [8]. It results from a loss-of-function mutation in the mitochondrial genome [9] and causes a dysfunction of the respiratory metabolism and an abnormal production of male gametes [10] [11]. The plants develop no or nonvital pollen but the female fertility is not affected. Three main types of CMS are known for maize [12]: CMS-T or Texas-cytoplasm [13] [14], CMS-S or USDA-cytoplasm [15] and CMS-C or Charrua-cytoplasm [16]. Their differences are based on different restorer of fertility genes (*Rf*). Even though the determination of CMS occurs in extranuclear, these nuclear *Rf* genes can compensate the CMS effect of the cytoplasm [8]. As a result tassels can be partly restored or even become fertile with more or less vital pollen in the first-generation progeny (F1).

Irrespective of the internal interactions between mitochondrial and nuclear genes, fertility of many CMS plant species can also be restored by environmental impacts like heavy rain, extreme heat etc. [17]-[21]. As can be expected, our field studies showed that the climate had an impact on the performance of several CMS maize hybrid genotypes [1]. In a trial year with consistently lower temperatures than the longstanding mean and high precipitation rates, the growth of the maize stands was unequal and partly delayed. Later developed plants (delay of ca. 10 days) arose due to cold and wet weather conditions earlier in the season. These plants bloomed in round about 2°C lower temperatures and much more precipitation (+127 mm above the longstanding mean for August). They developed fertile tassels with a large amount of pollen while most of the earlier flowers of the same hybrid were sterile. These observations raised the question whether the air temperature and water availability might have an impact on the CMS trait stability. The influence of air temperature on the stability of the CMS trait was already demonstrated [22], but the underlying mechanisms are not completely understood.

The aim of this study was to investigate the influence of the level of irrigation on the CMS stability of maize hybrids under defined greenhouse conditions. Additionally the interaction of irrigation with two temperature regimes was studied. The results will provide more detailed information about the applicability of CMS maize hybrids under different environmental conditions as a tool for coexistence purposes.

2. Material and Methods

Two experiments (experiment 1 and 2) were carried out to test the influence of irrigation on the level of sterility. In a third experiment (experiment 3) the interaction between irrigation level and air temperature was tested. In all experiments differentiated irrigation was applied until the end of flowering.

Experiment 1 was sown in April 2013 and flowering ended in the third quarter of July. Experiment 2 started at the beginning of September 2013. The flowering period ended beginning of November. In both experiments the air temperature was not specified. The irrigation levels of experiment 1 and 2 were: "dry", "regular" and "wet". In experiment 3 only "dry" and "wet" were tested. To ensure a correct irrigation the water content of the soil was tested every second day with a soil (PCE-SMM1, PCE Instruments, Meschede, Germany). The pots were irrigated to free soil water contents of 12% - 18% in "dry" conditions (just before plant decrease), 20% - 25% in "regular" conditions (sufficient water for plant growth) and >50% in "wet" conditions (water saturated soil).

The experiment 3 was carried out to test possible interactions of irrigation levels and air temperatures. Two different air temperature regimes ("cold", "hot") were applied, each in one separate greenhouse chamber (**Table 1**). The regime "cold" started with defined temperatures of 21°C at night and 25°C at day time to provide positive initial conditions. The temperatures were reduced stepwise to 16°C at night and 21°C at daytime. In "hot" temperature conditions sowing was performed at 15°C at night and 25°C at daytime, and temperatures were increased stepwise to reach 22°C at night and 35°C at daytime after 35 days. To control the defined temperature regimes permanent measurements of air temperature and humidity in the green house chambers were carried out.

During daytime the plants were illuminated 16 hours by 5000 Lux. Ten plants per CMS hybrid and irrigation level were cultivated in 30 l plastic pots filled with a standard greenhouse soil based on crumble peat. The CMS maize hybrids tested in the greenhouse were previously used in field trials for testing their potential as a biological confinement tool [1]: Torres and Zidane (Kleinwanzlebener Saatzucht AG, Germany) both belong to the

Table 1. Temperature regimes of the greenhouse experiment 3.

Temperature regimes	Setting	Night/day
cold	sowing	21°C/25°C
	after 14 days	18°C/25°C
	after 21 days	16°C/21°C
hot	sowing	15°C/20°C
	after 14 days	15°C/25°C
	after 21 days	18°C/28°C
	after 35 days	22°C/35°C

CMS-S cytoplasm and DSP2 (Delley Seeds and Plants Company, DSP, Switzerland) belongs to the CMS-T cytoplasm.

In all experiments the tassels were assessed whether they stayed sterile, were partly restored or even fertile (**Figure 1**). The results of the flower assessment were weighted as follows: sterile tassel = 1; weak partly restored tassel = 2; partly restored tassel = 3; strong partly restored tassel = 4; complete restored tassel = 5. The pollen production was weighted as follows: no pollen = 1; very few pollen = 2; few pollen = 3; many pollen = 4. The values for tassel and pollen development were added and gave the value for the CMS stability ("loss of CMS").

Pollen vitality was tested in experiment 3. Therefore, all plants were self-pollinated by hand and developed kernels were counted. Statistics about flower assessment and kernel number per ear were done by 2-way-anova (Origin software, version 8.1, Heise, Germany).

3. Results

Experiment 1 and 2 were carried out in different seasons. Consequently, different air temperatures of 5°C on average were measured in the green house chamber from sowing until the end of flowering (not shown). The period for 14 days before and until the end of flowering is most important for the fertility of inflorescence and the maturation of pollen. In experiment 1 this period was reached in June and July. Accordingly, the air temperatures of 25°C on average reached up to 37°C maximum (**Figure 2**). The lowest temperature was 17°C. In experiment 2 the flowering period was in October and November. Average temperatures were 22°C with a peak value of 37°C and the lowest temperature of 17°C as well.

In both irrigation experiments (experiment 1 and 2) Torres was the CMS hybrid with the highest level of sterility (**Figure 3**) independent of the irrigation quantity. "Loss of CMS" values were between 3.6 and 4.6 (**Table 2**). Compared to Torres, Zidane restored more to fertility in both experiments and reached "loss of CMS" values of 6 to 9. An impact of the irrigation level could not be detected. DSP2 expressed different levels of "loss of CMS" in experiment 1 and 2. It stayed close to sterility—similar to Torres—in experiment 1 ("Loss of CMS" values 4 - 4.8) and developed fertility ("Loss of CMS" value 9) in experiment 2, independently from irrigation. Generally, on a level of $p = 0.05\%$ the CMS maize hybrids differed significantly from each other (**Table 3**). Experiment 1 showed significant differences between Zidane and Torres as well as DSP2. In experiment 2 all three CMS maize hybrids showed significant differences of the stability level at $p = 0.05\%$.

Experiment 3 was more complex depending on the defined temperature regimes (**Table 1**). The temperature measurements during the experiment (**Figure 4**) showed a minimal deviation of ± 1°C from the temperature defined. In the variant "hot" the average temperature was 27°C with a maximum of 37°C and a minimum of 17°C. The variant "cold" had an average temperature of 19°C with maximum and minimum temperatures of 37°C and 15°C, respectively.

The CMS maize hybrid DSP2 stayed completely sterile in hot temperature conditions (**Figure 5**—value 2). In a cold environment all plants were fertile (value 9). The results were independent from the irrigation level (**Figure 5, Table 4**). Torres stayed nearly sterile in hot temperatures (22°C/35°C—"Loss of CMS" value 2.4 and 2.5). Some plants developed a very low amount of anthers which produced nearly no pollen or even stayed sterile. In low temperatures a few plants restored to fertility and developed some anthers which shed a small amount

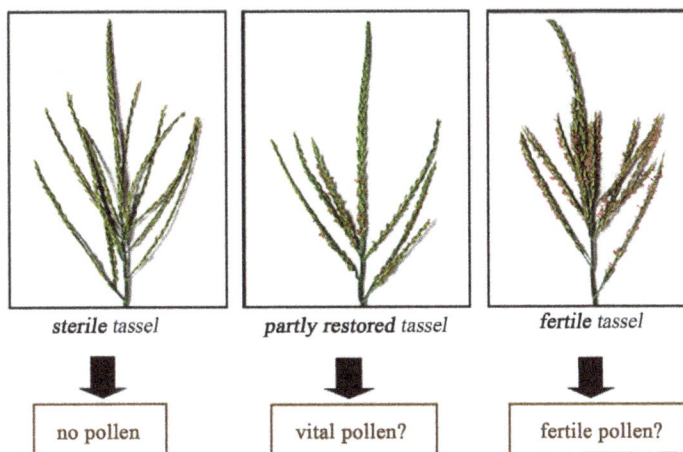

Figure 1. Tassel characteristics of CMS maize plants.

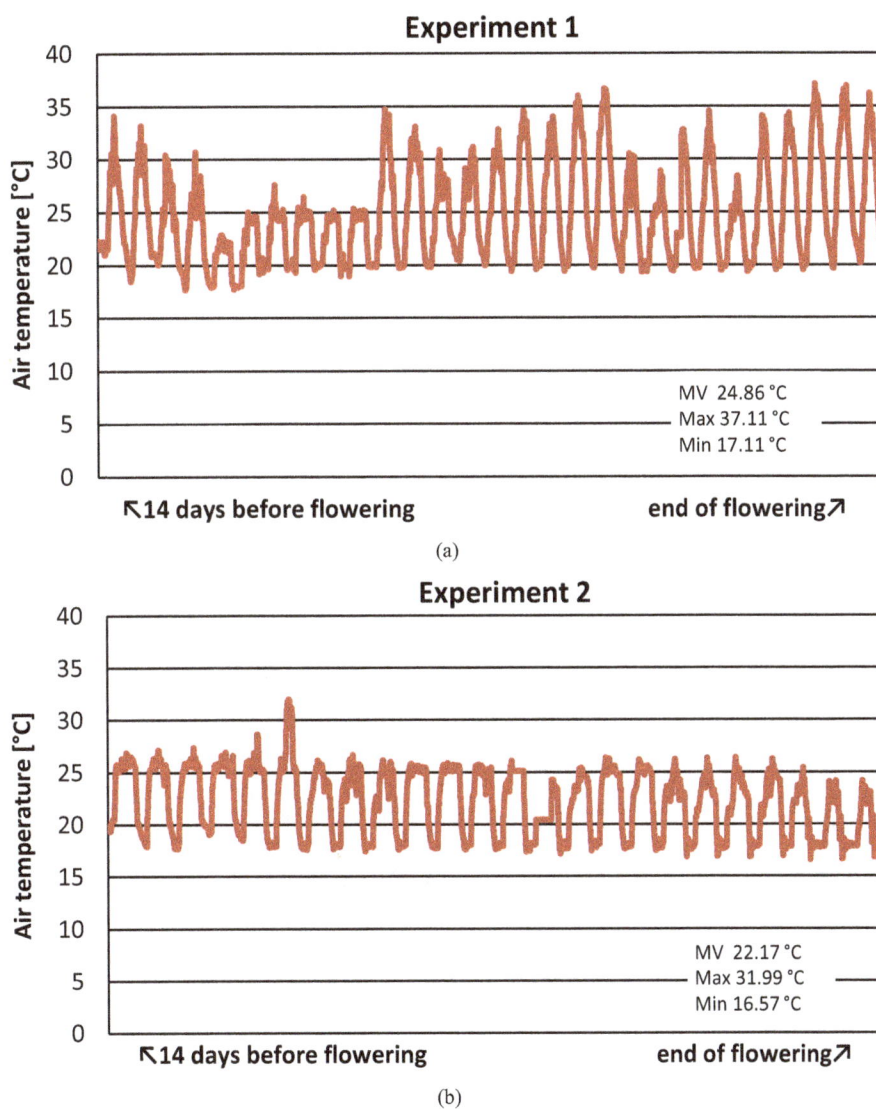

(a)

(b)

Figure 2. Air temperatures in Experiment 1 and 2 from 14 days before flowering until the end of flowering.

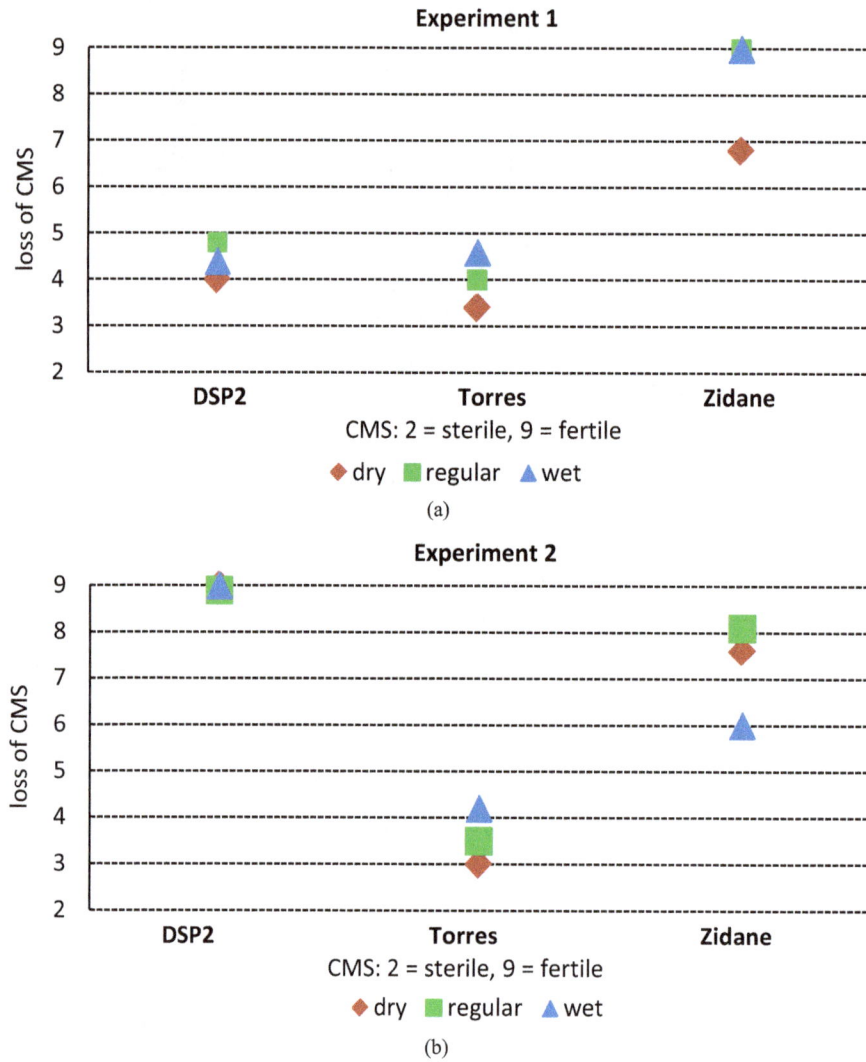

Figure 3. Stability of the sterility trait of three CMS maize hybrids in dependence of irrigation level in Experiment 1 and 2.

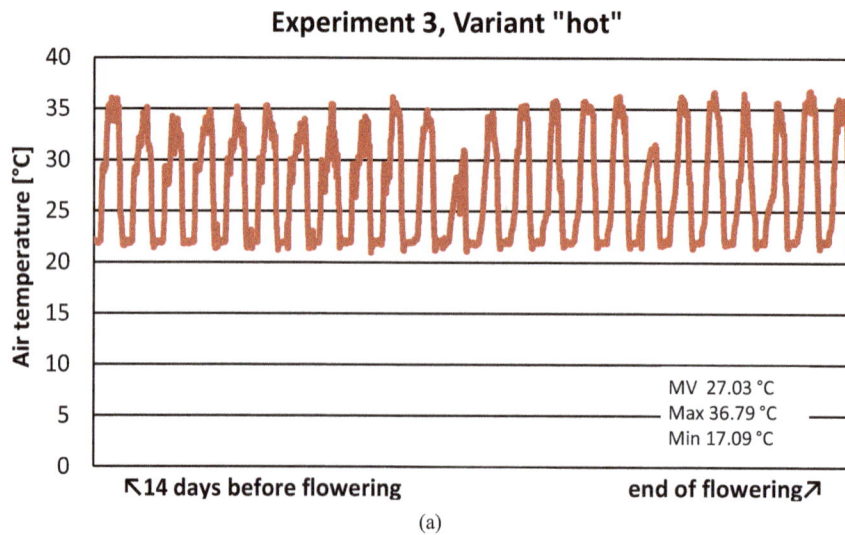

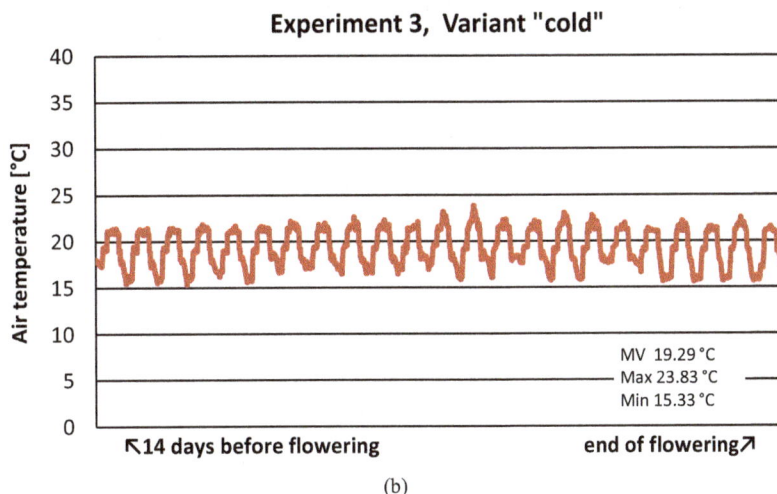

(b)

Figure 4. Measured air temperatures in variant "hot" and "cold" from 14 days before flowering until the end of flowering—Experiment 3.

Table 2. Loss of the stability of CMS and standard deviations of three CMS maize hybrids in different irrigation levels—Experiment 1 and 2.

Experiment 1 Irrigation level	DSP2	Torres	Zidane
dry	4.0	3.4	6.8
Standard deviation	1.76	0.97	1.23
regular	4.8	4.0	9.0
Standard deviation	1.93	0.00	0.00
wet	4.4	4.6	9.0
Standard deviation	2.07	0.97	0.00

Experiment 2 Irrigation level	DSP2	Torres	Zidane
dry	9.0	3.0	7.6
Standard deviation	0.00	0.00	0.84
regular	8.9	3.5	8.1
Standard deviation	0.32	0.71	0.32
wet	9.0	4.2	6.0
Standard deviation	0.00	0.42	0.94

2 = sterile, 9 = fertile.

Table 3. Significance of irrigation level and genotype influences on the level of stability of the CMS trait, 2-way anova, $p = 0.05$ (Experiment 1 and 2). Different letters mark significant differences.

Experiment 1	Significance level $p = 0.05$		
	DSP2	Torres	Zidane
Irrigation level	a	a	b
	dry	regular	wet
CMS maize hybrid	a	b	b

Experiment 2			
	DSP2	Torres	Zidane
Irrigation level	a	b	c
	dry	regular	wet
CMS maize hybrid	ab	a	b

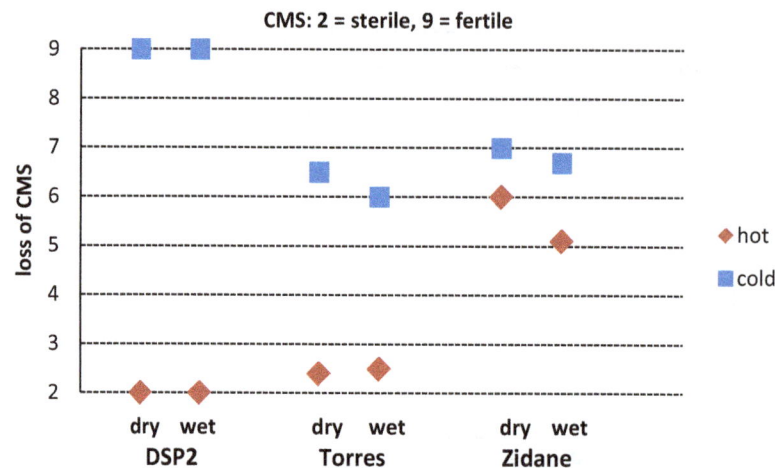

Figure 5. Stability of the sterility trait of three CMS maize hybrids in dependence of air temperature and irrigation level—Experiment 3.

Table 4. Loss of the stability of CMS and standard deviations of three CMS maize hybrids in different air temperature regimes and irrigation levels (dry, wet—Experiment 3).

Temperature regime	DSP2		Torres		Zidane	
	dry	wet	dry	wet	dry	wet
cold (21°C/16°C)	9	9	6.5	6	7	6.7
Standard deviation	0	0	0.53	0.53	0.53	0.32
hot (35°C/22°C)	2	2	2.4	2.5	6	5.1
Standard deviation	0	0	0.84	0.85	0	1.37

2 = sterile, 9 = fertile.

of pollen and reached "loss of CMS" values of 6 to 6.5 (**Figure 5**, **Table 4**). The level of irrigation did not affect these results. In contrast to Torres and DSP2, Zidane showed partly restored tassels and the production of some pollen (values of 5.1 to 6) in hot air temperatures (22°C/35°C). In colder temperatures the "loss of CMS" was higher (value 6.7 to 7) because more tassels of Zidane restored to fertility and produced more pollen compared to "hot" temperature conditions. The irrigation level had no major affect on the tassel and pollen production of Zidane. Experiment 3 showed significant differences of the level of sterility ($p = 0.05\%$) between the CMS maize hybrids depending on the temperature (**Table 5**) while the irrigation level resulted in a different level of sterility between Torres and Zidane. Generally, the temperature regime had a significant influence on the level of CMS sterility while the level of irrigation had not.

In experiment 3, the development of kernels after self pollination was measured. Generally, no CMS maize hybrid developed kernels in variant "hot" (**Figure 6**) but in variant "cold" substantial kernel development occurs. Corresponding to the "loss of CMS" in cold conditions DSP2 plants developed the largest number of kernels per ear (on average 335 in dry and 391 in wet soil) followed by Zidane (on average 226, dry, and 213, wet—**Figure 6**, **Table 6**) and Torres (on average 15, dry, and 58, wet). No significant differences of kernel development were measured between "wet" and "dry" but between "hot" and "cold" variants (**Table 7**). In summary, the level of sterility of the CMS maize hybrids tested was depending on the temperature regime and the extent of the loss of sterility was depending on the genotype.

4. Discussion

The objective of this study was to verify whether the irrigation level or soil water availability, the air temperature or the interaction of both can affect the stability of the CMS trait of maize hybrids. The results will help to understand the climatic impacts on the fertility restoration of CMS maize hybrids and to better define conditions

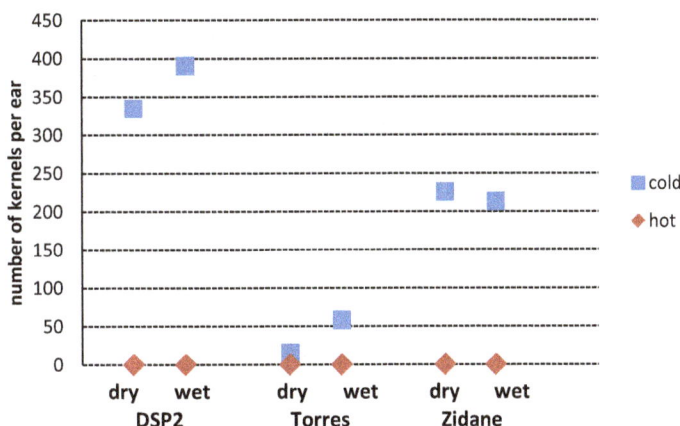

Figure 6. Kernels per ear after self-pollination of three CMS maize hybrids in different air temperature regimes and irrigation levels—Experiment 3.

Table 5. Significance of air temperature, irrigation level and genotype influences on the level of stability of CMS, 2-way anova, $p = 0.05$, Experiment 3. Different letters mark significant differences.

	Significance level $p = 0.05$		
	DSP2	Torres	Zidane
Temperature regime	a	b	c
	DSP2	Torres	Zidane
Irrigation level	a	ab	ac
	hot (35°C/22°C)	cold (21°C/16°C)	
CMS maize hybrid	a	b	
	dry	wet	
CMS maize hybrid	a	a	

Table 6. Kernels per ear after self-pollination and standard deviations of three CMS maize hybrids in different air temperature regimes and irrigation levels (dry, wet—Experiment 3).

Temperature regime	DSP2		Torres		Zidane	
	dry	wet	dry	wet	dry	wet
cold (21°C/16°C)	335	391	15	58	226	213
Standard deviation	65	107	17	74	126	141
hot (35°C/22°C)	0	0	0	0	0	0
Standard deviation	0	0	0	0	0	0

Table 7. Significance of air temperature, irrigation level and genotype influences on the amount of kernels produced, 2-way anova, $p = 0.05$, Experiment 3. Different letters mark significant differences.

	Significance level $p = 0.05$		
Temperature regime	DSP2	Torres	Zidane
	a	b	c
Irrigation level	DSP2	Torres	Zidane
	a	b	a
CMS maize hybrid	hot (35°C/22°C)	cold (21°C/16°C)	
	a	b	
CMS maize hybrid	dry	wet	
	a	a	

for the applicability of CMS maize hybrids for coexistence purposes. Therefore, three CMS maize hybrids were cultivated with different irrigation levels and different air temperature regimes in greenhouse conditions.

In all experiments, no significant difference between the levels of sterility developed in different irrigation regimes were observed with the exception of the effect of low irrigation on Zidane. An explanation for this effect might be that the amount of water could have been too low to ensure a sufficient plant development. The expectation that the amount of available soil water affected the stability of the CMS sterility trait following observations in field trials [1] could not be confirmed. In one of these trial years with high precipitation rates, the growth of the maize stands was unequal and partly delayed. Later developed plants (delay of app. 10 days) were detected which blossomed in round about 2°C lower temperatures and much more precipitation (+127 mm above the longstanding mean for August). The plants developed fertile tassels with a large amount of pollen while most of the earlier flowers of the same hybrid were sterile [1].

Zidane had always a high "loss of CMS" value while Torres had always low "loss of CMS" values which match the findings of Bueckmann et al. (2014) for the same varieties. The sterility level of DSP2 showed sensitivities to the air temperatures. With regard to the temperature curves in the green house cabins of experiment 1 and 2 the plants grew in a temperature environment of average 3°C difference. The cooler conditions caused higher "loss of CMS" values of DSP2 while Torres and Zidane were not strongly affected. The results concerning the impact of temperature on the stability of CMS were confirmed in experiment 3. High air temperatures forced the sterility of all three CMS maize hybrids and low temperatures forced the "loss of CMS". These results circumstantiate the findings of Bueckmann et al. (2014) about the influence of air temperature on the CMS trait which might be explained by the sensitivity of Rf gene activation or inactivation to environmental impacts just before and during anthesis [21]. The amount of kernels detected at each ear in experiment 3 matched the "loss of CMS" data: as higher the value of "loss of CMS" as more kernels were produced.

Especially for DSP2 the results of kernel production per cob underline the results found by Bueckmann et al. (2014) and Weider et al. (2009) for the same CMS maize hybrid. The latter authors tested DSP2 among other CMS maize hybrids of the three CMS cytoplasms. DSP2 belongs to the T-cytoplasm which is known as providing-with exceptions—the most reliable male sterility [21]. Restoration to fertility can only happen if two dominant Rf genes act together, $Rf1$ and $Rf2$. Solely in the presence of $Rf2$ the $Rf1$ allel affects the transcription of the responsible mitochondrial gene (T-urf13) and causes pollen fertility [23]. The influence of the $Rf1$ gene can be compensated by other dominant Rf genes like $Rf8$ [24]. These genes are rarely present in T-cytoplasm types. Their expression is environmentally sensitive and can result in a few fertile or partly restored plants in one population [21] [24]. Obviously, the slight dependency of DSP2 male sterility on weather conditions which is paralleled by pollen dispersal and kernel production is depending on the sporophytic T-cytoplasm as described by Dill et al. (1997). Concerning the kernel production Bueckmann et al. (2014) demonstrated that the kernel development of DSP2 increases with decreasing temperatures occurring shortly before the onset of anthesis (anther and pollen formation).

Torres and Zidane both belong to the CMS S type. This cytoplasm is the most unstable one [25] [26] and the restoration to fertility is more complex compared to the T-cytoplasm. It is probably depending on one dominant and not environment-sensitive restorer gene ($Rf3$) and a large number of spontaneously occurring Rf genes and more than 60 restoring allele mutations [25] [27] [28]). Gabay-Laughnan et al. (2009) [29] identified the $Rf9$ gene which is stronger expressed under moderate compared to higher temperatures. According to the greenhouse results, an increasing level of $Rf9$ gene expression with decreasing temperatures could be expected for Zidane, resulting in partly restored or fertile tassels and, therefore, a higher "loss of CMS" value. In the case of Torres, the partly restored tassels produced only a very small amount of pollen and reached a low value of "loss of CMS stability".

Apart from the type of the CMS cytoplasm the influence of the air temperature on the CMS trait was obvious and might be explained by the sensitivity of rf gene activation or inactivation just before anthesis. Weider et al. (2009) tested twenty two CMS maize hybrids in seventeen environments. They found an influence of climate on CMS restoration, but a clear correlation of climatic parameters and fertility restorations could not be found. This also matches the field trial results of Bueckmann et al. (2013) which were collected in three different German environments in 2009 and 2010. There, a higher level of sterility was measured in 2010, depending on weather conditions: In the first half of July 2010 the average daily air temperatures increased by 3°C to 3.5°C above the longstanding average temperature at all trial locations. In general, less pollen and, consequently, fewer kernels were produced by all CMS maize hybrids tested. This indicates an influence of air temperature just before anthesis on the restoration of fertility on both T-cytoplasm and S-cytoplasm. Duvick (1956) [30] and Tracy *et al.* (1991) [31] already assumed that cool and humid conditions cooperate in restoration of fertility, whereas hot and dry conditions maintain sterility. Hot and dry weather causes the development of fewer anthers and pollen can lose its viability or is non-competitive compared with pollen of fertile plants [32].

The influence of high temperatures on the level of sterility was described for other crops as well. Sarvella (1966) described in experiments with cotton that lower temperatures combined with higher humidity were associated with an increased restoration of fertility of CMS plants, but the temperature was the most important factor. Marschall *et al.* (1974) found that temperatures above 33°C were required for the consistent expression of the CMS trait in cotton. This might not only be linked to *Rf* expressions but might be caused by toxin production [33] which affected the CMS trait. Apart from the influence of air temperature on the stability of the CMS trait, other environmental impacts were discussed in the literature, e.g. water limiting conditions for sorghum [34].

CMS is a proven coexistence tool for maize. Cultivation of GM CMS maize hybrids in the frame of coexistence regimes requires an improved knowledge about the influence of the environment on the CMS trait and a sufficient pollination of the GM CMS plants by admixing a male-fertile and non-GM pollen donor [35] [36]. If the CMS maize hybrid and the pollinator plant provide a different genetic background, yield can be significantly increased [37]-[39]. The so-called Plus-Hybrid-Effect [35] [40] combines the potential benefits of CMS and a Xenia effect. CMS hybrids have a "female advantage" over their male-fertile counterparts, which may be caused by increased female fertility related to the reallocation of resources unused in male function or by greater seed vitality by avoiding self-pollination [11]. Hence, when growing GM CMS maize hybrids, pollen containment will be combined with yield increase.

Acknowledgements

The authors thank the European Commission for funding the PRICE project (Practical Implementation of Coexistence in Europe, http://price-coexistence.com/, FP7/2007-2013, grant agreement No. KBB-2011-5-289157).

References

[1] Bueckmann, H., Huesken, A. and Schiemann, J. (2013) Applicability of Cytoplasmic Male Sterility (CMS) as a Reliable Biological Confinement Method for the Cultivation of Genetically Modified Maize in Germany. *Journal of Agricultural Science and Technology A*, **3**, 385-403.

[2] Westgate, M., Lizaso, J. and Batchelor, W. (2003) Quantitative Relationship between Pollen-Shed Density and Grain Yield in Maize. *Crop Science*, **43**, 934-942. http://dx.doi.org/10.2135/cropsci2003.9340

[3] Uribelarrea, M., Cárcova, J., Otegui, M. and Westgate, M. (2002) Pollen Production, Pollination Dynamics, and Kernel Set in Maize. *Crop Science*, **42**, 1910-1918. http://dx.doi.org/10.2135/cropsci2002.1910

[4] Poehlmann, J.M. and Sleper, D.A. (1995) Breeding Field Crops. Iowa State University Press, Ames, XV, 494 p.

[5] Brookes, G., Barfoot, P., Melé, E., Messeguer, J., Bénétrix, F., Bloc, D., Foueillassar, X., Fabié, A. and Poeydomenge, C. (2004) Genetically Modified Maize: Pollen Movement and Crop Coexistence. PG Economics Ltd., Dorchester, UK. http://www.pgeconomics.co.uk

[6] Messeguer, J., Peñas, G., Ballester, J., Bas, M., Serra, J. and Salvia, J. (2006) Pollen-Mediated Gene Flow in Maize in Real Situations of Coexistence. *Plant Biotechology Journal*, **4**, 633-645. http://dx.doi.org/10.1111/j.1467-7652.2006.00207.x

[7] Laser, K.D. and Lersten, N.R. (1972) Anatomy and Cytology of Microsporogeneses in Cytoplasmic Male Sterile Angiosperms. *The Botanical Review*, **38**, 425-454. http://dx.doi.org/10.1007/BF02860010

[8] Schnable, P.S. and Wise, R.P. (1998) The Molecular Basis of Cytoplasmic Male Sterility and Fertility Restoration. *Trends in Plant Science*, **3/5**, 175-180. http://dx.doi.org/10.1016/S1360-1385(98)01235-7

[9] Chase, C.D. and Gabay-Laughnan, S. (2004) Cytoplasmic Male Sterility and Fertility Restoration by Nuclear Genes. In: Daniell, H. and Chase, C.D., Eds., *Molecular Biology and Biotechnology of Plant Organelles*, Springer, New York, 593-622. http://dx.doi.org/10.1007/978-1-4020-3166-3_22

[10] Chase, C.D. (2006) Genetically Engineered Cytoplasmic Male Sterility. *Trends in Plant Science*, **11**, 7-9. http://dx.doi.org/10.1016/j.tplants.2005.11.003

[11] Budar, F., Touzet, P. and De Paepe, R. (2003) The Nucleo-Mitochondrial Conflict in Cytoplasmic Male Sterilities Revisited. *Genetica*, **117**, 3-16. http://dx.doi.org/10.1023/A:1022381016145

[12] Sofi, P.A., Rather, A.G. and Wani, S.A. (2007) Genetic and Molecular Basis of Cytoplasmic Male Sterility in Maize.

Communications in Biometry Crop Science, **2**, 49-60.

[13] Rogers, J.S. and Edwardson, J.R. (1952) The Utilization of Cytoplasmic Male-Sterile Inbreds in the Production of Corn Hybrids. *Agronomy Journal*, **44**, 8-13. http://dx.doi.org/10.2134/agronj1952.00021962004400010004x

[14] Levings, C.S. (1993) Thoughts on Cytoplasmic Male Sterility in *cms*-T Maize. *Plant Cell*, **5**, 1285-1290. http://dx.doi.org/10.2307/3869781

[15] Jones, D.F., Stinson, H.T.J. and Khoo, U. (1954) Pollen Restoring Genes. Connecticut Agricultural Experiment Station Bulletin, 610, New Haven.

[16] Beckett, J.B. (1971) Classification of Male-Sterile Cytoplasms in Maize. *Crop Science*, **11**, 724-727.

[17] Sarvella, P. (1966) Environmental Influences on Sterility in Cytoplasmic Male-Sterile Cottons. *Crop Science*, **6**, 361-364. http://dx.doi.org/10.2135/cropsci1966.0011183X000600040020x

[18] Marshall, D.R., Thomson, N.J., Nicholls, G.H. and Patrick, C.M. (1974) Effects of Temperature and Daylength on Cytoplasmic Male Sterility in Cotton (*Gossypium*). *Australian Journal of Agricultural Research*, **25**, 443-447. http://dx.doi.org/10.1071/AR9740443

[19] Fan, Z.G. and Stefansson, B.R. (1986) Influence of Temperature on Sterility of 2 Cytoplasmic Male-Sterility Systems in Rape (*Brassica napus* L.). *Canadian Journal of Plant Science*, **66**, 221-227. http://dx.doi.org/10.4141/cjps86-035

[20] Peterson, C.E. and Foskett, R.L. (1953) Occurrence of Pollen Sterility in Seed Fields of Scott County Globe Onions. *Proceedings of the American Society of Horticultural Science*, **62**, 443-448.

[21] Weider, C., Stamp, P., Christov, N., Hüsken, A., Foueillassar, X., Camp, K.-H. and Munsch, M. (2009) Stability of Cytoplasmic Male Sterility in Maize under Different Environmental Conditions. *Crop Science*, **49**, 77-84. http://dx.doi.org/10.2135/cropsci2007.12.0694

[22] Bueckmann, H., Thiele, K., Huesken, A. and Schiemann, J. (2014) Influence of Air Temperature on the Stability of Cytoplasmic Male Sterility (CMS) in Maize (*Zea mays* L.). *AgBioForum*, **17**, 205-212

[23] Dewey, R.E, Timothy, D.H. and Levings, C.S. (1987) A Mitochondrial Protein Associated with Cytoplasmic Male Sterility in the T-Cytoplasm of Maize. *Proceedings of the National Academy of Science of the United States of America*, **84**, 5374-5378. http://dx.doi.org/10.1073/pnas.84.15.5374

[24] Dill, C.L., Wise, R. and Schnable, P.S. (1997) Rf8 and Rf* Mediate Unique T-urf13-transcript Accumulation, Revealing a Conserved Motif Associated with RNA Processing and Restoration of Pollen Fertility in T-Cytoplasm Maize. *Genetics*, **147**, 1367-1379.

[25] Gabay-Laughnan, S., Zabala, G. and Laughnan, J.R. (1995) S-Type Cytoplasmic Male Sterility in Maize. In: Levings, C.S. and Vasil, I.K., Eds., *The Molecular Biology of Plant Mitochondria*, Kluwer Academic Publishers, Dordrecht, 395-432. http://dx.doi.org/10.1007/978-94-011-0163-9_12

[26] Gabay-Laughnan, S., Chase, C.D., Ortega, V.M. and Zhao, L. (2004) Molecular-Genetic Characterization of CMS-S Restorer-of-Fertility Alleles Identified in Mexican Maize and Theosinte. *Genetics*, **166**, 959-970. http://dx.doi.org/10.1534/genetics.166.2.959

[27] Laghnan, J.R. and Gabay, S.J. (1978) Nuclear and Cytoplasmic Mutations to Fertility in S Male-Sterile Maize. In: Walden, D.B., Ed., *Maize Breeding and Genetics*, John Wiley & Sons, New York, 427-446.

[28] Gabay-Laughnan, S. (1997) Late Reversion Events Can Mimic Imprinting of Restorer-of-Fertility Genes in CMS-S Maize. *Maydica*, **42**, 163-172.

[29] Gabay-Laughnan, S., Kuzmin, E.V., Monroe, J., Roark, L. and Newton, K.J. (2009) Characterization of a Novel Thermosensitive Restorer of Fertility for Cytoplasmic Male Sterility in Maize. *Genetics*, **182**, 91-103. http://dx.doi.org/10.1534/genetics.108.099895

[30] Duvick, D.N. (1965) Cytoplasmic Pollen Sterility in Corn. *Advances in Genetics*, **13**, 1-56. http://dx.doi.org/10.1016/S0065-2660(08)60046-2

[31] Tracy, W.F., Everett, H.L. and Gracen, V.E. (1991) Inheritance, Environmental Effects and Partial Male Fertility in C-Type CMS in a Maize Inbred. *Journal of Heredity*, **82**, 343-346.

[32] Sotchenko, V., Gorbacheva, A. and Kosogorova, N. (2007) C-Type Cytoplasmic Male Sterility in Corn. *Russian Agricultural Sciences*, **33**, 83-86. http://dx.doi.org/10.3103/S1068367407020048

[33] Sakata, T., Oshino, T., Miura, S., Tomabechi, M., Tsunaga, Y., Higashitani, N., Miyazawa, Y., Takahashi, H., Watanabe, M. and Higashitani, A. (2010) Auxins Reverse Plant Male Sterility Caused by High Temperatures. *Proceedings of the National Academy of the United States of America*, **107**, 8569-8574. http://dx.doi.org/10.1073/pnas.1000869107

[34] Elkonin, L.A., Kozhemyakin, V.V. and Tsvetova, M.I. (2009) Epigenetic Control of the Expression of Fertility-Restoring Genes for the "9E" CMS-Inducing Cytoplasm of Sorghum. *Maydica*, **54**, 243-251.

[35] Feil, B., Weingartner, U. and Stamp, P. (2003) Controlling the Release of Pollen from Genetically Modified Maize and

Increasing Its Grain Yield by Growing Mixtures of Male-Sterile and Male-Fertile Plants. *Euphytica*, **130**, 163-165. http://dx.doi.org/10.1023/A:1022843504598

[36] Paterniani, E. and Stort, A.C. (1974) Effective Maize Pollen Dispersal in the Field. *Euphytica*, **23**, 129-134. http://dx.doi.org/10.1007/BF00032751

[37] Weingartner, U., Kaeser, O., Long, M. and Stamp, P. (2002) Combining Cytoplasmic Male Sterility and Xenia Increases Grain Yield of Maize Hybrids. *Crop Science*, **42**, 1848-1856. http://dx.doi.org/10.2135/cropsci2002.1848

[38] Munsch, M., Stamp, P., Christoph, N.K., Foueillassar, X., Hüsken, A., Camp, K.-H. and Weider, C. (2010) Grain Yield Increase and Pollen Containment by Plus-Hybrids Could Improve Acceptance of Transgenic Maize. *Crop Science*, **50**, 909-919. http://dx.doi.org/10.2135/cropsci2009.03.0117

[39] Stamp, P., Chowchong, S., Menzi, M., Weingartner, U. and Kaeser, O. (2000) Increase in the Yield of Cytoplasmic Male Sterile Maize Revisited. *Crop Science*, **40**, 1586-1587. http://dx.doi.org/10.2135/cropsci2000.4061586x

[40] Feil, B. and Stamp, P. (2002) The Pollen-Mediated Flow of Transgenes in Maize Can Already Be Controlled by Cytoplasmic Male Sterility. AgBiotechNet, 4. http://www.agbiotechnet.com/Reveiws

19

Determination of the Effect of the System of Rice Intensification (SRI) on Rice Yields and Water Saving in Mwea Irrigation Scheme, Kenya

Kepha G. Omwenga, Bancy M. Mati, Patrick G. Home

Biomechanical and Environmental Engineering Department, Jomo Kenyatta University of Agriculture and Technology, Nairobi, Kenya
Email: beed@jkuat.ac.ke

Abstract

Irrigated rice cultivation has long been associated with large amounts of water. Currently convectional rice production is faced with major challenges of water shortage as a result of increasing population sharing the same water resources, as well as global environmental changes. The System of Rice Intensification (SRI), as opposed to conventional rice production, involves alternate wetting and drying (AWD) of rice fields. The objective of this study was to determine the optimum drying days period of paddy fields that has a positive effect on rice yields and the corresponding water saving. The experimental design used was randomized complete block design (RCBD). Four treatments and the conventional rice irrigation method were used. The treatments were the dry days allowed after draining the paddy under SRI before flooding again. These were set as 0, 4, 8, 12 and 16 day-intervals. Yield parameters were monitored during the growth period of the crop where a number of tillers, panicles, panicle length and panicle filling were monitored. Amount of water utilized for crop growth for each treatment was measured. Average yield and corresponding water saving were determined for each treatment. The results obtained show that the 8 days drying period gave the highest yield of 7.13 tons/ha compared with the conventional method of growing rice which gave a yield of 4.87 tons/ha. This was an increase of 46.4% above the conventional method of growing rice. Water saving associated with this drying regime was 32.4%. This was taken as evidence that SRI improved yields with reduction in water use.

Keywords

Alternate Wetting and Drying, System of Rice Intensification, Water Saving, Rice Yield

1. Introduction

Water is the most important component of sustainable rice production in traditional rice growing areas of the world. The traditional method of growing rice has been continuously flooded conditions during vegetative growth of the crop with draining of the water during grain ripening stage. This method of continuous flooding is practiced in all the rice growing schemes in Kenya. This is because it is believed rice is an aquatic plant or at least a hydrophilic one [1]. This method is associated with high water use because of the continuous flooding of rice paddies and it is occasioned by losses through seepage, percolation and evaporation. The quantity of irrigation water used for example in Mwea Irrigation Scheme (MIS) has positive but insignificant effect on rice output, probably implying overuse of water [2].

With the decreasing availability of water for use in agriculture due to climate change and increased competition from other players like the industrial and urban sectors, water use in rice production systems has to be reduced in order to maintain production. At the same time, output per unit volume of irrigation water has to be increased so as to meet the increasing demand for rice in the world. There are various methods which have been used in reducing water use in rice production. One of the most tried methods is alternate wetting and drying (AWD) of rice paddy fields.

In a research done in lowland irrigated lands in Asia [3], alternate submergence-non submerged systems reduced water use up to 15% without affecting yields when the shallow ground water stays within about 0 - 30 cm. Another method of saving water use in rice production is saturated soil culture [4]. The soil is kept at around saturation part of or the entire growing season. But this resulted in lower yields than the traditional method of continuous flooding although the differences in yields were not statistically significant. It reduced the water input by 31% - 58%.

But these conclusions cannot be generalized as the AWD irrigation methods adopted in given areas may not transfer to other areas because of difference in climatic, soil and topographic conditions. Therefore, there is a need for area specific research to come up with recommendations for the best regime for AWD suited for such areas.

The system of rice intensification (SRI) is an AWD method that has been tried in a number of countries in the world. It is a methodology for increasing the productivity of irrigated rice by changing the management of plants, soil, water and nutrients. It involves transplanting young seedlings (8 days old) and planting singly at a wider spacing (20 cm × 20 cm - 40 cm × 40 cm). Compost is preferred for nutrients supply and the paddy is dried and wetted alternately. This method has been associated with water saving, because of alternate wetting and drying of rice paddies during the vegetative growth of the crop as opposed to the conventional continuous flooding. There are reported cases of increased yields in a number of countries where it has been tried [5].

2. Methodology

This study was carried out at the Mwea Irrigation Agricultural Development (MIAD) research centre at Mwea Irrigation Scheme (MIS) in Wamumu sub-location, Kirinyaga County of Kenya. The experimental design used was randomized complete block design (RCBD). Four treatments under SRI and conventional rice irrigation methods were used in three blocks each with five plots. Plots used were of size 3 m × 3 m. The variety of rice planted was Basmati 370.

For SRI plots, seedlings raised in a nursery were transplanted singly after 8 days when 2 leaves had developed. Spacing used for the seedlings was 20 cm × 20 cm. For conventional rice irrigation, 21 days old seedlings were randomly transplanted with 3 - 5 seedlings per hill at approximately 10 cm × 10 cm. After a week, when the SRI plots had dried, they were flooded to a depth of 3 cm and drained after two days. Flooding and draining of the plots was repeated for the four different treatments. The different treatments used were: flooding 4 days after draining (T1), flooding 8 days after draining (T2), flooding 12 days after draining (T3) and flooding 16 days after draining (T4). T5 was the control which comprised the conventional practice of growing rice in MIS which involves continuous flooding.

The parameters measured for each drying regime were the number of tillers, number of productive tillers (with panicles), length of panicles and percentage of grain filling. Yield for each drying regime was then determined. The yields were subjected to analysis of variance to determine their statistical significance.

In determining the water used for each treatment, the amount of water added and drained for each drying regime was measured using a 1 inch Parshall flume. This was installed at each plot inlet when adding water into the plot and at each plot outlet when draining the plot. The initial height of water at the inlet of the flume was recorded. Water was then allowed into the plot through the flume throat. Time taken to flood the plot to the 3 cm depth was recorded. Final height of water on the converging side was recorded at the end of the flooding process. This was repeated during draining where the initial height of water on the converging side of the Parshall flume and the final height after draining were recorded

The amount of water added into or drained out of each plot was calculated using the formula (Equation (1)):

$$Q = CH^{1.55} \tag{1}$$

where

Q = Flow rate of water into or out of the plot (cubic feet/sec);

C = Constant depending on the throat of the Parshall flume (0.338 for 1 inch throat used in the study);

H = Average of initial and final height of water in the converging zone of flume (feet).

The difference between the flooded and drained water was the amount utilized in the plot for crop growth. The water utilized for each drying regime was then compared to that used in the conventional method of growing rice to determine the water saving.

3. Results

SRI on average produced more tillers and panicles (55 and 51 respectively) compared to the conventional method of growing rice (25 and 21) as shown in **Figure 1**.

These were found to be statistically significant at 95% significance level. The calculated significance value (<0.001) for both cases is less than 0.05. This is shown in **Table 1** and **Table 2**.

The average panicle length was longer in SRI (25.4 cm) compared to the conventional method (22.6 cm) **Figure 2** shows the variation of panicle length for conventional irrigation and SRI.

The differences in panicle length were found to be statistically significant at 95% confidence level. The calculated significance value (<0.001) is less than 0.05 as shown in **Table 3**.

SRI recorded a higher average percentage of grain filling (81.5%) than conventional method (75.1%) as illustrated in **Table 4**.

But the percentage grain filling is not statistically significant at 95% significance level. The calculated significance value (0.663) is more than 0.05. This is shown in **Table 5**.

On average yields, the 8 days drying regime gave the highest yield of 7.13 tons/ha while conventional method yielded 4.87 tons/ha. This is an increase of 46.4% above conventional method of growing rice. This is shown in **Table 6**.

The average yields are statistically significant at 95% significance level. The calculated significant value (0.001) is less than 0.05 (**Table 7**).

The 12 days drying regime yielded slightly less than the 8 days' drying regime although the difference was not statistically significant. A comparison between means of yields conducted showed the significance value for the test of the contrast (0.053) to be larger than 0.05 (**Table 8**).

The water saving realized for the 8 and the 12 days' drying regimes was 32% and 42% respectively as shown in **Table 9**.

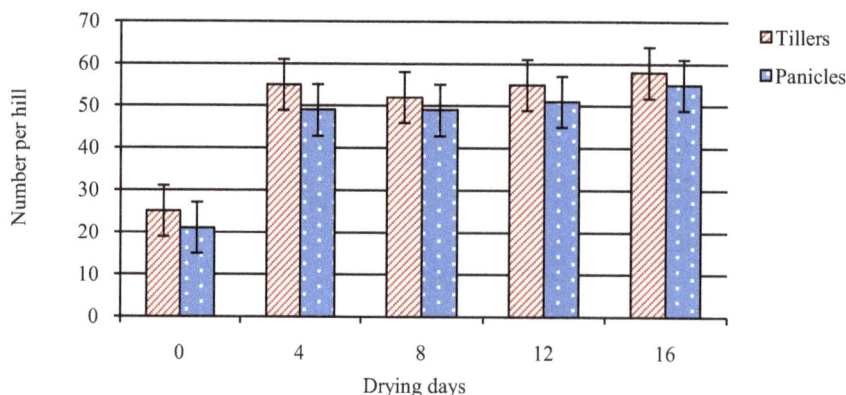

Figure 1. Average tillers and panicles.

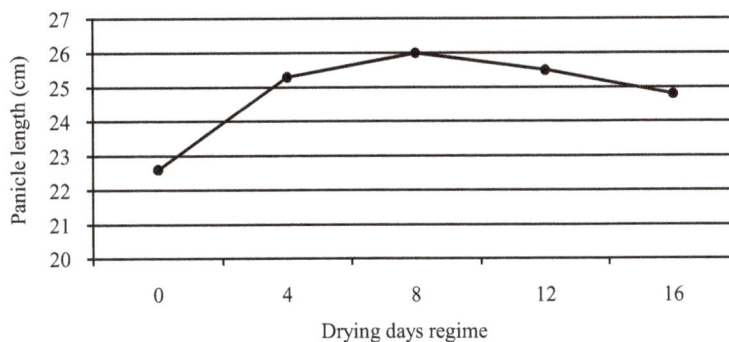

Figure 2. Average panicle lengths.

Table 1. ANOVA for tillers.

	Sum of Squares	df	Mean Square	F	Sig.
Between Groups	2209.733	4	552.433	19.316	0.000
Within Groups	286.000	10	28.600		
Total	2495.733	14			

Table 2. ANOVA for panicles.

	Sum of Squares	df	Mean Square	F	Sig.
Between Groups	2149.733	4	537.433	17.875	0.000
Within Groups	300.667	10	30.067		
Total	2450.400	14			

Table 3. ANOVA for panicle length.

	Sum of Squares	df	Mean Square	F	Sig.
Between Groups	20.329	4	5.082	27.823	0.000
Within Groups	1.827	10	0.183		
Total	22.156	14			

Table 4. Percent grain filling.

Drying Days	Percent Grain Filling	% Deviation from Conventional Irrigation
0	75.1	-
4	82.6	7.5
8	80.9	5.8
12	80.3	5.2
16	82.2	7.1

Table 5. ANOVA for grain filling in main crop.

	Sum of Squares	df	Mean Square	F	Sig.
Between Groups	110.837	4	27.709	0.612	0.663
Within Groups	452.560	10	45.256		
Total	563.397	14			

Table 6. Average yields.

Drying Days	Yield (t/ha)	Increase above Conventional Irrigation(t/ha)	Percent Increase
0	4.87	-	-
4	5.52	0.65	13.4
8	7.13	2.26	46.4
12	6.17	1.3	26.6
16	4.59	−0.28	−5.85

Table 7. ANOVA for yield in main crop.

	Sum of Squares	df	Mean Squares	F	Sig.
Between Groups	12.624	4	3.156	10.874	0.001
Within Groups	2.902	10	0.290		
Total	15.526	14			

Table 8. Contrast test for 8 and 12 days drying regimes.

	Contrast		Value of Contrast	Std. Error	t	df	Sig. (2-tailed)
Yield t/ha	Assume Equal Variances	1	0.9633	0.43986	2.190	10	0.053
	Does Not Assume Equal Variances	1	0.9633	0.52215	1.845	3.850	0.142

Table 9. Irrigation water use and saving.

Drying Days	Irrigation Water Use (×10^3 m^3/ha)	Percent Irrigation Water Saving
0	12.0	-
4	8.71	27
8	8.13	32
12	6.93	42
16	4.64	58

Average water productivity per unit volume of water for rice yield under conventional rice production and SRI is 0.4 kg/m^3 and 0.84 kg/m^3 respectively.

4. Discussion

Productivity of rice is a function of many parameters. Among the parameters which influence the yields are the number of tillers produced, the number of panicles formed, panicle length and percentage grain filling. There is a significant effect on these parameters brought about by SRI. The difference between SRI and conventional method of growing rice is evident in the number of tillers per hill and the number of panicles formed. But the different drying regimes had different effects on the other yield parameters. This in turn affected the individual yields for the drying regimes.

The better performance of the four, eight and twelve days drying regimes can be attributed to higher average number of tillers and panicles formed and longer average panicle length compared to the conventional irrigation method.

The number of tillers produced under SRI was more than those produced under conventional practice of growing rice. This is because the seedlings used in SRI were planted at an early age and tillers start developing as early as 14 days. The synergetic effect of young seedlings is seen in tiller formation, panicle length and percentage grain filling. These lead to higher yields [6]. Transplanting of young seedlings induces higher tiller production at an early stage. With increasing transplanting age of rice seedlings, emergence of tillers is delayed [7].

Seedlings under SRI are transplanted singly and spaced widely as opposed to conventional method of growing rice where they were transplanted 3 - 5 per hill at close spacing. This implied that there was no competition for nutrients between the seedlings under SRI. Wider spacing in SRI means that each plant had a wider area where it would draw it's nutrients from leading to more tillers, panicle length and number of filled grains [6]. This has a direct implication on the increased rice yield for SRI [8].

Wetting and drying of rice paddies has the beneficial effect of enhancing root growth. The rewetting facilitates nitrogen mineralization and this is made available to the plant for growth [8].

The main attracting attribute of SRI to farmers is what they will expect, in terms of yields, from adopting it. Generally SRI posted higher yields than conventional method of growing rice except for the sixteen days drying regime for the main crop. If water deficit is properly controlled especially during grain filling stage of rice, it enhances whole-plant senescence. This facilitates remobilization of carbon reserves, accelerate grain filling and increase grain yield [9].

From the comparison between means of yields conducted, since there are no statistically significant difference in yield between the 8 and 12 days drying regime, the two drying regimes can be used for the study area. This helps to support the notion that rice is not an aquatic plant but can be considered hydrophilic [1].

The lower yields in the 16 days drying regime for the main crop can be attributed to water stress suffered by the crop because of the prolonged drying period. Water stress imposed during the reproductive period can lead to reduced grain yield [10]. These moisture sensitive periods are flowering and head development stages [11]. Decrease in yield can be attributed to decrease in number of filled grains per panicle. This is manifested in the good average panicle length and low percentage grain filling. Conventional method performed marginally better than the 16 days drying regime because of presence of water during grain formation and filling.

Challenges facing rice production are to save on water use, increase water productivity and produce more rice with less water. The water saving associated with highest yields is 32.3%. This is for the eight days drying regime. Since the difference in yields between the eight and twelve days drying regimes are not statistically significant, the water saving for twelve drying days of 42.2% is also applicable.

The saving in irrigation water use under SRI is because of the remarked reduction in percolation, seepage and evaporation because of draining standing water from the plots. This saving should be sufficient reason for governments, international agencies and environmental organizations to promote the adoption of SRI [12] as water for agriculture use is becoming more and more limited.

5. Conclusion

The system of rice intensification is a rice production methodology that can be used by farmers to increase the water productivity in rice growing schemes. Drying of rice paddies for between 4 and 12 days under SRI has positive impacts on rice yields. This results in water saving of between 27% and 42%. This saving has an implication on increasing area under rice irrigation.

6. Recommendations

The system of rice intensification should be adopted in MIS as a means of increasing water productivity to address the diminishing water resources. A drying regime of eight days is ideal for MIS to help increase farmers' yield. But drying paddies for between 4 - 12 days can improve rice yields. Further research should be done to determine the effect of spacing on rice production for the 8 and 12 days drying regimes. Also studies on direct seeding should be done to determine the effect on rice yield.

References

[1]   Satyanarayana, A., Thiyagarajan, T.M. and Uphoff, N. (2007) Opportunities for Water Saving with Higher Yield from the System of Rice Intensification. *Irrigation Science*, **25**, 99-115. http://dx.doi.org/10.1007/s00271-006-0038-8

[2]   Obiero, O.B.P. (2011) Analysis of Economic Efficiency of Irrigation Water-Use in Mwea Irrigation Scheme, Kirinyaga District, Kenya. Kenyatta University, Nairobi Kenya.

[3]   Belder, P., Bouman, B.A.M., Cabangon, R., Lu, G., Qualing, E.J.P., Li, Y.H., Spiertz, J.H.J. and Tuong, T.P. (2004) Effect of Water-Saving Irrigation on Rice Yield and Water Use in Typical Lowland Conditions in Asia. *Agricultural Water Management*, **65**, 193-210. http://dx.doi.org/10.1016/j.agwat.2003.09.002

[4]   Tabbal, D.F, Bouman, B.A.M., Bhuiyan, S.I., Sibayan, E.B. and Sattar, M.A. (2002) On-Farm Strategies for Reducing Water Input in Irrigated Rice; Case Studies in the Philippines. *Agricultural Water Management*, **56**, 93-112.

http://dx.doi.org/10.1016/S0378-3774(02)00007-0

[5]   Uphoff, N. (2005) Features of SRI Apart from Increase in Yields. CIIFAD, New York.

[6]   Chapagain, T. and Yamaji, E. (2010) The Effect of Irrigation Method, Age of Seedling and Spacing on Crop Perfor-
      mance, Productivity and Water-Wise Rice Production in Japan. *Paddy and Water Environment Journal*, **8**, 81-90.
      http://dx.doi.org/10.1007/s10333-009-0187-5

[7]   Estela, P., Tubana, B., Bertheloot, J. and Lafarge, T. (2004) Impact of Early Transplanting on Tillering and Grain Yield
      in Irrigated Rice. 4*th International Crop Science Congress*, Brisbane, 26 September 2004.

[8]   Ceasey, M., Reid, W.S., Fernandes, E.C.M. and Uphoff, N.T. (2006). The Effects of Repeated Soil Wetting and Drying
      on Lowland Rice Yield with System of Rice Intensification (SRI) Methods. *International Journal of Agricultural Sus-
      tainability*, **4**, 5-14.

[9]   Yang, J.C., Zhang, J.H., Wang, Z.G., Liu, L.J. and Zhu, Q.S. (2001) Hormonal Changes in the Grains of Rice Sub-
      jected to Water Stress during Grain Filling. *Plant Physiology*, **127**, 315-323. http://dx.doi.org/10.1104/pp.127.1.315

[10] Lilley, J.M. and Fukai, S. (1994) Effect of Timing and Severity of Water Deficit on Four Diverse Rice Cultivars III.
      Phenological Development, Crop Growth and Grain Yield. *Field Crops Research*, **37**, 225-234.
      http://dx.doi.org/10.1016/0378-4290(94)90101-5

[11] Halil, S. and Neemi, B. (1999) The Effect of Water Stress on Grain and Total Biological Yield and Harvest Index in
      Rice (Oryzae sativae L.). In: Chataigner, J., Ed., *Future of Water Management for Rice in Mediterranean Climate
      Areas: Proceedings of the Workshops*, CIHEAM, Montpellier, 61-68.

[12] Uphoff, N. (2006) The System of Rice Intensification (SRI) as a Methodology for Reducing Water Requirements in Ir-
      rigated Rice Production. CIIFAD, New York.

20

Supply Response Analysis of Rice Growers in District Gujranwala, Pakistan

Sunair Junaid[1], Arif Ullah[2], Shaofeng Zheng[2], Syed Noor Muhammad Shah[3,4*], Shahid Ali[1], Munir Khan[1]

[1]Department of Agricultural and Applied Economics, The University of Agriculture, Peshawar, Pakistan
[2]College of Economics and Management, Northwest A & F University, Yangling, China
[3]Department of Horticulture, Faculty of Agriculture, Gomal University, D. I. Khan, Pakistan
[4]College of Horticulture, Northwest A & F University, Yangling, China
Email: *noor.aup@gmail.com

Abstract

The study was designed to estimate the restricted profit function in district Gujranwala, Punjab, Pakistan. Data were collected from 100 respondents using proportional allocation sampling technique. The analysis was done using SHAZAM software. The results indicate that the farmers are price-responsive. Rice own price elasticity was 1.873. The output supply elasticity of rice with respect to education, land, fertilizer price and irrigation cost were 0.0.169, 1.274, −0.873 and −0.953 respectively. Irrigation demand elasticity with respect to education, land, fertilizer price, irrigation cost and output price were 0.14, 1.14, −0.783, −1.84 and 1.78 respectively. Fertilizer demand elasticity with respect to education, land, fertilizer price, irrigation cost and output price was 0.023, 0.792, −1.65, −0.85 and 1.851 respectively. Lastly, the elasticity of profit with respect to education, land, fertilizer price, irrigation cost and output price was 0.20, 1.10, −0.83, −1.136 and 1.92 respectively. The study recommends that Government should provide consistent electricity with stable rates, so that, they irrigate their fields through electric tube wells and ultimately their cost of irrigation decreases. The study also suggests that government should stabilized fertilizer prices to encourage its application. Furthermore, government should raise procurement price of rice to encourage its supply this; it in turn will also increase the profit of the farmer.

Keywords

Rice, Supply Response, Translog Restricted Profit

---

*Corresponding author.

1. Introduction

Agriculture is back bone of economic growth and development in Pakistan. Being the dominant sector, it contributes 21.4% to GDP, employs 45% of the country's labour force and contributes in the growth of other sectors of the economy. The healthy expansion in agriculture stimulates domestic demand for industrial goods and other services and supplying raw material to agro-based industry notably cotton textile industry which is the largest subsector of manufacturing sector [1].

Rice is staple food worldwide, particularly in "East, South, Southeast Asia, the Middle East, Latin America, and the West Indies". It ranked second in production worldwide after maize crop. The world top ten producing countries are "China, India, Indonesia, Bangladesh, Vietnam, Thailand, Burma, Philippines, Brazil, and United States of America". The other minor producers of Rice are Japan, Cambodia, Pakistan, Republic of Korea, Madagascar, Nigeria, Egypt, and Sri Lanka [2]. Rice plays a significant role in agrarian economy of the Pakistan. Pakistan is considered as main rice exporter with share of world trade of 10%, which is about 2 million tons annually. Share of Pakistani basmati rice in export is 25% managed by private traders [3]. The main competitors of Pakistani rice are Thailand, India and Chad. The government of Pakistan has taken effective measures to increase the production and quality of export rice. That's why research efforts are continuing on developing high yielding basmati and IRRI varieties. Emphases are also being laid on agronomic research like fertilizer use, direct seedling etc. and on improved extension services. The flow of input and credits is also being substantially increased [4].

The importance of estimating valid elasticities of farm output supply and input demand can hardly be overemphasized [5]. Reliable estimates of the responsiveness of the supply of and demand for agricultural products to prices and other factors are fundamental to accurate economic forecasting and valid policy decision making. For example, own-price elasticities of demand indicate the extent to which buyers vary their purchases as the price of the product rises or falls. These variations are measured as movements along the demand curve. Cross-price elasticities of demand provide a framework for understanding the interactions in food and fibre choice decisions by consumers. These are reflected in shifts in the location of demand curves. This understanding is necessary for the accurate analysis of the response of consumers to changes in prices of products due to changes in their external environment [6].

The policy analyses which evaluate the impact of changes in single price or non-price variables or combinations of them become available. Due to resulted both the price elasticities and the elasticities with respect to several other variables that are usually considered as constraints on farm production for the elasticity estimates of input demand and output supply [7]. Technological change and positive price policy can play a significant role in stimulating agricultural production through the desired allocation of resources. At these stages, the policy planners face the challenges to formulate suitable agricultural policy by which the desired growth of agricultural output can be achieved. In order to formulate effective price policy, one needs reliable empirical knowledge about the degree of responsiveness of demand for factors and supply of products, to reliable prices and technological changes. The output supply and factor demand are closely interlinked to each other. Therefore, any change in factor and product prices affects the factor demand and output supply simultaneously. Rising cost of inputs discourages the input use and reduces the output supply. The decline in output supply raises food prices [8].

The literature suggests that production target achieved by providing very high support prices. As these high support prices may not practicable for Government, a more particular option is to bring more land area under rice crop and this area also allocated under modern rice varieties [9].

The study results are likely to be essential used for different reasons. For farm output supply reliable evaluation should be made by using input demand elasticities, are of enormous importance for the accurate prediction of the responsiveness of farmers to changes in input-output prices and government taxes and therefore used intensive programs to create the national requirement of development and food exports in agricultural sector. Results of the coefficient of elasticity may distribute a solid foundation in the development of effective policies applicable to the intrusion to support production, equity, efficiency and circulation revenues finally free in the agricultural sector of the economy.

The purpose of this paper is to apply the normalized restricted translog profit function and the corresponding system of derived demand equations to the farm-level data for rice from three villages of district Gujranwala, Pakistan in order to generate policy-relevant empirical estimates for rice supply and input demand function. The study addresses some of the issues that need to be confronted in determining what adjustments might be appropriate. In particular, the study explores the response of rice producers in district Gujranwala of Pakistan, to changes in prices and fixed inputs.

2. Methodology

2.1. Data Collection and Sampling

A multistage sampling technique was used for the selection of sampled respondents. Various stages are involved in collection of data through multistage sampling technique. In the first stage district Gujranwala was purposively selected as it is the second largest rice producing district of Punjab province, Pakistan. In stage 2, out of five tehsils (Gujranwala City, Gujranwala Sadar, Kamonki, NowshehraVirkan and Wazirabad) one tehsil (Gujranwala City) was randomly selected through simple random sampling technique. In third stage, three villages were randomly selected from a list of major rice producing villages of Gujranwala City. And then in the fourth and last stage a sample size of 100 respondents were randomly selected from selected villages through proportional allocation sampling technique. Data collected for this study openly from farmers through face to face discussion. The collected data were entered in tally sheets of computer software EXCEL and SHAZAM. Various analyses of the data were presented in the tabulated form.

2.2. Specification of Empirical Model

The Normalized Restricted Translog Profit Function Approach was used, as formulated by Diewert [10] and Christensen, *et al.* [11].

$$\ln \Pi_R^* = \alpha_0 + \alpha_i \sum_{i=1}^{2} \ln p_i^* + 1/2 \sum_{i=1}^{2} \sum_{j=1}^{2} \gamma_{ij} \ln p_i^* \ln p_j + \sum_{i=1}^{2} \sum_{k=1}^{2} \delta_{ik} \ln p_i^* \ln z_k$$

$$+ \sum_{k=1}^{2} \beta_k \ln z_k + 1/2 \sum_{k=1}^{2} \sum_{k=1}^{2} \theta_{kh} \ln z_k \ln z_h + \varepsilon$$

(1)

where

Π_R^* = Restricted profit, Π_R, normalized by the output price (P_R).

P_i^* = Price of ith input (P_i) normalized by the output price (P_R).

$i = j = 1$, Irrigation.

= 2, Fertilizer.

Z_k = Quantity of fixed input, k.

$k = h = 1$, Area under rice crop.

= 2, Average no. of schooling years per male family member above 13 years.

α_0, α_i, γ_{ij}, δ_{ik}, β_k and θ_{kh} are parameters to be estimated.

E = Random error.

For econometric estimation, irrigation and fertilizer were used as variable inputs and land under rice crop and education as fixed inputs. These variables are defined as below:

1) *Restricted profit*: It is calculated by excluding cost of irrigation plus fertilizer cost from total revenue and then normalized/divided by output price.

2) *Irrigation*: In profit function irrigation input is used consist of per acre irrigation cost and then normalized/divided by output price. It is represented by P_{Ig} in the model.

3) *Fertilizer*: Fertilizer input used in equation is calculated as per kilogram price in rupees per acre, it is actually the total cost of different fertilizer converted in kilograms and then normalized or divided by output price of rice.

4) *Land*: Land for rice crop is calculated as total land under rice crop it is represented by Z_1.

5) *Education*: Education was calculated as total education level of all male family members who were older than 13 and it is represented by "Z_2".

The corresponding share equations are expressed as,

$$S_i = P_i x_i / \Pi_R = -\partial \ln \Pi^* / \partial \ln P_i^* = -\alpha_i - \sum_{j=1}^{2} \gamma_{ij} \ln P_j^* - \sum_{k=1}^{2} \delta_{ik} Z_k$$

(2)

$$S_R = P_R x_R / \Pi_R = 1 + \partial \ln \Pi^* / \partial \ln P_R^* = 1 - \sum_{i=1}^{2} \alpha_i - \sum_{i=1}^{2} \sum_{j=1}^{2} \gamma_{ij} \ln p_j^* - \sum_{i=1}^{2} \sum_{k=1}^{2} \delta_{ik} \ln Z_k \qquad (3)$$

where S_i is the share of ith input, S_R is the share of output, X_i denotes the quantity of input i and y_R is the level of rice output.

Since the input and output shares come from a singular system of equations (since by definition $S_R - \Sigma S_i = 1$), one of the share equations, the output share, was dropped and the profit and factor demand equations were estimated as a simultaneous system.

2.3. Estimation of Elasticities

Production elasticities were calculated by using following formulae.

2.4. Input Demand Elasticities

The own price elasticity of demand for variable input i (η_{ii}), was estimated as:

$$\eta_{ii} = -S_i - \gamma_{ii} / S_i - 1 \qquad (4)$$

where S_i is the ith share equation, at the sample mean.

For the cross-price elasticity of demand for ith variable input with respect to the price of jth variable input (η_{ij}), the following expression was used.

$$\eta_{ij} = -S_j - \gamma_{ij} / S_i \quad \text{for } i \neq j \qquad (5)$$

The following equation was used for estimating the elasticity of demand for variable input with respect to output price, P_R (η_{iR})

$$\eta_{iR} = S_R + \sum_{j=1}^{2} \gamma_{ij} / S_i \qquad (6)$$

The elasticity of demand for variable input with respect to kth fixed factor, η_{ik}

$$\eta_{ik} = \beta_k + \delta_{ik} \ln \ln P_i^* + \theta_{kh} \ln Z_h - \delta / S_R \qquad (7)$$

2.5. Output Supply Elasticities

To compute the elasticity of output supply with respect to price of ith variable input (ϵ_{Ri}) the following equation was used.

$$\epsilon_{Ri} = -S_i - \gamma_{ji} / S_R \qquad (8)$$

The own price elasticity (ϵ_{RR}) was calculated using the following equation

$$\epsilon_{RR} = \sum_{i=1}^{2} S_i + \gamma_{ji} / S_R \qquad (9)$$

The elasticity of output supply with respect to fixed input k (ϵ_{RK}) was computed as:

$$(\epsilon_{RK}) = \beta_k + \sum_{i=1}^{2} \delta_{ik} \ln P_i^* + \theta_{kh} \ln Z_h + \sum_{i=1}^{2} \delta / S_R \qquad (10)$$

2.6. Profit Elasticities

These are defined as:

$$\partial \ln \Pi^* / \partial \ln P_i^* \qquad (11)$$

For the elasticity of profit with respect to changes in input prices and

$$\partial \ln \Pi^* / \partial \ln Z_k \qquad (12)$$

For the profit elasticity with respect to changes in fixed inputs.

3. Results and Discussion

3.1. Parameters Estimates of the Normalized Restricted Translog Profit Function

Estimated factors of the normalized restricted translog profit function and demand equations are entered in the following equations. The estimated parameters of translog profit function and input demand equations are used to calculate the elasticities of input demand and output supply in relation to the price of rice, the amount of fixed and changeable inputs prices.

The following model was estimated empirically:

$$\ln \Pi_R = \beta_0 + \beta_1 \ln p_{lg} + \beta_2 \ln P_F + \beta_3 \ln Z_1 + \beta_4 \ln Z_2 + 1/2\,\beta_5 \left(\ln P_{lg}\right)^2 + 1/2\,\beta_6 \left(\ln P_F\right)^2$$
$$+ \beta_7 \ln P_{lg} \ln P_F + \beta_8 \ln P_{lg} \ln Z_1 + \beta_9 \ln P_{lg} \ln Z_2 + \beta_{10} \ln P_F \ln Z_1$$
$$+ \beta_{11} \ln P_F \ln Z_2 + \beta_{12} \ln Z_1 \ln Z_2 + 1/2\,\beta_{13} \left(\ln Z_1\right)^2 + 1/2\,\beta_{14} \left(\ln Z_2\right)^2$$

$\ln \Pi_R^*$	Coefficients	Std. Errors	t-Ratios
β_0	7.255	(1.823)	(3.980)
$\ln P_{lg}^2$	−0.2968	(0.112)	(−2.650)
$\ln P_F$	− 0.1125	(0.018)	(−6.252)
$\ln Z_1$	+ 0.5206	(0.043)	(12.10)
$\ln Z_2 - 1/2$	+0.302	(0.158)	(1.917)
$\left(\ln P_{lg}\right)^2$	−1/2 0.152	(0.021)	(−7.198)
$\left(\ln P_F\right)^2$	−1/2 0.2406	(0.023)	(−10.19)
$\ln P_{lg} \ln P_F$	−0.1354	(0.039)	(−3.474)
$\ln P_{lg} \ln Z_1$	+0.1846	(0.056)	(3.297)
$\ln P_{lg} \ln Z_2$	−0.0514	(0.023)	(−2.238)
$\ln P_F \ln Z_1$	−0.0820	(0.014)	(−5.858)
$\ln P_F \ln Z_2$	+0.015	(0.009)	(1.771)
$\ln Z_1 \ln Z_2$	−0.1939	(0.217)	(−0.894)
$\left(\ln Z_1\right)^2$	−1/2 0.0271	(0.012)	(−2.262)
$\left(\ln Z_2\right)^2$	+1/2 0.426	(0.379)	(1.125)

$R^2 = 0.8478$; Adj. $R^2 = 0.8217$; F = 116.43; DW = 1.96.

The coefficient of determination value *i.e.* R^2 reveals that the independent variable explains 82.1% of the variation in the dependent variables. The value of F (116.43) shows that the model is good to fit. DW (1.96) shows that there is no auto co-relation problem [12]. The above model was used to calculate the equation of the demand for changeable/variable inputs *i.e.* irrigation and fertilizer.

3.2. Factor Demand Equations

$$D_{ig} = \beta_0 + \beta_1 \ln P_{lg} + \beta_2 \ln P_F + \beta_3 \ln Z_1 + \beta_4 \ln Z_2$$
$$D_f = \beta_0 + \beta_1 \ln P_{lg} + \beta_2 \ln P_F + \beta_3 \ln Z_1 + \beta_4 \ln Z_2$$

		Coefficients	Std. Errors	t−Ratios
	β_0	−0.2968	(0.112)	(−2.650)
	$\ln P_{Ig}$	− 0.152	(0.021))	(−7.198)
D_{ig}	$\ln P_F$	− 0.135	(0.039)	(−3.474)
	$\ln Z_1$	+ 0.184	(0.056)	(3.297)
	$\ln Z_2$	−0.051	(0.023)	(−2.238)
	β_0	−0.112	(0.018)	(−6.252)
	$\ln P_{Ig}$	− 0.135	(0.039)	(−3.474)
D_f	$\ln P_F$	− 0.240	(0.023)	(−10.194)
	$\ln Z_1$	− 0.082	(0.014)	(−5.858)
	$\ln Z_2$	+0.015	(0.009)	(1.77)

where D_{Ig} and D_f shows equation for demand of irrigation and fertilizer respectively. The intercept of both equations were negative. The relationship of irrigation demand with irrigation cost, fertilizer price and education is negative. But positive relationship for land the demand for fertilizer has negative relation with fertilizer price, irrigation cost and land while positive relationship is estimated with education.

3.3. Descriptive Statistics of the Variable Included in the Analysis

After detailed interview of the rice growers in the study area results were presented in tabular form (**Table 1**) which showed that per kg price of rice mean was computed Rs. 39.4/- and its Standard Deviation was Rs. 8.7/- and its per acre Irrigation cost mean was found Rs. 21276.7 while Standard Deviation was Rs. 2932.1/-. Furthermore, per Kg price of fertilizer mean was calculated Rs. 37.5/- having Standard Deviation Rs. 10.2/- was figures out. During survey Area under rice was calculated in Acre(s) its mean and Standard Deviation were estimated 8.4 and 5.6 respectively. In addition, for the production analysis Education was estimated in years. The mean value for education was 2.9 and Standard Deviation 1.46. At the end, Restricted Profit mean value was obtained Rs.38666.72 and Standard Deviation share was 10805.01.

3.4. Calculated Elasticities and Their Accusation

Table 2 demonstrates the elasticities of input demand and supply of output in relation to price of changeable inputs, rice price and quantity of fixed inputs. Following elasticities were calculated by using translog profit function and equations of input demand.

Farmer feedback for increasing price of rice is positive and elastic in the study area, Surge of price by 1% of rice boost the output supply of rice increase by 1.873 percent. This Surge in rice prices would also encourage the expansion of direct and significant demand for input variables/changeable inputs. In quantitative terms, the percent surge in demand for irrigation associated with an increased percent of price of rice was 1.780 percent and for fertilizer it was 1.851 percent. Rising price of rice, also increase the net return of farmers in the study area: 1% increase in rice price, surge profit/net return of 1.920 percent. These results are in conformity with the results obtained by Rahman [13].

The outcomes of the own-price elasticity of demand for changeable inputs are negative, as expected and price elastic. Irrigation demand owns price elasticity is −1.842. Which shows 1% rush in the cost of irrigation, decline demand for irrigation by 1.842%. While for fertilizer the own price elasticity is −1.650, shows that 1% increase, in the cost fertilizer, decline demand by 1.65%. Similarly the effect of surge in the price of changeable inputs has negative impact on the supply of rice. Increase, in 1% cost of irrigation, reduce the supply of rice by 0.953%, likewise irrigation, fertilizer costs decreases the supply by 0.873%. In the same way surge in the prices of changeable inputs also adversely affect profit. As the fertilizer elasticity and irrigation in relation to profit is negative. Cost of irrigation surge by 1%, decline profit by 1.136% and cost of fertilizer rise by 1% decline profit by 0.832%. This is also consistent with the findings of Ullah [14].

Table 1. Descriptive statistics of the variables.

Variable	Unit	Mean	Standard deviation
Price of rice	Rs/Kg	39.4	8.7
Irrigation cost	Rs/acre	21276.7	2932.1
Price of fertilizer	Rs/Kg	37.5	10.2
Area under rice	Acre(s)	8.4	5.6
Education	Years	2.9	1.46
Restricted profit	Rs.	38666.72	10805.01

Source: Survey Data, 2011.

Table 2. Calculated elasticities and their accusation.

With respect to	Output price	Irrigation cost	Fertilizer price	Land	Education
Output	1.873	−0.953	−0.873	1.274	0.162
Irrigation	1.780	−1.842	−0.783	1.142	0.144
Fertilizer	1.851	−0.851	−1.650	0.792	0.023
Profit	1.920	−1.136	−0.832	1.101	0.200

Source: Survey Data, 2011.

Through overall estimation, the results reveal that changes in market prices whether output or input prices effects farmer profit, rice supply and resource use. The response of farmer with changing in input or output prices was significant in the study area.

Irrigation demand elasticity in relation to fertilizer price is −0.783, which reveals that 1% raise in the cost of fertilizer, irrigation demand decrease by 0.783%. Similarly the fertilizer demand elasticity in relation to irrigation cost is also negative −0.851, 1% rises in irrigation cost, decline 0.851% demand for fertilizer. These negative elasticities show that irrigation and fertilizer are complementary inputs. Their combined application surges production synergistically.

Irrigation elasticity in relation to fixed input land is positive 1.142, which shows demand for irrigation, surge by 1.142% by increasing 1% land area. Fertilizer elasticity over the area under rice is 0.792 shows that 1% surge in the area under rice, surge demand for fertilizer by 0.792%. Similarly elasticity of irrigation and fertilizer with relation to education is 0.144 and 0.023 correspondingly, which shows that the demand for the fertilizer and irrigation increases with the surge of education level of hose hold. So improving education level of the hose members, surge the demand for changeable inputs. The profit elasticity in relation to education is 0.200, reveals that the surge of 1 % of the education levels of farmers, profit increased by 0.200%.

Rice supply elasticity in relation to fixed inputs land and education are positive *i.e.* 1.274 and 0.162 respectively. The effect of land on rice supply is significant and expansion in fixed input land, not only surge supply of rice but also profit of the farmer surges significantly. On the other hand the effect of education on rice supply and profit is positive but inelastic. This is also in agreement with the estimation of Farooq *et al.* [9].

3.5. Concluding Remarks

The analysis shows that farmers are price sensitive; that prevalence ensures equitable output and that input prices is essential for preserving the incentive for farmers ultimately surged rice production. Furthermore, application of fertilizers has surged the demand for irrigation affects positively. Surge in rice acreage significantly increases the supply of rice and profits and stimulates the demand for changeable inputs. In terms of education on the other hand, they also have positive but inelastic effects on the supply of rice, net return of the farmer and the demand for changeable inputs.

On the other hand, education level of the farming household also has a positive but inelastic effect on the supply of rice, the profit of the farmer and the demand for variable inputs. The results of the study suggest that the government should subsidize fertilizer prices to encourage its application. Increasing fertilizer application will not only increase output but also will encourage on-farm employment. Similarly, bringing more area under rice cultivation will increase production and the rate of the farm employment.

References

[1]   Farooq, O. (2013) Agriculture. Ministry of Finance, Government of Pakistan, Islamabad.

[2]   FAO (2013) Food and Agriculture Organization. Cereal Crop Statistics.
        http://www.fao.org/tc/tci/newsandmeetings/en/s

[3]   Anonymous (2014) Basmati Rice. http://www.pakissan.com/english/allabout/crop/rice/

[4]   Shaikh, F.M. and Shah, M.A. (2008) Dynamic Supply Response Analysis of Pakistani Rice Growers. *Pakistan Journal of Commerce and Social Sciences*, **1**, 48-55.

[5]   Chaudhary, M.A., Khan, M.A. and Naqvi, K.H. (1998) Estimates of Farm Output Supply and Input Demand Elasticities: The Translog Profit Function Approach. *The Pakistan Development Review*, **37**, 1031-1050.

[6]   Griffith, G., I'Anson, K., Hill, D. and Lubett, R. (2001) Previous Supply Elasticity Estimates for Australian Broadacre Agriculture.

[7]   Sidhu, S.S. and Baanante, C.A. (1981) Estimating Farm-Level Input Demand and Wheat Supply in the Indian Punjab Using a Translog Profit Function. *American Journal of Agricultural Economics*, **63**, 237-246.
        http://dx.doi.org/10.2307/1239559

[8]   Thakare, S.S., Shende, N.V. and Shinde, K.J. (2012) Mathematical Modeling for Demand and Supply Estimation for Cotton in Maharashtra. *International Journal of Scientific and Research Publications*, **2**, 1-5.

[9]   Farooq, U., Young, T., Russell, N. and Iqbal, M. (2001) The Supply Response of Basmati Rice Growers in Punjab, Pakistan: Price and Non-Price Determinants. *Journal of International Development*, **13**, 227-237.
        http://dx.doi.org/10.1002/jid.728

[10]   Diewert, W.E. (1974) () Applications of Duality Theory. In: Intrlligator, M.D. and Kendrick, D.A., Eds., Frontiers of Qumtitatiue Economics, Stanford Institute for Mathematical Studies in the Social Sciences, North-Holland Publishing Company, Amsterdam.

[11]   Christensen, L.R., Jorgenson, D.W. and Lau, L.J. (1973) Transcendental Logarithmic Production Frontiers. *The Review of Economics and Statistics*, **55**, 28-45. http://dx.doi.org/10.2307/1927992

[12]   Gujarati, D.N. (2012) Basic Econometrics. Tata McGraw-Hill Education, Noida.

[13]   Rahman, S. (2003) Profit Efficiency among Bangladeshi Rice Farmers. *Food Policy*, **28**, 487-503.
        http://dx.doi.org/10.1016/j.foodpol.2003.10.001

[14]   Ullah, R., Ali, S., Safi, Q.S., Shah, J. and Khan, K.H. (2012) Supply Response Analysis of Wheat Growers in District Peshawar: Pakistan. *International Journal of Latest Trends in Agriculture and Food Sciences*, **2**, 33-38.

Permissions

List of Contributors

Chiraz Masmoudi-Charfi
Institution of Agriculture Research & High Education, Tunis, Tunisia

Hamadi Habaieb
National Institute of Agronomy, Tunis, Tunisia

Daffa Alla M. Abdel
Agricultural Engineering Section, Research and Development Department, Kenana Sugar Company, Rabak City, Sudan

Wahab Ali M. Adeeb
Irrigation and Water Management Institute, Gezira University, Wad Medani, Sudan

Noor Amin and Dawood Ibrar
Department of Chemistry, Abdul Wali Khan University, Mardan, Pakistan

Sultan Alam
Department of Chemistry, University of Malakand, Chakdara, Pakistan

Nithya Rajan
Texas A & M AgriLife Research and Extension Center, Vernon, Texas, USA

Stephan J. Maas
Texas Tech University and Texas A & M AgriLife Research Center, Lubbock, Texas, USA

Ali A. Alderfasi and Mostafa M. Selim
Department of Plant Production, Faculty of Food and Agriculture Science, King Saud University, Riyadh, Saudi Arabia

Bushra A. Alhammad
Faculty of Sciences and Humanities, Salman Bin Abdulaziz University, Alkharj, Saudi Arabia

El Sayed Ali El Abd
Geology Department, Desert Research Centre, Cairo, Egypt

Maged Mostafa El Osta
Geology Department, Faculty of Science, Damanhour University, Damanhour, Egypt

Zahra Paydar and Yun Chen
CSIRO Land and Water, Canberra, Australia

Riaz Bhanbhro
Department of Civil, Environmental and Natural Resources Engineering, Luleå University of Technology, Luleå, Sweden
Quaid-e-Awam University of Engineering, Science & Technology, Nawabshah, Pakistan

Nadhir Al-Ansari and Sven Knutsson
Department of Civil, Environmental and Natural Resources Engineering, Luleå University of Technology, Luleå, Sweden

Shabya Choudhary, Shobhana Ramteke, Keshaw Prakash Rajhans, Pravin Kumar Sahu, Suryakant Chakradhari and Khageshwar Singh Patel
School of Studies in Chemistry/Environmental Science, Pt. Ravishankar Shukla University, Raipur, India

Laurent Matini
Department of Exact Sciences, E.N.S., Marien Ngouabi University, Brazzaville, Congo

Mohamed Kamel Fattah
Environmental Studies and Researches Institute, Sadat City University, Sadat, Egypt

S. A. E. Abdelrazek
Soil, Water and Environment Research Institute, Agriculture Research Center, Cairo, Egypt

Ch. Ramesh Naidu
GVP College of Engineering (Autonomous), Visakhapatnam, India

M. V. S. S. Giridhar
Jawaharlal Nehru Technological University, Hyderabad, India

Mohamed R. El Tahlawi, Mohamed Abo-El Kassem, Gamal. Y. Baghdadi and Hussein A. Saleem
Mining and Metallurgical Engineering Department, Faculty of Engineering, Assiut University, Assiut, Egypt

M. Arif
Department of Chemistry, Banasthali University, Niwai, District-Tonk, India

J. Hussain
National River Water Quality Laboratory, Central Water Commission, New Delhi, India

I. Hussain
Public Health Engineering Department (PHED) Laboratory, Bhilwara, India

S. Kumar and G. Bhati
Department of Space, Regional Remote Sensing Centre (West), NRSC/ISRO, Jodhpur, India

Clinton C. Shock
Malheur Experiment Station, Oregon State University, Ontario, USA

André B. Pereira
Department of Soil Science and Agricultural Engineering, State University of Ponta Grossa, Ponta Grossa, Brazil

Erik B. G. Feibert
Oregon State University, Malheur Experiment Station, Ontario, USA

Cedric A. Shock
Oregon State University, Malheur Experiment Station, Ontario, USA

Ali Ibrahim Akin
Ankara Nuclear Agriculture and Animal Research Center, Ankara, Turkey

Levent Abdullah Unlenen
General Directorate of Agricultural Researches, Potato Research Institute, Ministry of Agriculture and Rural Affairs, Nigde, Turkey

Sivarajah Mylevaganam and Chittaranjan Ray
University of Nebraska-Lincoln, Lincoln, NE, USA

Mohamed M. Hussein
Water Relations and Irrigation Department, National Research Centre, Cairo, Egypt

Ashok K. Alva
USDA-ARS, Prosser, WA, USA

Abdelraouf Ramadan Eid
Water Relations & Field Irrigation Department, National Research Centre, Cairo, Egypt

Essam Mohamed Hoballah
Agricultural Microbiology Department, National Research Centre, Cairo, Egypt

Sahar E. A. Mosa
Bio-Engineering Department, Agricultural Engineering Research Institute (AEnRI), Agricultural Research Center (ARC), Ministry of Agriculture, Giza, Egypt

Heidrun Bueckmann, Katja Thiele and Joachim Schiemann
Institute for Biosafety in Plant Biotechnology, Julius Kühn-Institut, Federal Research Centre for Cultivated Plants (JKI), Quedlinburg, Germany

Kepha G. Omwenga, Bancy M. Mati and Patrick G. Home
Biomechanical and Environmental Engineering Department, Jomo Kenyatta University of Agriculture and Technology, Nairobi, Kenya

Sunair Junaid, Shahid Ali and Munir Khan
Department of Agricultural and Applied Economics, The University of Agriculture, Peshawar, Pakistan

Arif Ullah and Shaofeng Zheng
College of Economics and Management, Northwest A & F University, Yangling, China

Syed Noor Muhammad Shah
Department of Horticulture, Faculty of Agriculture, Gomal University, D. I. Khan, Pakistan
College of Horticulture, Northwest A & F University, Yangling, China

www.ingramcontent.com/pod-product-compliance
Lightning Source LLC
Chambersburg PA
CBHW080507200326

41458CB00012B/4123